"ONE OF THE MOST IMPORTANT BOOKS ON EARTH'S ROOTS EVER WRITTEN."

EAST-WEST MAGAZINE

- How did the Nefilim—gold-seekers from a distant, alien planet—use cloning to create beings in their own image on Earth?
- Why did these "gods" seek the destruction of humankind through the Great Flood 13,000 years ago?
- What happens when their planet returns to Earth's vicinity every thirty-six centuries?
- Do Bible and Science conflict?
- Are were alone?

THE 12TH PLANET

"Heavyweight scholarship . . .
For thousands of years priests, poets,
and scientists have tried
to explain how life began . . .
Now a recognized scholar has come forth
with a theory that is
the most astonishing of all."

United Press International

Avon Books by
Zecharia Sitchin

THE EARTH CHRONICLES

ZECHARIA SITCHIN

THE 12TH PLANET

BOOK I OF THE EARTH CHRONICLES

HARPER

An Imprint of HarperCollinsPublishers

ACKNOWLEDGMENTS

The author wishes to express his gratitude to the many scholars who, over a span of more than a century, have uncovered, deciphered, translated, and explained the textual and artistic relics of the Ancient Near East, and to the many institutions and their staffs by whose excellence and courtesies the texts and pictorial evidence on which this book is based were made available to the author.

The author wishes especially to thank the New York Public Library and its Oriental Division; the Research Library (Reading Room and Oriental Students Room) of the British Museum, London; the Research Library of The Jewish Theological Seminary, New York; and for pictorial assistance to the Trustees of the British Museum and the keeper of the Assyrian and Egyptian Antiquities; the director of the Vorderasiatisches Museum, Staatliche Museen, East Berlin; the University Museum, Philadelphia; la Réunion des Musées Nationaux, France (Musées du Louvre); the Curator, Museum of Antiquities, Aleppo; the U.S. National Aeronautics and Space Administration.

HARPER

An Imprint of HarperCollins*Publishers*
10 East 53rd Street
New York, New York 10022-5299

Copyright © 1976 by Zecharia Sitchin
Author's Note copyright © 2007 by Zecharia Sitchin
Excerpt from *The End of Days* copyright © 2007 by Zecharia Sitchin
ISBN: 978-0-06-137913-0
ISBN-10: 0-06-137913-1

First Harper paperback printing: April 2007
First Avon Books paperback printing: July 1978

HarperCollins® and Harper® are trademarks of HarperCollins Publishers.

Printed in the United States of America

Visit Harper paperbacks on the World Wide Web at www.harpercollins.com

10 9

CONTENTS

•

AUTHOR'S NOTE

•

THIS EDITION of *The Twelfth Planet* is offered to the reader in celebration of a double milestone. First, it coincides with the publication of the seventh and concluding volume of *The Earth Chronicles* series; and, second, it marks the **thirtieth anniversary** of the initial publication of *The Twelfth Planet,* the first book in the series.

In the annals of publishing, especially of nonfiction, few books stay so long in print. *The Twelfth Planet* has managed not only to do that, but to do so very actively: This volume represents the **forty-fifth** printing of the original Avon Books paperback edition—a record!—and coincides with the book's translation into its twenty-first language (Korean). Published and republished in English and other various languages in hardcover, softcover, paperback, pocketbook, mass market, and other formats (including Braille for the blind and Books-on-Tape), the book has reached millions of readers, has been quoted (or misquoted . . .) in scores of other books, and has been reviewed and discussed in countless articles and on radio, TV, and more than half a million websites. It is, by all standards, a "classic."

When I set out to write it, I had no idea that the book would attain such heights, or that it would spawn a seven-book series (plus four "companion" books). In fact, I had no idea that it would include "planet" in its title. All I wanted to find out was who were the biblical *Nefilim*. The realization that they were the Sumerian *Anunnaki* ("Those Who from Heaven to Earth Came") opened up for me new vistas of research and understanding. The greatest breakthrough was the insight that the Sumerian/Babylonian Epic of Creation was a sophisticated scientific document. The resulting conclusions regarding the existence of one more planet in our solar system, the origins of life on Earth, and past space travel presaged a host of subsequent discoveries in astronomy, genetics, and other scientific fields that were unthinkable thirty years ago.

The book has shown that there is no conflict between Bible and Evolution. It endures, I believe, because it tells us the truth about our own origins—and that WE ARE NOT ALONE.

Zecharia Sitchin
New York, October 2006

PROLOGUE: GENESIS

•

GENESIS

THE OLD TESTAMENT has filled my life from childhood. When the seed for this book was planted, nearly fifty years ago, I was totally unaware of the then raging Evolution versus Bible debates. But as a young schoolboy studying *Genesis* in its original Hebrew, I created a confrontation of my own. We were reading one day in Chapter VI that when God resolved to destroy Mankind by the Great Flood, "the sons of the deities", who married the daughters of men, were upon the Earth. The Hebrew original named them *Nefilim;* the teacher explained it meant "giants"; but I objected: didn't it mean literally "Those Who Were Cast Down", who had descended to Earth? I was reprimanded and told to accept the traditional interpretation.

In the ensuing years, as I have learned the languages and history and archaeology of the ancient Near East, the *Nefilim* became an obsession. Archaeological finds and the deciphering of Sumerian, Babylonian, Assyrian, Hittite, Canaanite and other ancient texts and epic tales increasingly confirmed the accuracy of the biblical references to the kingdoms, cities, rulers, places, temples, trade routes, artifacts, tools and customs of antiquity. Is it not now time, therefore, to accept the word of these same ancient records regarding the *Nefilim* as visitors to Earth from the heavens?

The Old Testament repeatedly asserted: "The throne of Yahweh is in heaven"—"from heaven did the Lord behold the Earth". The New Testament spoke of "Our Father, which art in Heaven". But the credibility of the Bible was shaken by the advent and general acceptance of Evolution. If Man evolved, then surely he could not have been created all at once by a Deity who, premeditating, had suggested "Let us make Adam in our image and after our likeness". All the ancient peoples believed in gods who had descended to Earth from the heavens and who could at will soar heavenwards. But these tales were never given credibility, having been branded by scholars from the very beginning as myths.

The writings of the ancient Near East, which include a profusion of astronomical texts, clearly speak of a planet from which these astronauts or "gods" had come. However, when scholars, fifty and one hundred years ago, deciphered and translated the ancient lists of celestial bodies, our astronomers were not yet aware of Pluto (which was only located in 1930). How then could they be expected to accept the evidence of yet one more member of our solar system? But now that we too, like the ancients, are aware of the planets beyond Saturn, why not accept that ancient evidence for the existence of the Twelfth Planet?

As we ourselves venture into space, a fresh look and an acceptance of the ancient scriptures is more than timely. Now that astronauts have landed on the Moon, and unmanned spacecraft explore other planets, it is no longer impossible to believe that a civilization on another planet more advanced than ours was capable of landing its astronauts on the planet Earth some time in the past.

Indeed, a number of popular writers have speculated that ancient artifacts such as the pyramids and giant stone sculptures must have been fashioned by advanced visitors from another planet—for surely primitive man could not have possessed by himself the required technology? How was it, for another example, that the civilization of Sumer seemed to flower so suddenly nearly 6,000 years ago without a precursor? But since these writers usually fail to show *when, how* and, above all, *from where* such ancient astronauts did come—their intriguing questions remain unanswered speculations.

It has taken thirty years of research, of going back to the ancient sources, of accepting them literally, to re-create in my own mind a continuous and plausible scenario of prehistoric events. *The Twelfth Planet*, therefore, seeks to provide the reader with a narrative giving *answers* to the specific questions of When, How, Why and Wherefrom.

The evidence I adduce consists primarily of the ancient texts and pictures themselves.

In *The Twelfth Planet* I have sought to decipher a sophisticated cosmogony which explains, perhaps as well as modern scientific theories, how the solar system could have been formed, an invading planet caught into solar orbit, and Earth and other parts of the solar system brought into being.

The evidence I offer includes celestial maps dealing with space flight to Earth from that Planet, the Twelfth. Then, in sequence, follow the dramatic establishment of the first settlements on Earth by the *Nefilim:* their leaders were named; their relationships, loves, jealousies, achievements and struggles described; the nature of their "immortality" explained.

Above all, *The Twelfth Planet* aims to trace the momentous events that led to the creation of Man, and the advanced methods by which this was accomplished.

It then suggests the tangled relationship between Man and his lords, and throws fresh light on the meaning of the events in the Garden of Eden, of the Tower of Babel, of the great Flood. Finally, Man—endowed by his makers biologically and materially—ends up crowding his gods off the Earth.

This book suggests that we are not alone in our solar system. Yet it may enhance rather than diminish the faith in a universal Almighty. For, if the *Nefilim* created Man on Earth, they may have only been fulfilling a vaster Master Plan.

Z. SITCHIN

New York, February 1977

ABOUT THE SOURCES

•

THE PRIME SOURCE for the biblical verses quoted in *The Twelfth Planet* is the Old Testament in its original Hebrew text. It must be borne in mind that all the translations consulted—of which the principal ones are listed at the end of the book—are just that: translations or interpretations. In the final analysis, what counts is what the original Hebrew says.

In the final version quoted in *The Twelfth Planet*, I have compared the available translations against each other and against the Hebrew source and the parallel Sumerian and Akkadian texts/tales, to come up with what I believe is the most accurate rendering.

The rendering of Sumerian, Assyrian, Babylonian, and Hittite texts has engaged a legion of scholars for more than a century. Decipherment of script and language was followed by transcribing, transliterating, and finally translating. In many instances, it was possible to choose between differing translations or interpretations only by verifying the much earlier transcriptions and transliterations. In other instances, a late insight by a contemporary scholar could throw new light on an early translation.

The list of sources for Near Eastern texts, given at the end of this book, thus ranges from the oldest to the newest, and is followed by the scholarly publications in which valuable contributions to the understanding of the texts were found.

1

THE ENDLESS BEGINNING

OF THE EVIDENCE that we have amassed to support our conclusions, exhibit number one is Man himself. In many ways, modern man—*Homo sapiens*—is a stranger to Earth.

Ever since Charles Darwin shocked the scholars and theologians of his time with the evidence of evolution, life on Earth has been traced through Man and the primates, mammals, and vertebrates, and backward through ever-lower life forms to the point, billions of years ago, at which life is presumed to have begun.

But having reached these beginnings and having begun to contemplate the probabilities of life elsewhere in our solar system and beyond, the scholars have become uneasy about life on Earth: Somehow, it does not belong here. If it began through a series of spontaneous chemical reactions, why does life on Earth have but a single source, and not a multitude of chance sources? And why does all living matter on Earth contain too little of the chemical elements that abound on Earth, and too much of those that are rare on our planet?

Was life, then, imported to Earth from elsewhere?

Man's position in the evolutionary chain has compounded the puzzle. Finding a broken skull here, a jaw there, scholars at first believed that Man originated in Asia some 500,000 years ago. But as older fossils were found, it became evident that the mills of evolution grind much, much slower. Man's ancestor apes are now placed at a staggering 25,000,000 years ago. Discoveries in East Africa reveal a transition to manlike apes (hominids) some 14,000,000 years ago. It was about 11,000,000 years later that the first ape-man worthy of the classification *Homo* appeared there.

The first being considered to be truly manlike—"Advanced Australopithecus"—existed in the same parts of Africa some 2,000,000 years ago. It took yet another million years to produce *Homo erectus*. Finally, after another 900,000 years, the first primitive Man appeared; he is named Neanderthal after the site where his remains were first found.

In spite of the passage of more than 2,000,000 years between Advanced Australopithecus and Neanderthal, the tools of these two groups—sharp stones—were virtually alike; and the groups themselves (as they are believed to have looked) were hardly distinguishable. (Fig. 1)

 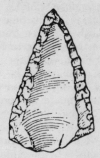

Fig. 1

Then, suddenly and inexplicably, some 35,000 years ago, a new race of Men—*Homo sapiens* ("thinking Man")— appeared as if from nowhere, and swept Neanderthal Man from the face of Earth. These modern Men—named Cro-Magnon—looked so much like us that, if dressed like us in modern clothes, they would be lost in the crowds of any European or American city. Because of the magnificent cave art which they created, they were at first called "cavemen." In fact, they roamed Earth freely, for they knew how to build shelters and homes of stones and animal skins wherever they went.

For millions of years, Man's tools had been simply stones of useful shapes. Cro-Magnon Man, however, made specialized tools and weapons of wood and bones. He was no longer a "naked ape," for he used skins for clothing. His society was organized; he lived in clans with a patriarchal hegemony. His cave drawings bespeak artistry and depth of feeling; his drawings and sculptures evidence some form of "religion," apparent in the worship of a Mother Goddess, who was sometimes depicted with the sign of the Moon's crescent. He buried his dead, and must therefore have had some philosophies regarding life, death, and perhaps even an afterlife.

As mysterious and unexplained as the appearance of Cro-Magnon Man has been, the puzzle is still more complicated. For, as other remains of modern Man were discovered (at sites including Swanscombe, Steinheim, and Montmaria), it became apparent that Cro-Magnon Man stemmed from an even earlier *Homo sapiens* who lived in western Asia and North Africa some 250,000 years before Cro-Magnon Man.

The appearance of modern Man a mere 700,000 years after *Homo erectus* and some 200,000 years before Neanderthal Man is absolutely implausible. It is also clear that *Homo sapiens* represents such an extreme departure from the slow evolutionary process that many of our features, such as the ability to speak, are totally unrelated to the earlier primates.

An outstanding authority on the subject, Professor Theodosius Dobzhansky *(Mankind Evolving)*, was especially puzzled by the fact that this development took place during a period when Earth was going through an ice age, a most unpropitious time for evolutionary advance. Pointing out that *Homo sapiens* lacks completely some of the peculiarities of the previously known types, and has some that never appeared before, he concluded: "Modern man has many fossil collateral relatives but no progenitors; the derivation of *Homo sapiens*, then, becomes a puzzle."

How, then, did the ancestors of modern Man appear some 300,000 years ago—instead of 2,000,000 or 3,000,000 years in the future, following further evolutionary development? Were we imported to Earth from elsewhere, or were we, as the Old Testament and other ancient sources claim, created by the gods?

We now know where civilization began and how it developed, once it began. The unanswered question is: *Why*—why did civilization come about at all? For, as most scholars now admit in frustration, by all data Man should still be without civilization. There is no obvious reason that we should be any more civilized than the primitive tribes of the Amazon jungles or the inaccessible parts of New Guinea.

But, we are told, these tribesmen still live as if in the Stone Age because they have been isolated. But isolated from what? If they have been living on the same Earth as we, why have they not acquired the same knowledge of sciences and technologies on their own as we supposedly have?

The real puzzle, however, is not the backwardness of the Bushmen, but our advancement; for it is now recognized that in the normal course of evolution Man should still be typified by the Bushmen and not by us. It took Man some 2,000,000 years to advance in his "tool industries" from the use of stones as he found them to the realization that he could chip and shape stones to better suit his purposes. Why not another 2,000,000 years to learn the use of other materials, and another 10,000,000 years to master mathematics and engineering and astronomy? Yet here we are, less than 50,000 years from Neanderthal Man, landing astronauts on the Moon.

The obvious question, then, is this: Did we and our Mediterranean ancestors really acquire this advanced civilization on our own?

Though Cro-Magnon Man did not build skyscrapers nor use metals, there is no doubt that his was a sudden and revolutionary civilization. His mobility, ability to build shelters, his desire to clothe himself, his manufactured tools, his art—all were a sudden high civilization breaking an endless beginning of Man's culture that stretched over millions of years and advanced at a painfully slow pace.

Though our scholars cannot explain the appearance of *Homo sapiens* and the civilization of Cro-Magnon Man, there is by now no doubt regarding this civilization's place of origin: the Near East. The uplands and mountain ranges that extend in a semiarc from the Zagros Mountains in the east (where present-day Iran and Iraq border on each other), through the Ararat and Taurus ranges in the north, then down, westward and southward, to the hill lands of Syria, Lebanon, and Israel, are replete with caves where the evidence of prehistoric but modern Man has been preserved. (Fig. 2)

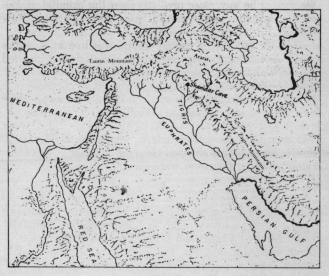

Fig. 2

One of these caves, Shanidar, is located in the north-eastern part of the semiarc of civilization. Nowadays, fierce Kurdish tribesmen seek shelter in the area's caves for themselves and their flocks during the cold winter months. So it was, one wintry night 44,000 years ago, when a family of seven (one of whom was a baby) sought shelter in the cave of Shanidar.

Their remains—they were evidently crushed to death by a rockfall—were discovered in 1957 by a startled Ralph

Solecki, who went to the area in search of evidence of early Man.* What he found was more than he expected. As layer upon layer of debris was removed, it became apparent that the cave preserved a clear record of Man's habitation in the area from about 100,000 to some 13,000 years ago.

What this record showed was as surprising as the find itself. Man's culture has shown not a progression but a regression. Starting from a certain standard, the following generations showed not more advanced but less advanced standards of civilized life. And from about 27,000 B.C. to 11,000 B.C., the regressing and dwindling population reached the point of an almost complete absence of habitation. For reasons that are assumed to have been climatic, Man was almost completely gone from the whole area for some 16,000 years.

And then, circa 11,000 B.C., "thinking Man" reappeared with new vigor and on an inexplicably higher cultural level.

It was as if an unseen coach, watching the faltering human game, dispatched to the field a fresh and better-trained team to take over from the exhausted one.

•

Throughout the many millions of years of his endless beginning, Man was nature's child; he subsisted by gathering the foods that grew wild, by hunting the wild animals, by catching wild birds and fishes. But just as Man's settlements were thinning out, just as he was abandoning his abodes, when his material and artistic achievements were disappearing—just then, suddenly, with no apparent reason and without any prior known period of gradual preparation —Man became a farmer.

Summarizing the work of many eminent authorities on the subject, R. J. Braidwood and B. Howe (*Prehistoric Investigations in Iraqi Kurdistan*) concluded that genetic studies confirm the archaeological finds and leave no doubt that agriculture began exactly where Thinking Man had emerged earlier with his first crude civilization: in the Near East. There is no doubt by now that agriculture

---

* Professor Solecki has told me that nine skeletons were found, of which only four were crushed by rockfall.

spread all over the world from the Near Eastern arc of mountains and highlands.

Employing sophisticated methods of radiocarbon dating and plant genetics, many scholars from various fields of science concur in the conclusion that Man's first farming venture was the cultivation of wheat and barley, probably through the domestication of a wild variety of emmer. Assuming that, somehow, Man did undergo a gradual process of teaching himself how to domesticate, grow, and farm a wild plant, the scholars remain baffled by the profusion of other plants and cereals basic to human survival and advancement that kept coming out of the Near East. These included, in rapid succession, millet, rye, and spelt, among the edible cereals; flax, which provided fibers and edible oil; and a variety of fruit-bearing shrubs and trees.

In every instance, the plant was undoubtedly domesticated in the Near East for millennia before it reached Europe. It was as though the Near East were some kind of genetic-botanical laboratory, guided by an unseen hand, producing every so often a newly domesticated plant.

The scholars who have studied the origins of the grapevine have concluded that its cultivation began in the mountains around northern Mesopotamia and in Syria and Palestine. No wonder. The Old Testament tells us that Noah "planted a vineyard" (and even got drunk on its wine) after his ark rested on Mount Ararat as the waters of the Deluge receded. The Bible, like the scholars, thus places the start of vine cultivation in the mountains of northern Mesopotamia.

Apples, pears, olives, figs, almonds, pistachios, walnuts —all originated in the Near East and spread from there to Europe and other parts of the world. Indeed, we cannot help recalling that the Old Testament preceded our scholars by several millennia in identifying the very same area as the world's first orchard: "And the Lord God planted an orchard in Eden, in the east. . . . And the Lord God caused to grow, out of the ground, every tree that is pleasant to behold and that is good for eating."

The general location of "Eden" was certainly known to the biblical generations. It was "in the east"—east of the Land of Israel. It was in a land watered by four major rivers, two of which are the Tigris and the Euphrates.

There can be no doubt that the Book of Genesis located the first orchard in the highlands where these rivers originated, in northeastern Mesopotamia. Bible and science are in full agreement.

As a matter of fact, if we read the original Hebrew text of the Book of Genesis not as a theological but as a scientific text, we find that it also accurately describes the process of plant domestication. Science tells us that the process went from wild grasses to wild cereals to cultivated cereals, followed by fruit-bearing shrubs and trees. This is exactly the process detailed in the first chapter of the Book of Genesis.

And the Lord said:
"Let the Earth bring forth grasses;
cereals that by seeds produce seeds;
fruit trees that bear fruit by species,
      which contain the seed within themselves."

And it was so:
The Earth brought forth grass;
cereals that by seed produce seed, by species;
and trees that bear fruit, which contain
      the seed within themselves, by species.

The Book of Genesis goes on to tell us that Man, expelled from the orchard of Eden, had to toil hard to grow his food. "By the sweat of thy brow shalt thou eat bread," the Lord said to Adam. It was after that that "Abel was a keeper of herds and Cain was a tiller of the soil." Man, the Bible tells us, became a shepherd soon after he became a farmer.

Scholars are in full agreement with this biblical sequence of events. Analyzing the various theories regarding animal domestication, F. E. Zeuner (*Domestication of Animals*) stresses that Man could not have "acquired the habit of keeping animals in captivity or domestication before he reached the stage of living in social units of some size." Such settled communities, a prerequisite for animal domestication, followed the changeover to agriculture.

The first animal to be domesticated was the dog, and not necessarily as Man's best friend but probably also for food. This, it is believed, took place circa 9500 B.C. The first skeletal remains of dogs have been found in Iran, Iraq, and Israel.

Sheep were domesticated at about the same time; the Shanidar cave contains remains of sheep from circa 9000 B.C., showing that a large part of each year's young were killed for food and skins. Goats, which also provided milk, soon followed; and pigs, horned cattle, and hornless cattle were next to be domesticated.

In every instance, the domestication began in the Near East.

The abrupt change in the course of human events that occurred circa 11,000 B.C. in the Near East (and some 2,000 years later in Europe) has led scholars to describe that time as the clear end of the Old Stone Age (the Paleolithic) and the beginning of a new cultural era, the Middle Stone Age (Mesolithic).

The name is appropriate only if one considers Man's principal raw material—which continued to be stone. His dwellings in the mountainous areas were still built of stone; his communities were protected by stone walls; his first agricultural implement—the sickle—was made of stone. He honored or protected his dead by covering and adorning their graves with stones; and he used stone to make images of the supreme beings, or "gods," whose benign intervention he sought. One such image, found in northern Israel and dated to the ninth millennium B.C., shows the carved head of a "god" shielded by a striped helmet and wearing some kind of "goggles." (Fig. 3)

From an overall point of view, however, it would be more appropriate to call the age that began circa 11,000 B.C. not the Middle Stone Age but the Age of Domestication. Within the span of a mere 3,600 years—overnight in terms of the endless beginning—Man became a farmer, and wild plants and animals were domesticated. Then, a new age clearly followed. Our scholars call it the New Stone Age (Neolithic); but the term is totally inadequate, for the main change that had taken place circa 7500 B.C. was the appearance of pottery.

For reasons that still elude our scholars—but which will become clear as we unfold our tale of prehistoric events— Man's march toward civilization was confined, for the first

Fig. 3

several millennia after 11,000 B.C., to the highlands of
the Near East. The discovery of the many uses to which
clay could be put was contemporary with Man's descent
from his mountain abodes toward the lower, mud-filled
valleys.

By the seventh millennium B.C., the Near Eastern arc
of civilization was teeming with clay or pottery cultures,
which produced great numbers of utensils, ornaments, and
statuettes. By 5000 B.C., the Near East was producing clay
and pottery objects of superb quality and fantastic design.

But once again progress slowed, and by 4500 B.C.,
archaeological evidence indicates, regression was all

around. Pottery became simpler. Stone utensils—a relic of the Stone Age—again became predominant. Inhabited sites reveal fewer remains. Some sites that had been centers of pottery and clay industries began to be abandoned, and distinct clay manufacturing disappeared. "There was a general impoverishment of culture," according to James Melaart (*Earliest Civilizations of the Near East);* some sites clearly bear the marks of "the new poverty-stricken phase."

Man and his culture were clearly on the decline.

Then—suddenly, unexpectedly, inexplicably—the Near East witnessed the blossoming of the greatest civilization imaginable, a civilization in which our own is firmly rooted.

A mysterious hand once more picked Man out of his decline and raised him to an even higher level of culture, knowledge, and civilization.

2

THE SUDDEN CIVILIZATION

FOR A LONG TIME, Western man believed that his civilization was the gift of Rome and Greece. But the Greek philosophers themselves wrote repeatedly that they had drawn on even earlier sources. Later on, travelers returning to Europe reported the existence in Egypt of imposing pyramids and temple-cities half-buried in the sands, guarded by strange stone beasts called sphinxes.

When Napoleon arrived in Egypt in 1799, he took with him scholars to study and explain these ancient monuments. One of his officers found near Rosetta a stone slab on which was carved a proclamation from 196 B.C. written in the ancient Egyptian pictographic writing (hieroglyphic) as well as in two other scripts.

The decipherment of the ancient Egyptian script and language, and the archaeological efforts that followed, revealed to Western man that a high civilization had existed in Egypt well before the advent of the Greek civilization. Egyptian records spoke of royal dynasties that began circa 3100 B.C.—two full millennia before the beginning of Hellenic civilization. Reaching its maturity in the fifth and fourth centuries B.C., Greece was a latecomer rather than an originator.

Was the origin of our civilization, then, in Egypt?

As logical as that conclusion would have seemed, the facts militated against it. Greek scholars did describe visits to Egypt, but the ancient sources of knowledge of which they spoke were found elsewhere. The pre-Hellenic cultures of the Aegean Sea—the Minoan on the island of Crete and the Mycenaean on the Greek mainland—revealed evidence that the Near Eastern, not the Egyptian, culture had been adopted. Syria and Anatolia, not Egypt, were the principal avenues through which an earlier civilization became available to the Greeks.

Noting that the Dorian invasion of Greece and the Israelite invasion of Canaan following the Exodus from Egypt took place at about the same time (circa the thirteenth century B.C.), scholars have been fascinated to discover a growing number of similarities between the Semitic and Hellenic civilizations. Professor Cyrus H. Gordon *(Forgotten Scripts; Evidence for the Minoan Language)* opened up a new field of study by showing that an early Minoan script, called Linear A, represented a Semitic language. He concluded that "the pattern (as distinct from the content) of the Hebrew and Minoan civilizations is the same to a remarkable extent," and pointed out that the island's name, Crete, spelled in Minoan *Ke-re-ta,* was the same as the Hebrew word *Ke-re-et* ("walled city") and had a counterpart in a Semitic tale of a king of Keret.

Even the Hellenic alphabet, from which the Latin and our own alphabets derive, came from the Near East. The ancient Greek historians themselves wrote that a Phoenician named Kadmus ("ancient") brought them the alphabet, comprising the same number of letters, in the same order, as in Hebrew; it was the only Greek alphabet when the Trojan War took place. The number of letters was raised to twenty-six by the poet Simonides of Ceos in the fifth century B.C.

That Greek and Latin writing, and thus the whole foundation of our Western culture, were adopted from the Near East can easily be demonstrated by comparing the order, names, signs, and even numerical values of the original Near Eastern alphabet with the much later ancient Greek and the more recent Latin. (Fig. 4)

The scholars were aware, of course, of Greek contacts with the Near East in the first millennium B.C., culminating with the defeat of the Persians by Alexander the Macedonian in 331 B.C. Greek records contained much information about these Persians and their lands (which roughly paralleled today's Iran). Judging by the names of their kings—Cyrus, Darius, Xerxes—and the names of their deities, which appear to belong to the Indo-European linguistic stem, scholars reached the conclusion that they were part of the Aryan ("lordly") people that appeared from somewhere near the Caspian Sea toward the end of the second millennium B.C. and spread westward to Asia Minor, eastward to India, and southward to what the Old Testament called the "lands of the Medes and Parsees."

Yet all was not that simple. In spite of the assumed foreign origin of these invaders, the Old Testament treated them as part and parcel of biblical events. Cyrus, for example, was considered to be an "Anointed of Yahweh" —quite an unusual relationship between the Hebrew God and a non-Hebrew. According to the biblical Book of Ezra, Cyrus acknowledged his mission to rebuild the Temple in Jerusalem, and stated that he was acting upon orders given by Yahweh, whom he called "God of Heaven."

Cyrus and the other kings of his dynasty called themselves Achaemenids—after the title adopted by the founder of the dynasty, which was Hacham-Anish. It was not an Aryan but a perfect Semitic title, which meant "wise man." By and large, scholars have neglected to investigate the many leads that may point to similarities between the Hebrew God Yahweh and the deity Achaemenids called "Wise Lord," whom they depicted as hovering in the skies within a Winged Globe, as shown on the royal seal of Darius. (Fig. 5)

It has been established by now that the cultural, religious, and historic roots of these Old Persians go back to the earlier empires of Babylon and Assyria, whose extent and fall is recorded in the Old Testament. The symbols that make up the script that appeared on the Achaemenid monuments and seals were at first considered to be decorative designs. Engelbert Kampfer, who visited Persepolis, the Old Persian capital, in 1686, described the signs as "cuneates," or wedge-shaped impressions. The script has since been known as cuneiform.

As efforts began to decipher the Achaemenid inscriptions, it became clear that they were written in the same script as inscriptions found on ancient artifacts and tablets in Mesopotamia, the plains and highlands that lay between the Tigris and Euphrates rivers. Intrigued by the scattered finds, Paul Emile Botta set out in 1843 to conduct the first major purposeful excavation. He selected a site in northern Mesopotamia, near present-day Mosul, now

Fig. 4

Hebrew name	CANAANITE-PHOENICIAN	EARLY GREEK	LATER GREEK	Greek name	LATIN
Aleph	ⴽ ⴹ	Δ	A	Alpha	A
Beth	9 9	S ẞ	B	Beta	B
Gimel	⅂	⅂	Γ	Gamma	C G
Daleth	◿ ◿	Δ	Δ	Delta	D
He	ⴺⴺ	ⴺ	E	E(psilon)	E
Vau	Y	Y	ⴼ	Vau	F V
Zayin	ⵣ ⵣ	I	I	Zeta	
Heth (1)	ⴱ ⴱ	ⴱ	ⴱ	(H)eta	H
Teth	⊗	⊗	⊗	Theta	
Yod	ⵥ	ⵥ	ⵥ	Iota	I
Khaph	ⴿ ⵢⵢ	ⴽ	K	Kappa	
Lamed	ⵍ ⵍ	�v⵿⅂	L ⴷ	Lambda	L
Mem	ⵯⵯ ⵯⵯ	ⴹ	ⴹ	Mu	M
Nun	⵩ ⵩	ⴺ	N	Nu	N
Samekh	ⵣ ⵣⵣ	ⵣ	ⵣ	Xi	X
Ayin	◌◌	◌	◌	O(nicron)	O
Pe	ⵯⵯ ⵥ	ⵯ	Γ	Pi	P
Şade (2)	ⵯⵯⵯ	M	M	San	
Koph	φφφ	Φ	ⵕ	Koppa	Q
Resh	ⵕ	ⵕ	Ρ	Rho	R
Shin	W	ⵥ	ⵥ	Sigma	S
Tav	ⵝ	T	T	Tau	T

(1) "Ḥ", commonly transliterated as "H" for
    simplicity, is pronounced in the Sumerian
    and Semitic languages as "CH" in the
    Scottish or German "loch".

(2) "Ṣ", commonly transliterated at "S" for
    simplicity, is pronounced in the Sumerian
    and Semitic languages as "TS".

Fig. 5

called Khorsabad. Botta was soon able to establish that the cuneiform inscriptions named the place Dur Sharru Kin. They were Semitic inscriptions, in a sister language of Hebrew, and the name meant "walled city of the righteous king." Our textbooks call this king Sargon II.

This capital of the Assyrian king had as its center a magnificent royal palace whose walls were lined with sculptured bas-reliefs, which, if placed end to end, would stretch for over a mile. Commanding the city and the royal compound was a step pyramid called a ziggurat; it served as a "stairway to Heaven" for the gods. (Fig. 6)

The layout of the city and the sculptures depicted a way of life on a grand scale. The palaces, temples, houses, stables, warehouses, walls, gates, columns, decorations, statues, artworks, towers, ramparts, terraces, gardens—all were completed in just five years. According to Georges Contenau (*La Vie Quotidienne à Babylone et en Assyrie*), "the imagination reels before the potential strength of an empire which could accomplish so much in such a short space of time," some 3,000 years ago.

Not to be outdone by the French, the English appeared on the scene in the person of Sir Austen Henry Layard,

Fig. 6

who selected as his site a place some ten miles down the Tigris River from Khorsabad. The natives called it Kuyunjik; it turned out to be the Assyrian capital of Nineveh.

Biblical names and events had begun to come to life. Nineveh was the royal capital of Assyria under its last three great rulers: Sennacherib, Esarhaddon, and Ashurbanipal. "Now, in the fourteenth year of king Hezekiah, did Sennacherib king of Assyria come up against all the walled cities of Judah," relates the Old Testament (II Kings 18:13), and when the Angel of the Lord smote his army, "Sennacherib departed and went back, and dwelt in Nineveh."

The mounds where Nineveh was built by Sennacherib and Ashurbanipal revealed palaces, temples, and works of art that surpassed those of Sargon. The area where the remains of Esarhaddon's palaces are believed to lie cannot be excavated, for it is now the site of a Muslim mosque erected over the purported burial place of the prophet Jonah, who was swallowed by a whale when he refused to bring Yahweh's message to Nineveh.

Layard had read in ancient Greek records that an officer in Alexander's army saw a "place of pyramids and remains of an ancient city"—a city that was already buried in Alexander's time! Layard dug it up, too, and it turned out to be Nimrud, Assyria's military center. It was there that Shalmaneser II set up an obelisk to record his military expeditions and conquests. Now on exhibit at the British Museum, the obelisk lists, among the kings who were made to pay tribute, "Jehu, son of Omri, king of Israel."

Again, the Mesopotamian inscriptions and biblical texts supported each other!

Astounded by increasingly frequent corroboration of the biblical narratives by archaeological finds, the Assyriologists, as these scholars came to be called, turned to the tenth chapter of the Book of Genesis. There Nimrod—"a mighty hunter by the grace of Yahweh"—was described as the founder of all the kingdoms of Mesopotamia.

And the beginning of his kingdom:
Babel and Erech and Akkad, all in the Land of Shin'ar.
Out of that Land there emanated Ashur where
    Nineveh was built, a city of wide streets;
and Khalah, and Ressen—the great city
    which is between Nineveh and Khalah.

There were indeed mounds the natives called Calah, lying between Nineveh and Nimrud. When teams under W. Andrae excavated the area from 1903 to 1914, they uncovered the ruins of Ashur, the Assyrian religious center and its earliest capital. Of all the Assyrian cities mentioned in the Bible, only Ressen remains to be found. The name means "horse's bridle"; perhaps it was the location of the royal stables of Assyria.

At about the same time as Ashur was being excavated, teams under R. Koldewey were completing the excavation of Babylon, the biblical Babel—a vast place of palaces, temples, hanging gardens, and the inevitable ziggurat. Before long, artifacts and inscriptions unveiled the history of the two competing empires of Mesopotamia: Babylonia and Assyria, the one centered in the south, the other in the north.

Rising and falling, fighting and coexisting, the two constituted a high civilization that encompassed some 1,500 years, both rising circa 1900 B.C.. Ashur and Nineveh were finally captured and destroyed by the Babylonians in 614 and 612 B.C., respectively. As predicted by the biblical prophets, Babylon itself came to an inglorious end when Cyrus the Achaemenid conquered it in 539 B.C.

Though they were rivals throughout their history, one would be hard put to find any significant differences between Assyria and Babylonia in cultural or material matters. Even though Assyria called its chief deity Ashur ("all-seeing") and Babylonia hailed Marduk ("son of the pure mound"), the pantheons were otherwise virtually alike.

Many of the world's museums count among their prize exhibits the ceremonial gates, winged bulls, bas-reliefs, chariots, tools, utensils, jewelry, statues, and other objects made of every conceivable material that have been dug out of the mounds of Assyria and Babylonia. But the true treasures of these kingdoms were their written records: thousands upon thousands of inscriptions in the cuneiform script, including cosmologic tales, epic poems, histories of kings, temple records, commercial contracts, marriage and divorce records, astronomical tables, astrological forecasts, mathematical formulas, geographic lists, grammar and vocabulary school texts, and, not least of all, texts dealing with the names, genealogies, epithets, deeds, powers, and duties of the gods.

The common language that formed the cultural, historical, and religious bond between Assyria and Babylonia was Akkadian. It was the first known Semitic language, akin to but predating Hebrew, Aramaic, Phoenician, and Canaanite. But the Assyrians and Babylonians laid no claim to having invented the language or its script; indeed, many of their tablets bore the postscript that they had been copied from earlier originals.

Who, then, invented the cuneiform script and developed the language, its precise grammar and rich vocabulary? Who wrote the "earlier originals"? And why did the Assyrians and Babylonians call the language Akkadian?

Attention once more focuses on the Book of Genesis. "And the beginning of his kingdom: Babel and Erech and Akkad." Akkad—could there really have been such a royal capital, preceding Babylon and Nineveh?

The ruins of Mesopotamia have provided conclusive evidence that once upon a time there indeed existed a kingdom by the name of Akkad, established by a much earlier ruler, who called himself a *sharrukin* ("righteous ruler"). He claimed in his inscriptions that his empire stretched, by the grace of his god Enlil, from the Lower Sea (the Persian Gulf) to the Upper Sea (believed to be the Mediterranean). He boasted that "at the wharf of Akkad, he made moor ships" from many distant lands.

The scholars stood awed: They had come upon a Mesopotamian empire in the third millennium B.C.! There was a leap—backward—of some 2,000 years from the Assyrian Sargon of Dur Sharrukin to Sargon of Akkad. And yet the mounds that were dug up brought to light literature and art, science and politics, commerce and communications—a full-fledged civilization—long before the appearance of Babylonia and Assyria. Moreover, it was obviously the predecessor and the source of the later Mesopotamian civilizations; Assyria and Babylonia were only branches off the Akkadian trunk.

The mystery of such an early Mesopotamian civilization deepened, however, as inscriptions recording the achievements and genealogy of Sargon of Akkad were found. They stated that his full title was "King of Akkad, King of Kish"; they explained that before he assumed the throne, he had been a counselor to the "rulers of Kish." Was there, then—the scholars asked themselves—an even earlier kingdom, that of Kish, which preceded Akkad?

Once again, the biblical verses gained in significance.

> And Kush begot Nimrod;
> He was first to be a Hero in the Land. . . .
> And the beginning of his kingdom:
> Babel and Erech and Akkad.

Many scholars have speculated that Sargon of Akkad was the biblical Nimrod. If one reads "Kish" for "Kush" in the above biblical verses, it would seem Nimrud was indeed preceded by Kish, as claimed by Sargon. The scholars then began to accept literally the rest of his inscriptions: "He defeated Uruk and tore down its wall . . . he was victorious in the battle with the inhabitants of Ur . . . he defeated the entire territory from Lagash as far as the sea."

Was the biblical Erech identical with the Uruk of Sargon's inscriptions? As the site now called Warka was unearthed, that was found to be the case. And the Ur referred to by Sargon was none other than the biblical Ur, the Mesopotamian birthplace of Abraham.

Not only did the archaeological discoveries vindicate the biblical records; it also appeared certain that there must have been kingdoms and cities and civilizations in Mesopotamia even before the third millennium B.C. The only question was: How far back did one have to go to find the *first* civilized kingdom?

The key that unlocked the puzzle was yet another language.

•

Scholars quickly realized that names had a meaning not only in Hebrew and in the Old Testament but throughout the ancient Near East. All the Akkadian, Babylonian, and Assyrian names of persons and places had a meaning. But the names of rulers that preceded Sargon of Akkad did not make sense at all: The king at whose court Sargon was a counselor was called Urzababa; the king who reigned in Erech was named Lugalzagesi; and so on.

Lecturing before the Royal Asiatic Society in 1853, Sir Henry Rawlinson pointed out that such names were neither Semitic nor Indo-European; indeed, "they seemed to belong to no known group of languages or peoples." But if names had a meaning, what was the mysterious language in which they had the meaning?

Scholars took another look at the Akkadian inscriptions. Basically, the Akkadian cuneiform script was syllabic: Each sign stood for a complete syllable (*ab, ba, bat,* etc.). Yet the script made extensive use of signs that were not phonetic syllables but conveyed the meanings "god," "city," "country," or "life," "exalted," and the like. The only possible explanation for this phenomenon was that these signs were remains of an earlier writing method which used pictographs. Akkadian, then, must have been preceded by another language that used a writing method akin to the Egyptian hieroglyphs.

It was soon obvious that an earlier language, and not just an earlier form of writing, was involved here. Scholars found that Akkadian inscriptions and texts made extensive use of loanwords—words borrowed intact from another language (in the same way that a modern Frenchman would borrow the English word *weekend*). This was especially true where scientific or technical terminology was involved, and also in matters dealing with the gods and the heavens.

One of the greatest finds of Akkadian texts was the ruins of a library assembled in Nineveh by Ashurbanipal; Layard and his colleagues carted away from the site 25,000 tablets, many of which were described by the ancient scribes as copies of "olden texts." A group of twenty-three tablets ended with the statement: "23rd tablet: language of Shumer not changed." Another text bore an enigmatic statement by Ashurbanipal himself:

> The god of scribes has bestowed on me the gift of
> the knowledge of his art.
> I have been initiated into the secrets of writing.
> I can even read the intricate tablets in Shumerian;
> I understand the enigmatic words in the stone
> carvings from the days before the Flood.

The claim by Ashurbanipal that he could read intricate tablets in "Shumerian" and understand the words written on tablets from "the days before the Flood" only increased the mystery. But in January 1869 Jules Oppert suggested to the French Society of Numismatics and Archaeology that recognition be given to the existence of a pre-Akkadian language and people. Pointing out that the early rulers of Mesopotamia proclaimed their legitimacy by taking the title "King of Sumer and Akkad," he suggested that the people be called "Sumerians," and their land, "Sumer."

Except for mispronouncing the name—it should have been *Sh*umer, not Sumer—Oppert was right. Sumer was not a mysterious, distant land, but the early name for southern Mesopotamia, just as the Book of Genesis had clearly stated: The royal cities of Babylon and Akkad and Erech were in "the Land of Shin'ar." (Shinar was the biblical name for Shumer.)

Once the scholars had accepted these conclusions, the

flood gates were opened. The Akkadian references to the "olden texts" became meaningful, and scholars soon realized that tablets with long columns of words were in fact Akkadian-Sumerian lexicons and dictionaries, prepared in Assyria and Babylonia for their own study of the first written language, Sumerian.

Without these dictionaries from long ago, we would still be far from being able to read Sumerian. With their aid, a vast literary and cultural treasure opened up. It also became clear that the Sumerian script, originally pictographic and carved in stone in vertical columns, was then turned horizontally and, later on, stylized for wedge writing on soft clay tablets to become the cuneiform writing that was adopted by the Akkadians, Babylonians, Assyrians, and other nations of the ancient Near East. (Fig. 7)

The decipherment of the Sumerian language and script, and the realization that the Sumerians and their culture were the fountainhead of the Akkadian–Babylonian–Assyrian achievements, spurred archaeological searches in southern Mesopotamia. All the evidence now indicated that the beginning was there.

The first significant excavation of a Sumerian site was begun in 1877 by French archaeologists; and the finds from this single site were so extensive that others continued to dig there until 1933 without completing the job.

Called by the natives Telloh ("mound"), the site proved to be an early Sumerian city, the very Lagash of whose conquest Sargon of Akkad had boasted. It was indeed a royal city whose rulers bore the same title Sargon had adopted, except that it was in the Sumerian language: EN.SI ("righteous ruler"). Their dynasty had started circa 2900 B.C. and lasted for nearly 650 years. During this time, forty-three *ensi*'s reigned without interruption in Lagash: Their names, genealogies, and lengths of rule were all neatly recorded.

The inscriptions provided much information. Appeals to the gods "to cause the grain sprouts to grow for harvest . . . to cause the watered plant to yield grain," attest to the existence of agriculture and irrigation. A cup inscribed in honor of a goddess by "the overseer of the granary" indicated that grains were stored, measured, and traded. (Fig. 8)

SUMERIAN			CUNEIFORM		Pronun-	Meaning
Original	Turned	Archaic	Common	Assyrian	ciation	
					KI	Earth Land
					KUR	Mountain
					LU	Domestic Man
					SAL MUNUZ	Vulva Woman
					SAG	Head
					A	Water
					NAG	Drink
					DU	Go
					HA	Fish
					GUD	Ox Bull Strong
					SHE	Barley

Fig. 7

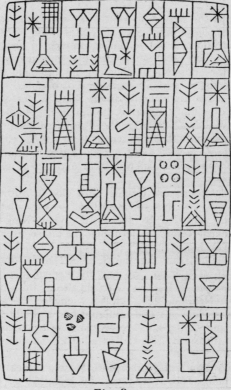

Fig. 8

An *ensi* named Eannatum left an inscription on a clay
brick which makes it clear that these Sumerian rulers could
assume the throne only with the approval of the gods. He
also recorded the conquest of another city, revealing to us
the existence of other city-states in Sumer at the beginning
of the third millennium B.C.

Eannatum's successor, Entemena, wrote of building a
temple and adorning it with gold and silver, planting

gardens, enlarging brick-lined wells. He boasted of building a fortress with watchtowers and facilities for docking
ships.

One of the better-known rulers of Lagash was Gudea.
He had a large number of statuettes made of himself, all
showing him in a votive stance, praying to his gods. This
stance was no pretense: Gudea had indeed devoted himself to the adoration of Ningirsu, his principal deity, and
to the construction and rebuilding of temples.

His many inscriptions reveal that, in the search for
exquisite building materials, he obtained gold from Africa
and Anatolia, silver from the Taurus Mountains, cedars
from Lebanon, other rare woods from Ararat, copper from
the Zagros range, diorite from Egypt, carnelian from
Ethiopia, and other materials from lands as yet unidentified
by scholars.

When Moses built for the Lord God a "Residence" in
the desert, he did so according to very detailed instructions provided by the Lord. When King Solomon
built the first Temple in Jerusalem, he did so after the Lord
had "given him wisdom." The prophet Ezekiel was shown
very detailed plans for the Second Temple "in a Godly
vision" by a "person who had the appearance of bronze
and who held in his hand a flaxen string and a measuring
rod." Ur-Nammu, ruler of Ur, depicted in an earlier
millennium how his god, ordering him to build for him a
temple and giving him the pertinent instructions, handed
him the measuring rod and rolled string for the job. (Fig.
9)

Twelve hundred years before Moses, Gudea made the
same claim. The instructions, he recorded in one very
long inscription, were given to him in a vision. "A man
that shone like the heaven," by whose side stood "a
divine bird," "commanded me to build his temple." This
"man," who "from the crown on his head was obviously a
god," was later identified as the god Ningirsu. With him
was a goddess who "held the tablet of her favorable star
of the heavens"; her other hand "held a holy stylus," with
which she indicated to Gudea "the favorable planet." A
third man, also a god, held in his hand a tablet of
precious stone; "the plan of a temple it contained." One of
Gudea's statues shows him seated, with this tablet on his

knees; on the tablet the divine drawing can clearly be seen. (Fig. 10)

Wise as he was, Gudea was baffled by these architectural instructions, and he sought the advice of a goddess who could interpret divine messages. She explained to him the meaning of the instructions, the plan's measurements, and the size and shape of the bricks to be used. Gudea then employed a male "diviner, maker of decisions" and a female "searcher of secrets" to locate the site, on the city's outskirts, where the god wished his temple to be built. He then recruited 216,000 people for the construction job.

Gudea's bafflement can readily be understood, for the simple-looking "floor plan" supposedly gave him the necessary information to build a complex ziggurat, rising high by seven stages. Writing in *Der Alte Orient* in 1900, A. Billerbeck was able to decipher at least part of the divine architectural instructions. The ancient drawing, even on the partly damaged statue, is accompanied at the top by groups of vertical lines whose number diminishes as the space between them increases. The divine architects, it appears, were able to provide, with a single floor plan, accompanied by seven varying scales, the complete instructions for the construction of a seven-stage high-rise temple. (Fig. 11)

It has been said that war spurs Man to scientific and material breakthroughs. In ancient Sumer, it seems, temple construction spurred the people and their rulers into greater technological achievements. The ability to carry out major construction work according to prepared architectural plans, to organize and feed a huge labor force, to flatten land and raise mounds, to mold bricks and transport stones, to bring rare metals and other materials from afar, to cast metal and shape utensils and oranaments—all clearly speak of a high civilization, already in full bloom in the third millennium B.C.

•

As masterful as even the earliest Sumerian temples were, they represented but the tip of the iceberg of the scope and richness of the material achievements of the first great civilization known to Man.

In addition to the invention and development of writ-

Fig. 9

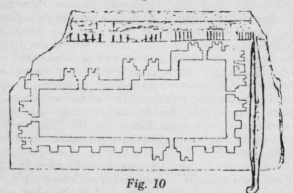

Fig. 10

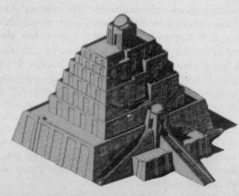

Fig. 11

ing, without which a high civilization could not have come about, the Sumerians should also be credited with the invention of printing. Millennia before Johann Gutenberg "invented" printing by using movable type, Sumerian scribes used ready-made "type" of the various pictographic signs, which they used as we now use rubber stamps to impress the desired sequence of signs in the wet clay.

They also invented the forerunner of our rotary presses —the cylinder seal. Made of extremely hard stone, it was a small cylinder into which the message or design had been engraved in reverse; whenever the seal was rolled on the wet clay, the imprint created a "positive" impression on the clay. The seal also enabled one to assure the authenticity of documents; a new impression could be made at once to compare it with the old impression on the document. (Fig. 12)

Many Sumerian and Mesopotamian written records concerned themselves not necessarily with the divine or spiritual but with such daily tasks as recording crops, measuring fields, and calculating prices. Indeed, no high civilization would have been possible without a parallel advanced system of mathematics.

The Sumerian system, called sexagesimal, combined a mundane 10 with a "celestial" 6 to obtain the base figure 60. This system is in some respects superior to our present one; in any case, it is unquestionably superior to later Greek and Roman systems. It enabled the Sumerians to divide into fractions and multiply into the millions, to calculate roots or raise numbers several powers. This was not only the first-known mathematical system but also one that gave us the "place" concept: Just as, in the decimal system, 2 can be 2 or 20 or 200, depending on the digit's place, so could a Sumerian 2 mean 2 or 120 (2×60), and so on, depending on the "place." (Fig. 13)

The 360-degree circle, the foot and its 12 inches, and the "dozen" as a unit are but a few examples of the vestiges of Sumerian mathematics still evident in our daily life. Their concomitant achievements in astronomy, the establishment of a calendar, and similar mathematical-celestial feats will receive much closer study in coming chapters.

Just as our own economic and social system—our books,

Fig. 12

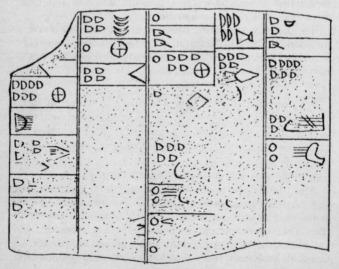

Fig. 13

court and tax records, commercial contracts, marriage certificates, and so on—depends on paper, Sumerian/ Mesopotamian life depended on clay. Temples, courts, and trading houses had their scribes ready with tablets of wet clay on which to inscribe decisions, agreements, letters, or calculate prices, wages, the area of a field, or the number of bricks required in a construction.

Clay was also a crucial raw material for the manufacture of utensils for daily use and containers for storage and transportation of goods. It was also used to make bricks— another Sumerian "first," which made possible the building of houses for the people, palaces for the kings, and imposing temples for the gods.

The Sumerians are credited with two technological breakthroughs that made it possible to combine lightness with tensile strength for all clay products: reinforcing and firing. Modern architects have discovered that reinforced concrete, an extremely strong building material, can be created by pouring cement into molds containing iron rods; long ago, the Sumerians gave their bricks great strength by mixing the wet clay with chopped reeds or straw. They also knew that clay products could be given tensile strength and durability by firing them in a kiln. The world's first high-rise buildings and archways, as well as durable ceramic wares, were made possible by these technological breakthroughs.

•

The invention of the kiln—a furnace in which intense but controllable temperatures could be attained without the risk of contaminating products with dust or ashes— made possible an even greater technological advance: the Age of Metals.

It has been assumed that man discovered that he could hammer "soft stones"—naturally occurring nuggets of gold as well as copper and silver compounds—into useful or pleasing shapes, sometime about 6000 B.C. The first hammered-metal artifacts were found in the highlands of the Zagros and Taurus mountains. However, as R. J. Forbes (*The Birthplace of Old World Metallurgy*) pointed out, "in the ancient Near East, the supply of native copper was quickly exhausted, and the miner had to turn to ores." This required the knowledge and ability to find and extract the ores, crush them, then smelt and refine them—processes that could not have been carried out without kiln-type furnaces and a generally advanced technology.

The art of metallurgy soon encompassed the ability to alloy copper with other metals, resulting in a castable, hard, but malleable metal we call bronze. The Bronze Age, our first metallurgical age, was also a Mesopotamian contribution to modern civilization. Much of ancient commerce was devoted to the metals trade; it also formed the basis for the development in Mesopotamia of banking and the first money—the silver *shekel* ("weighed ingot").

The many varieties of metals and alloys for which Sumerian and Akkadian names have been found and the extensive technological terminology attest to the high level of metallurgy in ancient Mesopotamia. For a while this puzzled the scholars because Sumer, as such, was devoid of metal ores, yet metallurgy most definitely began there.

The answer is energy. Smelting, refining, and alloying, as well as casting, could not be done without ample supplies of fuels to fire the kilns, crucibles, and furnaces. Mesopotamia may have lacked ores, but it had fuels in abundance. So the ores were brought to the fuels, which explains many early inscriptions describing the bringing of metal ores from afar.

The fuels that made Sumer technologically supreme were bitumens and asphalts, petroleum products that naturally seeped up to the surface in many places in Mesopotamia. R. J. Forbes (*Bitumen and Petroleum in Antiquity*) shows that the surface deposits of Mesopotamia were the ancient world's prime source of fuels from the earliest times to the Roman era. His conclusion is that the technological use of these petroleum products began in Sumer circa 3500 B.C.; indeed, he shows that the use and knowledge of the fuels and their properties were greater in Sumerian times than in later civilizations.

So extensive was the Sumerian use of these petroleum products—not only as fuel but also as road-building materials, for waterproofing, caulking, painting, cementing, and molding—that when archaeologists searched for ancient Ur they found it buried in a mound that the local

Arabs called "Mound of Bitumen." Forbes shows that the Sumerian language had terms for every genus and variant of the bituminous substances found in Mesopotamia. Indeed, the names of bituminous and petroleum materials in other languages—Akkadian, Hebrew, Egyptian, Coptic, Greek, Latin, and Sanskrit—can clearly be traced to the Sumerian origins; for example, the most common word for petroleum—*naphta*—derives from *napatu* ("stones that flare up").

The Sumerian use of petroleum products was also basic to an advanced chemistry. We can judge the high level of Sumerian knowledge not only by the variety of paints and pigments used and such processes as glazing but also by the remarkable artificial production of semiprecious stones, including a substitute for lapis lazuli.

•

Bitumens were also used in Sumerian medicine, another field where the standards were impressively high. The hundreds of Akkadian texts that have been found employ Sumerian medical terms and phrases extensively, pointing to the Sumerian origin of all Mesopotamian medicine.

The library of Ashurbanipal in Nineveh included a medical section. The texts were divided into three groups—*bultitu* ("therapy"), *shipir bel imti* ("surgery") and *urti mashmashshe* ("commands and incantations"). Early law codes included sections dealing with fees payable to surgeons for successful operations, and penalties to be imposed on them in case of failure: A surgeon, using a lancet to open a patient's temple, was to lose his hand if he accidentally destroyed the patient's eye.

Some skeletons found in Mesopotamian graves bore unmistakable marks of brain surgery. A partially broken medical text speaks of the surgical removal of a "shadow covering a man's eye," probably a cataract; another text mentions the use of a cutting instrument, stating that "if the sickness has reached the inside of the bone, you shall scrape and remove."

Sick persons in Sumerian times could choose between an A.ZU ("water physician") and an IA.ZU ("oil physician"). A tablet excavated in Ur, nearly 5,000 years old, names a medical practitioner as "Lulu, the doctor." There were also veterinarians—known either as "doctors of oxen" or as "doctors of asses."

A pair of surgical tongs is depicted on a very early cylinder seal, found at Lagash, that belonged to "Urlugaledina, the doctor." The seal also shows the serpent on a tree—the symbol of medicine to this day. (Fig. 14) An instrument that was used by midwives to cut the umbilical cord was also frequently depicted.

Fig. 14

Sumerian medical texts deal into diagnosis and prescriptions. They leave no doubt that the Sumerian physician did not resort to magic or sorcery. He recommended cleaning and washing; soaking in baths of hot water and mineral solvents; application of vegetable derivatives; rubbing with petroleum compounds.

Medicines were made from plant and mineral compounds and were mixed with liquids or solvents appropriate to the method of application. If taken by mouth, the powders were mixed into wine, beer, or honey; if "poured through the rectum"—administered in an enema—they were mixed with plant or vegetable oils. Alcohol, which plays such an important role in surgical disinfection and as a base for many medicines, reached our languages through the Arabic *kohl*, from the Akkadian *kuhlu*.

Models of livers indicate that medicine was taught at medical schools with the aid of clay models of human organs. Anatomy must have been an advanced science, for temple rituals called for elaborate dissections of sacrificial animals—only a step removed from comparable knowledge of human anatomy.

Several depictions on cylinder seals or clay tablets show people lying on some kind of surgical table, surrounded by teams of gods or people. We know from epics and other heroic texts that the Sumerians and their successors in Mesopotamia were concerned with matters of life, sickness, and death. Men like Gilgamesh, a king of Erech, sought the "Tree of Life" or some mineral (a "stone") that could provide eternal youth. There were also references to efforts to resurrect the dead, especially if they happened to be gods:

> Upon the corpse, hung from the pole,
> they directed the Pulse and the Radiance;
> Sixty times the Water of Life,
> Sixty times the Food of Life,
> they sprinkled upon it;
> And Inanna arose.

Were some ultramodern methods, about which we can only speculate, known and used in such revival attempts? That radioactive materials were known and used to treat certain ailments is certainly suggested by a scene of medical treatment depicted on a cylinder seal dating to the very beginning of Sumerian civilization. It shows, without question, a man lying on a special bed; his face is protected by a mask, and he is being subjected to some kind of radiation. (Fig. 15)

One of Sumer's earliest material achievements was the development of textile and clothing industries.

Our own Industrial Revolution is considered to have commenced with the introduction of spinning and weaving machines in England in the 1760s. Most developing nations have aspired ever since to develop a textile industry as the first step toward industrialization. The evidence shows that

Fig. 15

this has been the process not only since the eighteenth
century but ever since man's first great civilization. Man
could not have made woven fabrics before the advent of
agriculture, which provided him with flax, and the domestication of animals, creating a source for wool. Grace M.
Crowfoot (*Textiles, Basketry and Mats in Antiquity*) expressed the scholastic consensus by stating that textile
weaving appeared first in Mesopotamia, around 3800 B.C.

Sumer, moreover, was renowned in ancient times not
only for its woven fabrics, but also for its apparel. The
Book of Joshua (7:21) reports that during the storming
of Jericho a certain person could not resist the temptation
to keep "one good coat of Shin'ar," which he had found
in the city, even though the penalty was death. So highly
prized were the garments of Shinar (Sumer), that people
were willing to risk their lives to obtain them.

A rich terminology already existed in Sumerian times to
describe both items of clothing and their makers. The basic
garment was called TUG—without doubt, the forerunner
in style as well as in name of the Roman toga. Such
garments were TUG.TU.SHE, which in Sumerian meant
"garment which is worn wrapped around." (Fig. 16)

The ancient depictions reveal not only an astonishing
variety and opulence in matters of clothing, but also ele-

gance, in which good taste and coordination among clothes, hairdos, headdresses, and jewelry prevailed. (Figs. 17, 18)

•

Another major Sumerian achievement was its agriculture. In a land with only seasonal rains, the rivers were enlisted to water year-round crops through a vast system of irrigation canals.

Mesopotamia—the Land Between the Rivers—was a veritable food basket in ancient times. The apricot tree, the Spanish word for which is *damasco* ("Damascus tree"), bears the Latin name *armeniaca*, a loanword from the Akkadian *armanu*. The cherry—*kerasos* in Greek, *Kirsche* in German—originates from the Akkadian *karshu*. All the evidence suggests that these and other fruits and vegetables reached Europe from Mesopotamia. So did many special seeds and spices: Our word *saffron* comes from the Akkadian *azupiranu*, *crocus* from *kurkanu* (via *krokos* in Greek), *cumin* from *kamanu*, *hyssop* from *zupu*, *myrrh* from *murru*. The list is long; in many instances, Greece provided the physical and etymological bridge by which these products of the land reached Europe. Onions, lentils, beans, cucumbers, cabbage, and lettuce were common ingredients of the Sumerian diet.

What is equally impressive is the extent and variety of the ancient Mesopotamian food-preparation methods, their cuisine. Texts and pictures confirm the Sumerian knowledge of converting the cereals they had grown into flour, from which they made a variety of leavened and unleavened breads, porridges, pastries, cakes, and biscuits. Barley was also fermented to produce beer; "technical manuals" for beer production have been found among the texts. Wine was obtained from grapes and from date palms. Milk was available from sheep, goats, and cows; it was used as a beverage, for cooking, and for converting into yogurt, butter, cream, and cheeses. Fish was a common part of the diet. Mutton was readily available, and the meat of pigs, which the Sumerians tended in large herds, was considered a true delicacy. Geese and ducks may have been reserved for the gods' tables.

The ancient texts leave no doubt that the haute cuisine

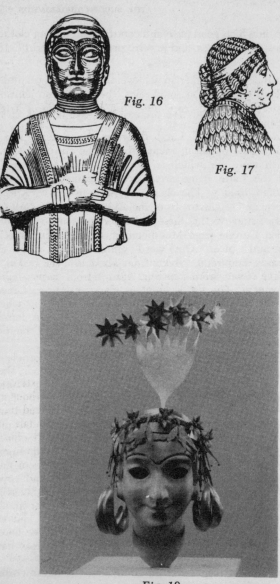

Fig. 16

Fig. 17

Fig. 18

of ancient Mesopotamia developed in the temples and in the service of the gods. One text prescribed the offering to the gods of "loaves of barley bread . . . loaves of emmer bread; a paste of honey and cream; dates, pastry . . . beer, wine, milk . . . cedar sap, cream." Roasted meat was offered with libations of "prime beer, wine, and milk." A specific cut of a bull was prepared according to a strict recipe, calling for "fine flour . . . made to a dough in water, prime beer, and wine," and mixed with animal fats, "aromatic ingredients made from hearts of plants," nuts, malt, and spices. Instructions for "the daily sacrifice to the gods of the city of Uruk" called for the serving of five different beverages with the meals, and specified what "the millers in the kitchen" and "the chef working at the kneading trough" should do.

Our admiration for the Sumerian culinary art certainly grows as we come across poems that sing the praises of fine foods. Indeed, what can one say when one reads a millennia-old recipe for "coq au vin":

> In the wine of drinking,
> In the scented water,
> In the oil of unction—
> This bird have I cooked,
> and have eaten.

A thriving economy, a society with such extensive material enterprises could not have developed without an efficient system of transportation. The Sumerians used their two great rivers and the artificial network of canals for waterborne transportation of people, goods, and cattle. Some of the earliest depictions show what were undoubtedly the world's first boats.

We know from many early texts that the Sumerians also engaged in deep-water seafaring, using a variety of ships to reach faraway lands in search of metals, rare woods and stones, and other materials unobtainable in Sumer proper. An Akkadian dictionary of the Sumerian language was found to contain a section on shipping, listing 105 Sumerian terms for various ships by their size, destination, or purpose (for cargo, for passengers, or for the exclusive use of certain gods). Another 69 Sumerian terms connected with the manning and construction of ships were translated into the Akkadian. Only a long seafaring tradition could have produced such specialized vessels and technical terminology.

For overland transportation, the wheel was first used in Sumer. Its invention and introduction into daily life made possible a variety of vehicles, from carts to chariots, and no doubt also granted Sumer the distinction of having been the first to employ "ox power" as well as "horse power" for locomotion. (Fig. 19)

•

In 1956 Professor Samuel N. Kramer, one of the great Sumerologists of our time, reviewed the literary legacy found beneath the mounds of Sumer. The table of contents of *From the Tablets of Sumer* is a gem in itself, for each one of the twenty-five chapters described a Sumerian "first," including the first schools, the first bicameral congress, the first historian, the first pharmacopoeia, the first "farmer's almanac," the first cosmogony and cosmology, the first "Job," the first proverbs and sayings, the first literary debates, the first "Noah," the first library catalogue; and Man's first Heroic Age, his first law codes and social reforms, his first medicine, agriculture, and search for world peace and harmony.

Fig. 19

This is no exaggeration.

The first schools were established in Sumer as a direct outgrowth of the invention and introduction of writing. The evidence (both archaeological, such as actual school buildings, and written, such as exercise tablets) indicates the existence of a formal system of education by the beginning of the third millennium B.C. There were literally thousands of scribes in Sumer, ranging from junior scribes to high scribes, royal scribes, temple scribes, and scribes who assumed high state office. Some acted as teachers at the schools, and we can still read their essays on the schools, their aims and goals, their curriculum and teaching methods.

The schools taught not only language and writing but also the sciences of the day—botany, zoology, geography, mathematics, and theology. Literary works of the past were studied and copied, and new ones were composed.

The schools were headed by the *ummia* ("expert professor"), and the faculty invariably included not only a "man in charge of drawing" and a "man in charge of Sumerian," but also a "man in charge of the whip." Apparently, discipline was strict; one school alumnus described on a clay tablet how he had been flogged for missing school, for insufficient neatness, for loitering, for not keeping silent, for misbehaving, and even for not having neat handwriting.

An epic poem dealing with the history of Erech concerns itself with the rivalry between Erech and the city-state of Kish. The epic text relates how the envoys of Kish proceeded to Erech, offering a peaceful settlement of their dispute. But the ruler of Erech at the time, Gilgamesh, preferred to fight rather than negotiate. What is interesting is that he had to put the matter to a vote in the Assembly of the Elders, the local "Senate":

> The lord Gilgamesh,
> Before the elders of his city put the matter,
> Seeks out the decision:
> "Let us not submit to the house of Kish,
> let us smite it with weapons."

The Assembly of the Elders was, however, for negotiations. Undaunted, Gilgamesh took the matter to the younger people, the Assembly of the Fighting Men, who voted for war. The significance of the tale lies in its disclosure that a Sumerian ruler had to submit the question of war or peace to the First Bicameral Congress, some 5,000 years ago.

The title of First Historian was bestowed by Kramer on Entemena, king of Lagash, who recorded on clay cylinders his war with neighboring Umma. While other texts were literary works or epic poems whose themes were historical events, the inscriptions by Entemena were straight prose, written solely as a factual record of history.

Because the inscriptions of Assyria and Babylonia were deciphered well before the Sumerian records, it was long believed that the first code of laws was compiled and decreed by the Babylonian king Hammurabi, circa 1900 B.C. But as Sumer's civilization was uncovered, it became clear that the "firsts" for a system of laws, for concepts of social order, and for the fair administration of justice belonged to Sumer.

Well before Hammurabi, a Sumerian ruler of the city-state of Eshnunna (northeast of Babylon) encoded laws that set maximum prices for foodstuffs and for the rental of wagons and boats so that the poor could not be oppressed. There were also laws dealing with offenses against person and property, and regulations pertaining to family matters and to master–servant relations.

Even earlier, a code was promulgated by Lipit-Ishtar, a ruler of Isin. The thirty-eight laws that remain legible on the partly preserved tablet (a copy of an original that was engraved on a stone stela) deal with real estate, slaves and servants, marriage and inheritance, the hiring of boats, the rental of oxen, and defaults on taxes. As was done by Hammurabi after him, Lipit-Ishtar explained in the prologue to his code that he acted on the instructions of "the great gods," who had ordered him "to bring well-being to the Sumerians and the Akkadians."

Yet even Lipit-Ishtar was not the first Sumerian law encoder. Fragments of clay tablets that have been found contain copies of laws encoded by Urnammu, a ruler of Ur circa 2350 B.C.—more than half a millennium before

Hammurabi. The laws, enacted on the authority of the god Nannar, were aimed at stopping and punishing "the grabbers of the citizens' oxen, sheep, and donkeys" so that "the orphan shall not fall prey to the wealthy, the widow shall not fall prey to the powerful, the man of one shekel shall not fall prey to a man of 60 shekels." Urnammu also decreed "honest and unchangeable weights and measurements."

But the Sumerian legal system, and the enforcement of justice, go back even farther in time.

By 2600 B.C. so much must already have happened in Sumer that the *ensi* Urukagina found it necessary to institute reforms. A long inscription by him has been called by scholars a precious record of man's first social reform based on a sense of freedom, equality, and justice—a "French Revolution" imposed by a king 4,400 years before July 14, 1789.

The reform decree of Urukagina listed the evils of his time first, then the reforms. The evils consisted primarily of the unfair use by supervisors of their powers to take the best for themselves; the abuse of official status; the extortion of high prices by monopolistic groups.

All such injustices, and many more, were prohibited by the reform decree. An official could no longer set his own price "for a good donkey or a house." A "big man" could no longer coerce a common citizen. The rights of the blind, poor, widowed, and orphaned were restated. A divorced woman—nearly 5,000 years ago—was granted the protection of the law.

How long had Sumerian civilization existed that it required a major reform? Clearly, a long time, for Urukagina claimed that it was his god Ningirsu who called upon him "to restore the decrees of former days." The clear implication is that a return to even older systems and earlier laws was called for.

The Sumerian laws were upheld by a court system in which the proceedings and judgments as well as contracts were meticulously recorded and preserved. The justices acted more like juries than judges; a court was usually made up of three or four judges, one of whom was a professional "royal judge" and the others drawn from a panel of thirty-six men.

While the Babylonians made rules and regulations, the Sumerians were concerned with justice, for they believed that the gods appointed the kings primarily to assure justice in the land.

More than one parellel can be drawn here with the concepts of justice and morality of the Old Testament. Even before the Hebrews had kings, they were governed by judges; kings were judged not by their conquests or wealth but by the extent to which they "did the righteous thing." In the Jewish religion, the New Year marks a ten-day period during which the deeds of men are weighed and evaluated to determine their fate in the coming year. It is probably more than a coincidence that the Sumerians believed that a deity named Nanshe annually judged Mankind in the same manner; after all, the first Hebrew patriarch —Abraham—came from the Sumerian city of Ur, the city of Ur-Nammu and his code.

The Sumerian concern with justice or its absence also found expression in what Kramer called "the first 'Job.'" Matching together fragments of clay tablets at the Istanbul Museum of Antiquities, Kramer was able to read a good part of a Sumerian poem which, like the biblical Book of Job, dealt with the complaint of a righteous man who, instead of being blessed by the gods, was made to suffer all manner of loss and disrespect. "My righteous word has been turned into a lie," he cried out in anguish.

In its second part, the anonymous sufferer petitions his god in a manner akin to some verses in the Hebrew Psalms:

> My god, you who are my father,
> who begot me—lift up my face. . . .
> How long will you neglect me,
> leave me unprotected . . .
> leave me without guidance?

Then follows a happy ending. "The righteous words, the pure words uttered by him, his god accepted; . . . his god withdrew his hand from the evil pronouncement."

Preceding the biblical Book of Ecclesiastes by some two millennia, Sumerian proverbs conveyed many of the same concepts and witticisms.

If we are doomed to die—let us spend;
If we shall live long—let us save.

When a poor man dies, do not try to revive him.

He who possesses much silver, may be happy;
He who possesses much barley, may be happy;
But who has nothing at all, can sleep!

Man: For his pleasure: Marriage;
On his thinking it over: Divorce.

It is not the heart which leads to enmity;
it is the tongue which leads to enmity.

In a city without watchdogs,
the fox is the overseer.

The material and spiritual achievements of the Sumerian civilization were also accompanied by an extensive development of the performing arts. A team of scholars from the University of California at Berkeley made news in March 1974 when they announced that they had deciphered the world's oldest song. What professors Richard L. Crocker, Anne D. Kilmer, and Robert R. Brown achieved was to read and actually play the musical notes written on a cuneiform tablet from circa 1800 B.C., found at Ugarit on the Mediterranean coast (now in Syria).

"We always knew," the Berkeley team explained, "that there was music in the earlier Assyrio-Babylonian civilization, but until this deciphering we did not know that it had the same heptatonic-diatonic scale that is characteristic of contemporary Western music, and of Greek music of the first millennium B.C." Until now it was thought that Western music originated in Greece; now it has been established that our music—as so much else of Western civilization—originated in Mesopotamia. This should not be surprising, for the Greek scholar Philo had already stated that the Mesopotamians were known to "seek world-wide harmony and unison through the musical tones."

There can be no doubt that music and song must also be

claimed as a Sumerian "first." Indeed, Professor Crocker could play the ancient tune only by constructing a lyre like those which had been found in the ruins of Ur. Texts from the second millennium B.C. indicate the existence of musical "key numbers" and a coherent musical theory; and Professor Kilmer herself wrote earlier (*The Strings of Musical Instruments: Their Names, Numbers and Significance*) that many Sumerian hymnal texts had "what appear to be musical notations in the margins." "The Sumerians and their successors had a full musical life," she concluded. No wonder, then, that we find a great variety of musical instruments—as well as of singers and dancers performing—depicted on cylinder seals and clay tablets. (Fig. 20)

Fig. 20

Like so many other Sumerian achievements, music and song also originated in the temples. But, beginning in the service of the gods, these performing arts soon were also prevalent outside the temples. Employing the favorite

Sumerian play on words, a popular saying commented on the fees charged by singers: "A singer whose voice is not sweet is a 'poor' singer indeed."

Many Sumerian love songs have been found; they were undoubtedly sung to musical accompaniment. Most touching, however, is a lullaby that a mother composed and sang to her sick child:

> Come sleep, come sleep, come to my son.
> Hurry sleep to my son;
> Put to sleep his restless eyes. . . .
>
> You are in pain, my son;
> I am troubled, I am struck dumb,
> I gaze up to the stars.
> The new moon shines down on your face;
> Your shadow will shed tears for you.
> Lie, lie in your sleep. . . .
>
> May the goddess of growth be your ally;
> May you have an eloquent guardian in heaven;
> May you achieve a reign of happy days. . . .
> May a wife be your support;
> May a son be your future lot.

What is striking about such music and songs is not only the conclusion that Sumer was the source of Western music in structure and harmonic composition. No less significant is the fact that as we hear the music and read the poems, they do not sound strange or alien at all, even in their depth of feeling and their sentiments. Indeed, as we contemplate the great Sumerian civilization, we find that not only are *our* morals and *our* sense of justice, *our* laws and architecture and arts and technology rooted in Sumer, but the Sumerian institutions are so familiar, so close. At heart, it would seem, we are all Sumerians.

•

After excavating at Lagash, the archaeologist's spade uncovered Nippur, the onetime religious center of Sumer and Akkad. Of the 30,000 texts found there, many remain unstudied to this day. At Shuruppak, schoolhouses dating to the third millennium B.C. were found. At Ur, scholars found magnificent vases, jewelry, weapons, chariots, helmets made of gold, silver, copper, and bronze, the remains of a weaving factory, court records—and a towering ziggurat whose ruins still dominate the landscape. At Eshnunna and Adab the archaeologists found temples and artful statues from pre-Sargonic times. Umma produced inscriptions speaking of early empires. At Kish monumental buildings and a ziggurat from at least 3000 B.C. were unearthed.

Uruk (Erech) took the archaeologists back into the fourth millennium B.C. There they found the first colored pottery baked in a kiln, and evidence of the first use of a potter's wheel. A pavement of limestone blocks is the oldest stone construction found to date. At Uruk the archaeologists also found the first ziggurat—a vast man-made mound, on top of which stood a white temple and a red temple. The world's first inscribed texts were also found there, as well as the first cylinder seals. Of the latter, Jack Finegan (*Light from the Ancient Past*) said, "The excellence of the seals upon their first appearance in the Uruk period is amazing." Other sites of the Uruk period bear evidence of the emergence of the Metal Age.

In 1919, H. R. Hall came upon ancient ruins at a village now called El-Ubaid. The site gave its name to what scholars now consider the first phase of the great Sumerian civilization. Sumerian cities of that period—ranging from northern Mesopotamia to the southern Zagros foothills—produced the first use of clay bricks, plastered walls, mosaic decorations, cemeteries with brick-lined graves, painted and decorated ceramic wares with geometric designs, copper mirrors, beads of imported turquoise, paint for eyelids, copper-headed "tomahawks," cloth, houses, and, above all, monumental temple buildings.

Farther south, the archaeologists found Eridu—the first Sumerian city, according to ancient texts. As the excavators dug deeper, they came upon a temple dedicated to Enki, Sumer's God of Knowledge, which appeared to have been built and rebuilt many times over. The strata clearly led the scholars back to the beginnings of Sumerian civilization: 2500 B.C., 2800 B.C., 3000 B.C., 3500 B.C.

Then the spades came upon the foundations of the first temple dedicated to Enki. Below that, there was virgin soil—nothing had been built before. The time was circa 3800 B.C. That is when civilization began.

It was not only the first civilization in the true sense of the term. It was a most extensive civilization, all-encompassing, in many ways more advanced than the other ancient cultures that had followed it. It was undoubtedly the civilization on which our own is based.

Having begun to use stones as tools some 2,000,000 years earlier, Man achieved this unprecedented civilization in Sumer circa 3800 B.C. And the perplexing fact about this is that to this very day the scholars have no inkling who the Sumerians were, where they came from, and how and why their civilization appeared.

For its appearance was sudden, unexpected, and out of nowhere.

H. Frankfort *(Tell Uqair)* called it "astonishing." Pierre Amiet *(Elam)* termed it "extraordinary." A. Parrot *(Sumer)* described it as "a flame which blazed up so suddenly." Leo Oppenheim *(Ancient Mesopotamia)* stressed "the astonishingly short period" within which this civilization had arisen. Joseph Campbell *(The Masks of God)* summed it up in this way: "With stunning abruptness . . . there appears in this little Sumerian mud garden . . . the whole cultural syndrome that has since constituted the germinal unit of all the high civilizations of the world."

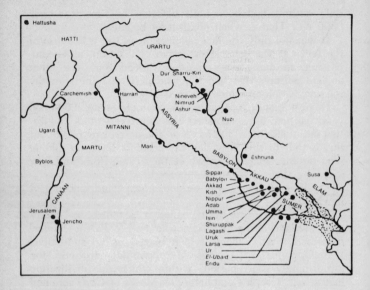

Napoleon conquers Europe
American Revolution

Columbus discovers America
Byzantine empire falls to Turks
Inca empire arises in South America
Aztec civilization in Mexico
Magna Carta granted by King John
Norman conquest of England

Charlemagne forms Holy Roman Empire

Muhammed proclaims Islam

Sack of Rome

Maya civilization in Central America

Jerusalem falls to Roman legions

Jesus of Nazareth

Hannibal challenges Rome
Great Wall begun in China
Alexander defeats Darius
Greek Classical age begins
Roman republic founded
Buddah rises in India
Cyrus captures Babylon
Fall of Nineveh

David king in Jerusalem
Dorian invasion of Greece
Israelite Exodus from Egypt

Mycenaean culture begins
Aryans migrate to India
Hittite empire rises
Abraham migrates from Ur
Hammurabi king in Babylon
Rise of Babylon & Assyria

Chinese civilization begins
Indus valley civilization
Hurrians arrive in Near East
Gudea rules in Lagash
Ur-Nammu rules in Ur

Sargon first king of Akkad
Minoan Civilization in Crete
Gilgamesh rules in Erech

Etana rules in Kish
Egyptian civilization begins

Kingship begins in Kish

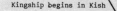
Sumerian civilization begins in Eridu

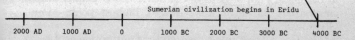

2000 AD   1000 AD   0   1000 BC   2000 BC   3000 BC   4000 BC

3
·
GODS OF HEAVEN AND EARTH

WHAT WAS IT that after hundreds of thousands and even millions of years of painfully slow human development abruptly changed everything so completely, and in a one–two–three punch—circa 11,000–7400–3800 B.C.—transformed primitive nomadic hunters and food gatherers into farmers and pottery makers, and then into builders of cities, engineers, mathematicians, astronomers, metallurgists, merchants, musicians, judges, doctors, authors, librarians, priests? One can go further and ask an even more basic question, so well stated by Professor Robert J. Braidwood *(Prehistoric Men)*: "Why did it happen at all? Why are all human beings not still living as the Maglemosians did?"

The Sumerians, the people through whom this high civilization so suddenly came into being, had a ready answer. It was summed up by one of the tens of thousands of ancient Mesopotamian inscriptions that have been uncovered: "Whatever seems beautiful, we made by the grace of the gods."

The gods of Sumer. Who were they?

Were the gods of the Sumerians like the Greek gods, who were described as living at a great court, feasting in the Great Hall of Zeus in the heavens—Olympus, whose counterpart on earth was Greece's highest peak, Mount Olympus?

The Greeks described their gods as anthropomorphic, as physically similar to mortal men and women, and human in character: They could be happy and angry and jealous;

they made love, quarreled, fought; and they procreated like humans, bringing forth offspring through sexual intercourse—with each other or with humans.

They were unreachable, and yet they were constantly mixed up in human affairs. They could travel at immense speeds, appear and disappear; they had weapons of immense and unusual power. Each had specific functions, and, as a result, a specific human activity could suffer or benefit by the attitude of the god in charge of that particular activity; therefore, rituals of worship and offerings to the gods were supposed to gain their favor.

The principal deity of the Greeks during their Hellenic civilization was Zeus, "Father of Gods and Men," "Master of the Celestial Fire." His chief weapon and symbol was the thunderbolt. He was a "king" upon earth who had descended from the heavens; a decision maker and the dispenser of good and evil to mortals, yet one whose original domain was in the skies.

He was neither the first god upon Earth nor the first deity to have been in the heavens. Mixing theology with cosmology to come up with what scholars treat as mythology, the Greeks believed that first there was Chaos; then Gaea (Earth) and her consort Uranus (the heavens) appeared. Gaea and Uranus brought forth the twelve Titans, six males and six females. Though their legendary deeds took place on Earth, it is assumed that they had astral counterparts.

Cronus, the youngest male Titan, emerged as the principal figure in Olympian mythology. He rose to supremacy among the Titans through usurpation, after castrating his father Uranus. Fearful of the other Titans, Cronus imprisoned and banished them. For that, he was cursed by his mother: He would suffer the same fate as his father, and be dethroned by one of his own sons.

Cronus consorted with his own sister Rhea, who bore him three sons and three daughters; Hades, Poseidon, and Zeus; Hestia, Demeter, and Hera. Once again, it was fated that the youngest son would be the one to depose his father, and the curse of Gaea came true when Zeus overthrew Cronus, his father.

The overthrow, it would seem, did not go smoothly. For many years battles between the gods and a host of

monstrous beings ensued. The decisive battle was between
Zeus and Typhon, a serpent-like deity. The fighting ranged
over wide areas, on Earth and in the skies. The final battle
took place at Mount Casius, near the boundary between
Egypt and Arabia—apparently somewhere in the Sinai
Peninsula. (Fig. 21)

Fig. 21

Having won the struggle, Zeus was recognized as the
supreme deity. Nevertheless, he had to share control with
his brothers. By choice (or, according to one version,
through the throwing of lots), Zeus was given control of
the skies, the eldest brother Hades was accorded the
Lower World, and the middle brother Poseidon was given
mastery of the seas.

Though in time Hades and his region became a synonym
for Hell, his original domain was a territory somewhere
"far below," encompassing marshlands, desolate areas, and
lands watered by mighty rivers. Hades was depicted as
"the unseen"—aloof, forbidding, stern; unmoved by prayer
or sacrifice. Poseidon, on the other hand, was frequently
seen holding up his symbol (the trident). Though ruler of
the seas, he was also master of the arts of metallurgy and
sculpting, as well as a crafty magician or conjurer. While

Zeus was depicted in Greek tradition and legend as strict with Mankind—even as one who at one point schemed to annihilate Mankind—Poseidon was considered a friend of Mankind and a god who went to great lengths to gain the praise of mortals.

The three brothers and their three sisters, all children of Cronus by his sister Rhea, made up the older part of the Olympian Circle, the group of Twelve Great Gods. The other six were all offspring of Zeus, and the Greek tales dealt mostly with their genealogies and relationships.

The male and female deities fathered by Zeus were mothered by different goddesses. Consorting at first with a goddess named Metis, Zeus had born to him a daughter, the great goddess Athena. She was in charge of common sense and handiwork, and was thus the Goddess of Wisdom. But as the only major deity to have stayed with Zeus during his combat with Typhon (all the other gods had fled), Athena acquired martial qualities and was also the Goddess of War. She was the "perfect maiden" and became no one's wife; but some tales link her frequently with her uncle Poseidon, and though his official consort was the goddess who was the Lady of the Labyrinth from the island of Crete, his niece Athena was his mistress.

Zeus then consorted with other goddesses, but their children did not qualify for the Olympian Circle. When Zeus got around to the serious business of producing a male heir, he turned to one of his own sisters. The eldest was Hestia. She was, by all accounts, a recluse—perhaps too old or too sick to be the object of matrimonial activities—and Zeus needed little excuse to turn his attentions to Demeter, the middle sister, the Goddess of Fruitfulness. But, instead of a son, she bore him a daughter, Persephone, who became wife to her uncle Hades and shared his dominion over the Lower World.

Disappointed that no son was born, Zeus turned to other goddesses for comfort and love. Of Harmonia he had nine daughters. Then Leto bore him a daughter and a son, Artemis and Apollo, who were at once drawn into the group of major deities.

Apollo, as firstborn son of Zeus, was one of the greatest gods of the Hellenic pantheon, feared by men and gods alike. He was the interpreter to mortals of the will of his father Zeus, and thus the authority in matters of religious law and temple worship. Representing moral and divine laws, he stood for purification and perfection, both spiritual and physical.

Zeus's second son, born of the goddess Maia, was Hermes, patron of shepherds, guardian of the flocks and herds. Less important and powerful than his brother Apollo, he was closer to human affairs; any stroke of good luck was attributed to him. As Giver of Good Things, he was the deity in charge of commerce, patron of merchants and travelers. But his main role in myth and epic was as herald of Zeus, Messenger of the Gods.

Impelled by certain dynastic traditions, Zeus still required a son by one of his sisters—and he turned to the youngest, Hera. Marrying her in the rites of a Sacred Marriage, Zeus proclaimed her Queen of the Gods, the Mother Goddess. Their marriage was blessed by a son, Ares, and two daughters, but rocked by constant infidelities on the part of Zeus, as well as a rumored infidelity on the part of Hera, which cast doubt on the true parentage of another son, Hephaestus.

Ares was at once incorporated into the Olympian Circle of twelve major gods and was made Zeus's chief lieutenant, a God of War. He was depicted as the Spirit of Carnage; yet he was far from being invincible—fighting at the battle of Troy, on the side of the Trojans, he suffered a wound which only Zeus could heal.

Hephaestus, on the other hand, had to fight his way into the Olympian summit. He was a God of Creativity; to him was attributed the fire of the forge and the art of metallurgy. He was a divine artificer, maker of both practical and magical objects for men and gods. The legends say that he was born lame and was therefore cast away in anger by his mother Hera. Another and more believable version has it that it was Zeus who banished Hephaestus —because of the doubt regarding his parentage—but Hephaestus used his magically creative powers to force Zeus to give him a seat among the Great Gods.

The legends also relate that Hephaestus once made an invisible net that would close over his wife's bed if it were warmed by an intruding lover. He may have needed such protection, for his wife and consort was Aphrodite, Goddess of Love and Beauty. It was only natural that many tales of love affairs would build up around her; in many of these the seducer was Ares, brother of Hephaestus. (One of the offspring of that illicit love affair was Eros, the God of Love.)

Aphrodite was included in the Olympian Circle of Twelve, and the circumstances of her inclusion shed light on our subject. She was neither a sister of Zeus nor his daughter, yet she could not be ignored. She had come from the Asian shores of the Mediterranean facing Greece (according to the Greek poet Hesiod, she arrived by way of Cyprus); and, claiming great antiquity, she ascribed her origin to the genitals of Uranus. She was thus genealogically one generation ahead of Zeus, being (so to say) a sister of his father, and the embodiment of the castrated Forefather of the Gods. (Fig. 22)

Aphrodite, then, had to be included among the Olympian gods. But their total number, twelve, apparently could not be exceeded. The solution was ingenious: Add one by dropping one. Since Hades was given domain over the Lower World and did not remain among the Great Gods on Mount Olympus, a vacancy was created, admirably handy for seating Aphrodite in the exclusive Circle of Twelve.

It also appears that the number twelve was a requirement that worked both ways: There could be no more than twelve Olympians, but no fewer than twelve, either. This becomes evident through the circumstances that led to the inclusion of Dionysus in the Olympian Circle. He was a son of Zeus, born when Zeus impregnated his own daughter, Semele. Dionysus, who had to be hidden from Hera's wrath, was sent to far-off lands (reaching even India), introducing vinegrowing and winemaking wherever he went. In the meantime, a vacancy became available on Olympus. Hestia, the oldest sister of Zeus, weaker and older, was dropped entirely from the Circle of Twelve. Dionysus then returned to Greece and was allowed to fill the vacancy. Once again, there were twelve Olympians.

Though Greek mythology was not clear regarding the origins of mankind, the legends and traditions claimed descent from the gods for heroes and kings. These semigods formed the link between the human destiny—daily toil, dependence on the elements, plagues, illness, death—and a golden past, when only the gods roamed Earth. And although so many of the gods were born on Earth, the select Circle of Twelve Olympians represented the celestial aspect of the gods. The original Olympus was described by the *Odyssey* as lying in the "pure upper air." The original Twelve Great Gods were Gods of Heaven who had come down to Eearth; and they represented the twelve celestial bodies in the "vault of Heaven."

The Latin names of the Great Gods, given them when the Romans adopted the Greek pantheon, clarify their astral associations: Gaea was Earth; Hermes, Mercury; Aphrodite, Venus; Ares, Mars; Cronus, Saturn; and Zeus, Jupiter. Continuing the Greek tradition, the Romans envisaged Jupiter as a thundering god whose weapon was the lightning bolt; like the Greeks, the Romans associated him with the bull. (Fig. 23)

There is now general agreement that the foundations of the distinct Greek civilization were laid on the island of Crete, where the Minoan culture flourished from circa 2700 B.C. to 1400 B.C. In Minoan myth and legend, the tale of the minotaur is prominent. This half-man, half-bull was the offspring of Pasiphaë, the wife of King Minos, and a bull. Archaeological finds have confirmed the extensive Minoan worship of the bull, and some cylinder seals depict the bull as a divine being accompanied by a cross symbol, which stood for some unidentified star or planet. It has therefore been surmised that the bull worshiped by the Minoans was not the common earthly creature but the Celestial Bull—the constellation Taurus—in commemoration of some events that had occurred when the Sun's spring equinox appeared in that constellation, circa 4000 B.C. (Fig. 24)

By Greek tradition, Zeus arrived on the Greek mainland via Crete, whence he had fled (by swimming the Mediterranean) after abducting Europa, the beautiful daughter of the king of the Phoenician city of Tyre. Indeed, when the earliest Minoan script was finally deciphered by Cyrus H. Gordon, it was shown to be "a Semitic dialect from the shores of the Eastern Mediterranean."

Fig. 22

Fig. 23

Fig. 24

The Greeks, in fact, never claimed that their Olympian gods came directly to Greece from the heavens. Zeus arrived from across the Mediterranean, via Crete. Aphrodite was said to have come by sea from the Near East, via Cyprus. Poseidon (Neptune to the Romans) brought the horse with him from Asia Minor. Athena brought "the olive, fertile and self-sown," to Greece from the lands of the Bible.

There is no doubt that the Greek traditions and religion arrived on the Greek mainland from the Near East, via Asia Minor and the Mediterranean islands. It is there that their pantheon had its roots; it is there that we should look for the origins of the Greek gods, and their astral relationship with the number twelve.

•

Hinduism, the ancient religion of India, considers the *Vedas*—compositions of hymns, sacrificial formulas, and other sayings pertaining to the gods—as sacred scriptures, "not of human origin." The gods themselves composed them, the Hindu traditions say, in the age that preceded the present one. But, as time went on, more and more of the original 100,000 verses, passed from generation to generation orally, were lost and confused. In the end, a sage wrote down the remaining verses, dividing them into four books and trusting four of his principal disciples to preserve one *Veda* each.

When, in the nineteenth century, scholars began to decipher and understand forgotten languages and trace the connections between them, they realized that the *Vedas* were written in a very ancient Indo-European language, the predecessor of the Indian root-tongue Sanskrit, of Greek, Latin, and other European languages. When they were finally able to read and analyze the *Vedas*, they were surprised to see the uncanny similarity between the Vedic tales of the gods and the Greek ones.

The gods, the *Vedas* told, were all members of one large, but not necessarily peaceful, family. Amid the tales of ascents to the heavens and descents to Earth, aerial battles, wondrous weapons, friendships and rivalries, marriages and infidelities, there appears to have existed a basic concern for genealogical record keeping—who fathered whom, and

who was the firstborn of whom. The gods on Earth originated in the heavens; and the principal deities, even on
Earth, continued to represent celestial bodies.

In primeval times, the Rishis ("primeval flowing ones")
"flowed" celestially, possessed of irresistible powers. Of
them, seven were the Great Progenitors. The gods Rahu
("demon") and Ketu ("disconnected") were once a single
celestial body that sought to join the gods without permission; but the God of Storms hurled his flaming weapon
at him, cutting him into two parts—Rahu, the "Dragon's
Head," which unceasingly traverses the heavens in search
of vengeance, and Ketu, the "Dragon's Tail." Mar-Ishi, the
progenitor of the Solar Dynasty, gave birth to Kash-Yapa
("he who is the throne"). The *Vedas* describe him as having
been quite prolific; but the dynastic succession was continued only through his ten children by Prit-Hivi ("heavenly
mother").

As dynastic head, Kash-Yapa was also chief of the devas
("shining ones") and bore the title Dyaus-Pitar ("shining
father"). Together with his consort and ten children, the
divine family made up the twelve Adityas, gods who were
each assigned a sign of the zodiac and a celestial body.
Kash-Yapa's celestial body was "the shining star"; Prit-Hivi
represented Earth. Then there were the gods whose celestial counterparts included the Sun, the Moon, Mars,
Mercury, Jupiter, Venus, and Saturn.

In time, the leadership of the pantheon of twelve passed
to Varuna, the God of the Heavenly Expanse. He was
omnipresent and all-seeing; one of the hymns to him reads
almost like a biblical psalm:

> It is he who makes the sun shine in the heavens,
> And the winds that blow are his breath.
> He has hollowed out the channels of the rivers;
> They flow at his command.
> He has made the depths of the sea.

His reign also came sooner or later to an end. Indra, the
god who slew the celestial "Dragon," claimed the throne
by slaying his father. He was the new Lord of the Skies
and God of Storms. Lightning and thunder were his
weapons, and his epithet was Lord of Hosts. He had, how-

ever, to share dominion with his two brothers. One was
Vivashvat, who was the progenitor of Manu, the first Man.
The other was Agni ("igniter"), who brought fire down
to Earth from the heavens, so that Mankind could use
it industrially.

•

The similarities between the Vedic and Greek pantheons
are obvious. The tales concerning the principal deities, as
well as the verses dealing with a multitude of other lesser
deities—sons, wives, daughters, mistresses—are clearly
duplicates (or originals?) of the Greek tales. There is no
doubt that Dyaus came to mean Zeus; Dyaus-Pitar, Jupiter;
Varuna, Uranus; and so on. And, in both instances, the
Circle of the Great Gods always stood at *twelve*, no matter
what changes took place in the divine succession.

How could such similarity arise in two areas so far apart,
geographically and in time?

Scholars believe that sometime in the second millennium
B.C. a people speaking an Indo-European language, and
centered in northern Iran or the Caucasus area, embarked
on great migrations. One group went southeast, to India.
The Hindus called them Aryans ("noble men"). They
brought with them the *Vedas* as oral tales, circa 1500 B.C.
Another wave of this Indo-European migration went westward, to Europe. Some circled the Black Sea and arrived
in Europe via the steppes of Russia. But the main route
by which these people and their traditions and religion
reached Greece was the shortest one: Asia Minor. Some
of the most ancient Greek cities, in fact, lie not on the
Greek mainland but at the western tip of Asia Minor.

But who were these Indo-Europeans who chose Anatolia
as their abode? Little in Western knowledge shed light
on the subject.

Once again, the only readily available—and reliable—
source proved to be the Old Testament. There the scholars
found several references to the "Hittites" as the people in-
habiting the mountains of Anatolia. Unlike the enmity
reflected in the Old Testament toward the Canaanites and
other neighbors whose customs were considered an
"abomination," the Hittites were regarded as friends and
allies to Israel. Bathsheba, whom King David coveted, was

the wife of Uriah the Hittite, an officer in King David's army. King Solomon, who forged alliances by marrying the daughters of foreign kings, took as wives the daughters both of an Egyptian pharaoh and of a Hittite king. At another time, an invading Syrian army fled upon hearing a rumor that "the king of Israel hath hired against us the kings of the Hittites and the kings of the Egyptians." These brief allusions to the Hittites reveal the high esteem in which their military abilities were held by other peoples of the ancient Near East.

With the decipherment of the Egyptian hieroglyphs—and, later on, of the Mesopotamian inscriptions—scholars have come across numerous references to a "Land of Hatti" as a large and powerful kingdom in Anatolia. Could such an important power have left no trace?

Forearmed with the clues provided in the Egyptian and Mesopotamian texts, the scholars embarked on excavations of ancient sites in Anatolia's hilly regions. The efforts paid off: They found Hittite cities, palaces, royal treasures, royal tombs, temples, religious objects, tools, weapons, art objects. Above all, they found many inscriptions—both in a pictographic script and in cuneiform. The biblical Hittites had come to life.

A unique monument bequeathed to us by the ancient Near East is a rock carving outside the ancient Hittite capital (the site is nowadays called Yazilikaya, which in Turkish means "inscribed rock"). After passing through gateways and sanctuaries, the ancient worshiper came into an open-air gallery, an opening among a semicircle of rocks, on which all the gods of the Hittites were depicted in procession.

Marching in from the left is a long procession of primarily male deities, clearly organized in "companies" of twelve. At the extreme left, and thus last to march in this amazing parade, are twelve deities who look identical, all carrying the same weapon. (Fig. 25)

The middle group of twelve marchers includes some deities who look older, some who bear diversified weapons, and two who are highlighted by a divine symbol. (Fig. 26)

The third (front) group of twelve is clearly made up of the more important male and female deities. Their weapons and emblems are more varied; four have the divine celestial symbol above them; two are winged. This group also includes nondivine participants: two bulls holding up a globe, and the king of the Hittites, wearing a skull cap and standing under the emblem of the Winged Disk. (Fig. 27)

Marching in from the right were two groups of female deities; the rock carvings are, however, too mutilated to ascertain their full original number. We will probably not be wrong in assuming that they, too, made up two "companies" of twelve each.

The two processions from the left and from the right met at a central panel which clearly depicted Great Gods, for they were all shown elevated, standing atop mountains, animals, birds, or even on the shoulders of divine attendants. (Fig. 28)

Much effort was invested by scholars (for example, E. Laroche, *Le Panthéon de Yazilikaya*) to determine from the depictions, the hieroglyphic symbols, as well as from partly legible texts and god names that were actually carved on the rocks, the names, titles, and roles of the deities included in the procession. But it is clear that the Hittite pantheon, too, was governed by the "Olympian" twelve. The lesser gods were organized in groups of twelve, and the Great Gods on Earth were associated with twelve celestial bodies.

That the pantheon was governed by the "sacred number" twelve is made additionally certain by yet another Hittite monument, a masonry shrine found near the present-day Beit-Zehir. It clearly depicts the divine couple, surrounded by ten other gods—making a total of twelve. (Fig. 29)

The archaeological finds showed conclusively that the Hittites worshiped gods that were "of Heaven and Earth," all interrelated and arranged into a genealogical hierarchy. Some were great and "olden" gods who were originally of the heavens. Their symbol—which in the Hittite pictographic writing meant "divine" or "heavenly god"—looked like a pair of eye goggles. (Fig. 30) It frequently appeared on round seals as part of a rocket-like object. (Fig. 31)

Other gods were actually present, not merely on Earth but among the Hittites, acting as supreme rulers of the land, appointing the human kings, and instructing the latter in matters of war, treaties, and other international affairs.

Fig. 25

Fig. 26

Fig. 27

Fig. 28

Fig. 29

Fig. 30

Fig. 31

Heading the physically present Hittite gods was a deity named Teshub, which meant "wind blower." He was thus what scholars call a Storm God, associated with winds, thunder, and lightning. He was also nicknamed Taru ("bull"). Like the Greeks, the Hittites depicted bull worship; like Jupiter after him, Teshub was depicted as the God of Thunder and Lightning, mounted upon a bull. (Fig. 32)

Hittite texts, like later Greek legends, relate how their chief deity had to battle a monster to consolidate his supremacy. A text named by the scholars "The Myth of the Slaying of the Dragon" identifies Teshub's adversary as the god Yanka. Failing to defeat him in battle, Teshub appealed to the other gods for help, but only one goddess came to his assistance, and disposed of Yanka by getting him drunk at a party.

Recognizing in such tales the origins of the legend of Saint George and the Dragon, scholars refer to the adversary smitten by the "good" god as "the dragon." But the fact is that Yanka meant "serpent," and that the ancient peoples depicted the "evil" god as such—as seen in this bas-relief from a Hittite site. (Fig. 33) Zeus, too, as we have shown, battled not a "dragon" but a serpent-god. As we shall show later on, there was deep meaning attached to these ancient traditions of a struggle between a god of winds and a serpent deity. Here, however, we can only stress that battles among the gods for the divine Kingship were reported in the ancient texts as events that had unquestionably taken place.

A long and well-preserved Hittite epic tale, entitled "Kingship in Heaven," deals with this very subject—the heavenly origin of the gods. The recounter of those premortal events first called upon twelve "mighty olden gods" to listen to his tale, and be witnesses to its accuracy:

> Let there listen the gods who are in Heaven,
> And those who are upon the dark-hued Earth!
> Let there listen, the mighty olden gods.

Thus establishing that the gods of old were both of Heaven and upon Earth, the epic lists the twelve "mighty olden ones," the forebears of the gods; and assuring their

Fig. 32

Fig. 33

attention, the recounter proceeded to tell how the god who was "king in Heaven" came to "dark-hued Earth":

Formerly, in the olden days, Alalu was king in Heaven;
He, Alalu, was seated on the throne.
Mighty Anu, the first among the gods, stood before him,
Bowed at his feet, set the drinking cup in his hand.
For nine counted periods, Alalu was king in Heaven.
In the ninth period, Anu gave battle against Alalu.
Alalu was defeated, he fled before Anu—
He descended to the dark-hued Earth.
Down to the dark-hued Earth he went;
On the throne sat Anu.

The epic thus attributed the arrival of a "king in Heaven" upon Earth to a usurpation of the throne: A god named Alalu was forcefully deposed from his throne (somewhere in the heavens), and, fleeing for his life, "descended to dark-hued Earth." But that was not the end. The text proceeded to recount how Anu, in turn, was also deposed by a god named Kumarbi (Anu's own brother, by some interpretations).

There is no doubt that this epic, written a thousand years before the Greek legends were composed, was the forerunner of the tale of the deposing of Uranus by Cronus and of Cronus by Zeus. Even the detail pertaining to the castration of Cronus by Zeus is found in the Hittite text, for that was exactly what Kumarbi did to Anu:

For nine counted periods Anu was king in Heaven;
In the ninth period, Anu had to do battle with Kumarbi.
Anu slipped out of Kumarbi's hold and fled—
Flee did Anu, rising up to the sky.
After him Kumarbi rushed, seized him by his feet;
He pulled him down from the skies.
He bit his loins; and the "Manhood" of Anu
with the insides of Kumarbi combined, fused as bronze.

According to this ancient tale, the battle did not result in a total victory. Though emasculated, Anu managed to fly back to his Heavenly Abode, leaving Kumarbi in control of Earth. Meanwhile, Anu's "Manhood" produced several deities within Kumarbi's insides, which he (like Cronus in the Greek legends) was forced to release. One of these was Teshub, the chief Hittite deity.

However, there was to be one more epic battle before Teshub could rule in peace.

Learning of the appearance of an heir to Anu in Kummiya ("heavenly abode"), Kumarbi devised a plan to "raise a rival to the God of Storms." "Into his hand he took his staff; upon his feet he put the shoes that are swift as winds"; and he went from his city Ur-Kish to the abode of the Lady of the Great Mountain. Reaching her,

> His desire was aroused;
> He slept with Lady Mountain;
> His manhood flowed into her.
> Five times he took her. . . .
> Ten times he took her.

Was Kumarbi simply lustful? We have reason to believe that much more was involved. Our guess would be that the succession rules of the gods were such that a son of Kumarbi by the Lady of the Great Mountain could have claimed to be the rightful heir to the Heavenly Throne; and that Kumarbi "took" the goddess five and ten times in order to make sure that she conceived, as indeed she did: she bore a son, whom Kumarbi symbolically named Ulli-Kummi ("suppressor of Kummiya"—Teshub's abode).

The battle for succession was foreseen by Kumarbi as one that would entail fighting in the heavens. Having destined his son to suppress the incumbents at Kummiya, Kumarbi further proclaimed for his son:

> Let him ascend to Heaven for kingship!
> Let him vanquish Kummiya, the beautiful city!
> Let him attack the God of Storms
> And tear him to pieces, like a mortal!
> Let him shoot down all the gods from the sky.

Did the particular battles fought by Teshub upon Earth and in the skies take place when the Age of Taurus commenced, circa 4000 B.C.? Was it for that reason that the winner was granted association with the bull? And were the events in any way connected with the beginning, at the very same time, of the sudden civilization of Sumer?

•

There can be no doubt that the Hittite pantheon and tales of the gods indeed had their roots in Sumer, its civilization, and its gods.

The tale of the challenge to the Divine Throne by Ulli-Kummi continues to relate heroic battles but of an indecisive nature. At one point, the failure of Teshub to defeat his adversary even caused his spouse, Hebat, to attempt suicide. Finally, an appeal was made to the gods to mediate the dispute, and an Assembly of the Gods was called. It was led by an "olden god" named Enlil, and another "olden god" named Ea, who was called upon to produce "the old tablets with the words of destiny"—some ancient records that could apparently help settle the dispute regarding the divine succession.

When these records failed to settle the dispute, Enlil advised another battle with the challenger, but with the help of some very ancient weapons. "Listen, ye olden gods, ye who know the olden words," Enlil said to his followers:

> Open ye the ancient storehouses
> Of the fathers and the forefathers!
> Bring forth the Olden Copper lance
> With which Heaven was separated from Earth;
> And let them sever the feet of Ulli-kummi.

Who were these "olden gods"? The answer is obvious, for all of them—Anu, Antu, Enlil, Ninlil, Ea, Ishkur—bear Sumerian names. Even the name of Teshub, as well as the names of other "Hittite" gods, were often written in Sumerian script to denote their identities. Also, some of the places named in the action were those of ancient Sumerian sites.

It dawned on the scholars that the Hittites in fact worshipped a pantheon of Sumerian origins, and that the arena of the tales of the "olden gods" was Sumer. This, however, was only part of a much wider discovery. Not only was the Hittite language found to be based on several Indo-European dialects, but it was also found to be subject to substantial Akkadian influence, both in speech and more so in writing. Since Akkadian was the international language of the ancient world in the second millennium B.C., its influence on Hittite could somehow be rationalized.

But there was cause for true astonishment when scholars discovered in the course of deciphering Hittite that it extensively employed Sumerian pictographic signs, syllables, and even whole words! Moreover, it became obvious that Sumerian was their language of high learning. The Sumerian language, in the words of O. R. Gurney *(The Hittites)*, "was intensively studied at Hattu-Shash [the capital city] and Sumerian-Hittite vocabularies were found there. . . . Many of the syllables associated with the cuneiform signs in the Hittite period are really Sumerian words of which the meaning had been forgotten [by the Hittites]. . . . In the Hittite texts the scribes often replaced common Hittite words by the corresponding Sumerian or Babylonian word."

Now, when the Hittites reached Babylon sometime after 1600 B.C., the Sumerians were already long gone from the Near Eastern scene. How was it, then, that their language, literature, and religion dominated another great kingdom in another millennium and in another part of Asia?

The bridge, scholars have recently discovered, were a people called the Hurrians.

Referred to in the Old Testament as the Horites ("free people"), they dominated the wide area between Sumer and Akkad in Mesopotamia and the Hittite kingdom in Anatolia. In the north their lands were the ancient "cedar lands" from which countries near and far obtained their best woods. In the east their centers embraced the present-day oil fields of Iraq; in one city alone, Nuzi, archaeologists found not only the usual structures and artifacts but also thousands of legal and social documents of great value. In the west, the Hurrians' rule and influence extended to the Mediterranean coast and encompassed such great ancient centers of trade, industry, and learning as Carchemish and Alalakh.

But the seats of their power, the main centers of the ancient trade routes, and the sites of the most venerated shrines were within the heartland that was "between the two rivers," the biblical Naharayim. Their most ancient capital (as yet undiscovered) was located somewhere on the Khabur River. Their greatest trading center, on the Balikh

River, was the biblical Ḫaran—the city where the family of the patriarch Abraham sojourned on their way from Ur in southern Mesopotamia to the Land of Canaan.

Egyptian and Mesopotamian royal documents referred to the Hurrian kingdom as Mitanni, and dealt with it on an equal footing—a strong power whose influence spread beyond its immediate borders. The Hittites called their Ḫurrian neighbors "Ḫurri." Some scholars pointed out, however, that the word could also be read "Har," and (like G. Contenau in *La Civilisation des Hittites et des Hurrites du Mitanni)* have raised the possibility that, in the name "Harri," "one sees the name 'Ary' or Aryans for these people."

There is no doubt that the Ḫurrians were Aryan or Indo-European in origin. Their inscriptions invoked several deities by their Vedic "Aryan" names, their kings bore Indo-European names, and their military and cavalry terminology derived from the Indo-European. B. Hrozny, who in the 1920s led an effort to unravel the Hittite and Ḫurrian records, even went so far as to call the Ḫurrians "the oldest Hindus."

These Ḫurrians dominated the Hittites culturally and religiously. The Hittite mythological texts were found to be of Ḫurrian provenance, and even epic tales of prehistoric, semidivine heroes were of Ḫurrian origin. There is no longer any doubt that the Hittites acquired their cosmology, their "myths," their gods, and their pantheon of twelve from the Hurrians.

The triple connection—between Aryan origins, Hittite worship, and the Ḫurrian sources of these beliefs—is remarkably well documented in a Hittite prayer by a woman for the life of her sick husband. Addressing her prayer to the goddess Hebat, Teshub's spouse, the woman intoned:

> Oh goddess of the Rising Disc of Arynna,
> My Lady, Mistress of the Hatti Lands,
> Queen of Heaven and Earth. . . .
> In the Hatti country, thy name is
> "Goddess of the Rising Disc of Arynna";
> But in the land that thou madest,
> In the Cedar Land,
> Thou bearest the name "Ḫebat."

With all that, the culture and religion adopted and transmitted by the Ḫurrians were not Indo-European. Even their language was not really Indo-European. There were undoubtedly Akkadian elements in the Ḫurrian language, culture, and traditions. The name of their capital, Washugeni, was a variant of the Semitic *resh-eni* ("where the waters begin"). The Tigris River was called Aranzakh, which (we believe) stemmed from the Akkadian words for "river of the pure cedars." The gods Shamash and Tashmetum became the Ḫurrian Shimiki and Tashimmetish—and so on.

But since the Akkadian culture and religion were only a development of the original Sumerian traditions and beliefs, the Ḫurrians, in fact, absorbed and transmitted the religion of Sumer. That this was so was also evident from the frequent use of the original Sumerian divine names, epithets, and writing signs.

The epic tales, it has become clear, were the tales of Sumer; the "dwelling places" of the olden gods were Sumerian cities; the "olden language" was the language of Sumer. Even the Ḫurrian art duplicated Sumerian art—its form, its themes, and its symbols.

When and how were the Ḫurrians "mutated" by the Sumerian "gene"?

Evidence suggests that the Ḫurrians, who were the northern neighbors of Sumer and Akkad in the second millennium B.C., had actually commingled with the Sumerians in the previous millennium. It is an established fact that Ḫurrians were present and active in Sumer in the third millennium B.C., that they held important positions in Sumer during its last period of glory, that of the third dynasty of Ur. There is evidence showing that the Ḫurrians managed and manned the garment industry for which Sumer (and especially Ur) was known in antiquity. The renowned merchants of Ur were probably Ḫurrians for the most part.

In the thirteenth century B.C., under the pressure of vast migrations and invasions (including the Israelite thrust from Egypt to Canaan), the Ḫurrians retreated to the northeastern portion of their kingdom. Establishing their new capital near Lake Van, they called their kingdom Urartu ("Ararat"). There they worshiped a pantheon headed by Tesheba (Teshub), depicting him as a vigorous god wearing a horned cap and standing upon his cult symbol, the bull. (Fig. 34) They called their main shrine Bitanu ("house of Anu") and dedicated themselves to making their kingdom "the fortress of the valley of Anu."

And Anu, as we shall see, was the Sumerian Father of the Gods.

•

What about the other avenue by which the tales and worship of the gods reached Greece—from the eastern shores of the Mediterranean, via Crete and Cyprus?

The lands that are today Israel, Lebanon, and southern Syria—which formed the southwestern band of the ancient Fertile Crescent—were then the habitat of peoples that can be grouped together as the Canaanites. Once again, all that was known of them until rather recently appeared in references (mostly adverse) in the Old Testament and scattered Phoenician inscriptions. Archaeologists were only beginning to understand the Canaanites when two discoveries came to light: certain Egyptian texts at Luxor and Saqqara, and, much more important, historical, literary, and religious texts unearthed at a major Canaanite center. The place, now called Ras Shamra, on the Syrian coast, was the ancient city of Ugarit.

The language of the Ugarit inscriptions, the Canaanite language, was what scholars call West Semitic, a branch of the group of languages that also includes the earliest Akkadian and present-day Hebrew. Indeed, anyone who knows Hebrew well can follow the Canaanite inscriptions with relative ease. The language, literary style, and terminology are reminiscent of the Old Testament.

The pantheon that unfolds from the Canaanite texts bears many similarities to the later Greek one. At the head of the Canaanite pantheon, too, there was a supreme deity called *El,* a word that was both the personal name of the god and the generic term meaning "lofty deity." He was the final authority in all affairs, human or divine. Ab Adam ("father of man") was his title; the Kindly, the Merciful was his epithet. He was the "creator of things created, and the one who alone could bestow kingship."

The Canaanite texts ("myths" to most scholars) depicted

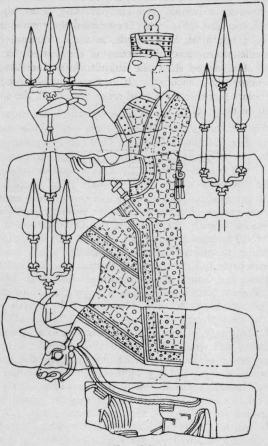

Fig. 34

El as a sage, elderly deity who stayed away from daily affairs. His abode was remote, at the "headwaters of the two rivers"—the Tigris and Euphrates. There he would sit on his throne, receive emissaries, and contemplate the problems and disputes the other gods brought before him.

A stela found in Palestine depicts an elderly deity sitting on a throne and being served a beverage by a younger deity. The seated deity wears a conical headdress adorned with horns—a mark of the gods, as we have seen, from prehistoric times—and the scene is dominated by the symbol of a winged star—the ubiquitous emblem that we shall increasingly encounter. It is generally accepted by the scholars that this sculptured relief depicts El, the senior Canaanite deity. (Fig. 35)

Fig. 35

El, however, was not always an olden lord. One of his epithets was Tor (meaning "bull"), signifying, scholars believe, his sexual prowess and his role as Father of the Gods. A Canaanite poem, called "Birth of the Gracious Gods," placed El at the seashore (probably naked), where two women were completely charmed by the size of his penis. While a bird was roasting on the beach, El had intercourse with the two women. Thus were the two gods

Shaḫar ("dawn") and Shalem ("completion" or "dusk") born.

These were not his only children nor his principal sons (of which he had, apparently, seven). His principal son was Baal—again the personal name of the deity, as well as the general term for "lord." As the Greeks did in their tales, the Canaanites spoke of the challenges by the son to the authority and rule of his father. Like El his father, Baal was what the scholars call a Storm God, a God of Thunder and Lightning. A nickname for Baal was Hadad ("sharp one"). His weapons were the battle-ax and the lightning-spear; his cult animal, like El's, was the bull, and, like El, he was depicted wearing the conical headdress adorned with a pair of horns.

Baal was also called Elyon ("supreme"); that is, the acknowledged prince, the heir apparent. But he had not come by this title without a struggle, first with his brother Yam ("prince of the sea"), and then with his brother Mot. A long and touching poem, pieced together from numerous fragmented tablets, begins with the summoning of the "Master Craftsman" to El's abode "at the sources of the waters, in the midst of the headwaters of the two rivers":

> Through the fields of El he comes
> He enters the pavilion of the Father of Years.
> At El's feet he bows, falls down,
> Prostrates himself, paying homage.

The Master Craftsman is ordered to erect a palace for Yam as the mark of his rise to power. Emboldened by this, Yam sends his messengers to the assembly of the gods, to ask for the surrender to him of Baal. Yam instructs his emissaries to be defiant, and the assembled gods do yield. Even El accepts the new lineup among his sons. "Ba'al is thy slave, O Yam," he declares.

The supremacy of Yam, however, was short-lived. Armed with two "divine weapons," Baal struggled with Yam and defeated him—only to be challenged by Mot (the name meant "smiter"). In this struggle, Baal was soon vanquished; but his sister Anat refused to accept this demise of Baal as final. "She seized Mot, the son of El, and with a blade she cleaved him."

The obliteration of Mot led, according to the Canaanite tale, to the miraculous resurrection of Baal. Scholars have attempted to rationalize the report by suggesting that the whole tale was only allegorical, representing no more than a tale of the annual struggle in the Near East between the hot, rainless summers that dry out the vegetation, and the coming of the rainy season in the autumn, which revives or "resurrects" the vegetation. But there is no doubt that the Canaanite tale intended no allegory, that it related what were then believed to be the true events: how the sons of the chief deity fought among themselves, and how one of them defied defeat to reappear and become the accepted heir, making El rejoice:

> El, the kindly one, the merciful, rejoices.
> His feet on the footstool he sets.
> He opens his throat and laughs;
> He raises his voice and cries out:
> "I shall sit and take my ease,
> The soul shall repose in my breast;
> For Ba'al the mighty is alive,
> For the Prince of Earth exists!"

Anat, according to Canaanite traditions, thus stood by her brother the Lord (Baal) in his life-and-death struggle with the evil Mot; and the parallel between this and the Greek tradition of the goddess Athena standing with the supreme god Zeus in his life-and-death struggle with Typhon is only too obvious. Athena, as we have seen, was called "the perfect maiden," yet had many illicit love affairs. Likewise, Canaanite traditions (which preceded the Greek ones) employed the epithet "the Maiden Anat," and, in spite of this, proceeded to report her various love affairs, especially with her own brother Baal. One text describes the arrival of Anat at Baal's abode on Mount Zaphon, and Baal's hurried dismissal of his wives. Then he sank by his sister's feet; they looked into each other's eyes; they anointed each other's "horns"—

> He seizes and holds her womb. . . .
> She seizes and holds his "stones.". . .
> The maiden Anat . . . is made to conceive and bear.

No wonder, then, that Anat was often depicted completely naked, to emphasize her sexual attributes—as in this seal impression, which illustrates a helmeted Baal battling another god. (Fig. 36)

Like the Greek religion and its direct forerunners, the Canaanite pantheon included a Mother Goddess, official consort of the chief deity. They called her Ashera; she paralleled the Greek Hera. Astarte (the biblical Ashtoreth) paralleled Aphrodite; her frequent consort was Athtar, who was associated with a bright planet, and who probably paralleled Ares, Aphrodite's brother. There were other young deities, male and female, whose astral or Greek parallels can easily be surmised.

But besides these young deities there were the "olden gods," aloof from mundane affairs but available when the gods themselves ran into serious trouble. Some of their sculptures, even in a partly damaged state, show them with commanding features, gods recognizable by their horned headgear. (Fig. 37)

Whence had the Canaanites, for their part, drawn their culture and religion?

The Old Testament considered them a part of the Hamitic family of nations, with roots in the hot (for that is what *ham* meant) lands of Africa, brothers of the Egyptians. The artifacts and written records unearthed by archaeologists confirm the close affinity between the two, as well as the many similarities between the Canaanite and Egyptian deities.

The many national and local gods, the multitude of their names and epithets, the diversity of their roles, emblems, and animal mascots at first cast the gods of Egypt as an unfathomable crowd of actors upon a strange stage. But a closer look reveals that they were essentially no different from those of the other lands of the ancient world.

The Egyptians believed in Gods of Heaven and Earth, Great Gods that were clearly distinguished from the multitudes of lesser deities. G. A. Wainwright (*The Sky-Religion in Egypt*) summed up the evidence, showing that the Egyptian belief in Gods of Heaven who descended to Earth from the skies was "extremely ancient." Some of the epithets of these Great Gods—Greatest God, Bull of Heaven, Lord/Lady of the Mountains—sound familiar.

Fig. 36

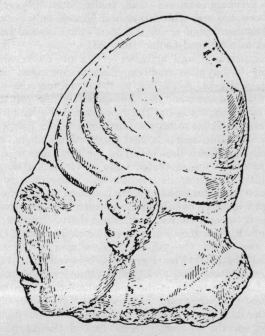

Fig. 37

Although the Egyptians counted by the decimal system, their religious affairs were governed by the Sumerian sexagesimal *sixty*, and celestial matters were subject to the divine number *twelve*. The heavens were divided into three parts, each comprising twelve celestial bodies. The afterworld was divided into twelve parts. Day and night were each divided into twelve hours. And all these divisions were paralleled by "companies" of gods, which in turn consisted of twelve gods each.

The head of the Egyptian pantheon was Ra ("creator"), who presided over an Assembly of the Gods that numbered twelve. He performed his wondrous works of creation in primeval times, bringing forth Geb ("Earth") and Nut ("sky"). Then he caused the plants to grow on Earth, and the creeping creatures—and, finally, Man. Ra was an unseen celestial god who manifested himself only periodically. His manifestation was the Aten—the Celestial Disc, depicted as a Winged Globe. (Fig. 38)

The appearance and activities of Ra on Earth were, according to Egyptian tradition, directly connected with kingship in Egypt. According to that tradition, the first rulers of Egypt were not men but gods, and the first god to rule over Egypt was Ra. He then divided the kingdom, giving Lower Egypt to his son Osiris and Upper Egypt to his son Seth. But Seth schemed to overthrow Osiris and eventually had Osiris drowned. Isis, the sister and wife of Osiris, retrieved the mutilated body of Osiris and resurrected him. Thereafter, he went through "the secret gates" and joined Ra in his celestial path; his place on the throne of Egypt was taken over by his son Horus, who was sometimes depicted as a winged and horned deity. (Fig. 39)

Though Ra was the loftiest in the heavens, upon Earth he was the son of the god Ptah ("developer," "one who fashioned things"). The Egyptians believed that Ptah actually raised the land of Egypt from under floodwaters by building dike works at the point where the Nile rises. This Great God, they said, had come to Egypt from elsewhere; he established not only Egypt but also "the mountain land and the far foreign land." Indeed, the Egyptians acknowledged, all their "olden gods" had come by boat from the south; and many prehistoric rock drawings have been found that show these olden gods—distinguished by

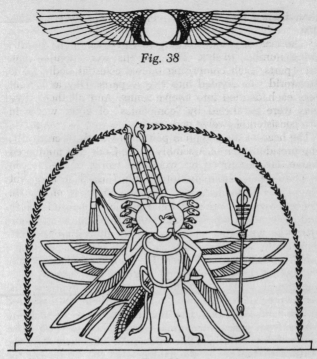

Fig. 38

Fig. 39

Fig. 40

their horned headdress—arriving in Egypt by boat. (Fig. 40)

The only sea route leading to Egypt from the south is the Red Sea, and it is significant that the Egyptian name for it was the Sea of Ur. Hieroglyphically, the sign for Ur meant "the far-foreign [land] in the east"; that it actually may also have referred to the Sumerian Ur, lying in that very direction, cannot be ruled out.

The Egyptian word for "divine being" or "god" was NTR, which meant "one who watches." Significantly, that is exactly the meaning of the name Shumer: the land of the "ones who watch."

The earlier notion that civilization may have begun in Egypt has been discarded by now. There is ample evidence now showing that the Egyptian-organized society and civilization, which began half a millennium and more *after* the Sumerian one, drew its culture, architecture, technology, art of writing, and many other aspects of a high civilization from Sumer. The weight of evidence also shows that the gods of Egypt originated in Sumer.

Cultural and blood kinsmen of the Egyptians, the Canaanites shared the same gods with them. But, situated in the land strip that was the bridge between Asia and Africa from time immemorial, the Canaanites also came under strong Semitic or Mesopotamian influences. Like the Hittites to the north, the Hurrians to the northeast, the Egyptians to the south, the Canaanites could not boast of an original pantheon. They, too, acquired their cosmogony, deities, and legendary tales from elsewhere. Their direct contacts with the Sumerian sources were the Amorites.

•

The land of the Amorites lay between Mesopotamia and the Mediterranean lands of western Asia. Their name derives from the Akkadian *amurru* and Sumerian *martu* ("westerners"). They were not treated as aliens but as related people who dwelt in the western provinces of Sumer and Akkad.

Persons bearing Amorite names were listed as temple functionaries in Sumer. When Ur fell to Elamite invaders circa 2000 B.C., a Martu named Ishbi-Irra reestablished Sumerian kingship at Larsa and made his first task the recapture of Ur and the restoration there of the great shrine to the god Sin. Amorite "chieftains" established the first independent dynasty in Assyria circa 1900 B.C. And Ḥammurabi, who brought greatness to Babylon circa 1800 B.C., was the sixth successor of the first dynasty of Babylon, which was Amorite.

In the 1930s archaeologists came upon the center and capital city of the Amorites, known as Mari. At a bend of the Euphrates, where the Syrian border now cuts the river, the diggers uncovered a major city whose buildings were erected and continuously reerected, between 3000 and 2000 B.C., on foundations that date to centuries earlier. These earliest remains included a step pyramid and temples to the Sumerian deities Inanna, Ninḥursag, and Enlil.

The palace of Mari alone occupied some five acres and included a throne room painted with most striking murals, three hundred various rooms, scribal chambers, and (most important to the historian) well over twenty thousand tablets in the cuneiform script, dealing with the economy, trade, politics, and social life of those times, with state and military matters, and, of course, with the religion of the land and its people. One of the wall paintings at the great palace of Mari depicts the investiture of the king Zimri-Lim by the goddess Inanna (whom the Amorites called Ishtar). (Fig. 41)

Fig. 41

As in the other pantheons, the chief deity physically present among the Amurru was a weather or storm god. They called him Adad—the equivalent of the Canaanite Baal ("lord")—and they nicknamed him Hadad. His symbol, as might be expected, was fork lightning.

In Canaanite texts, Baal is often called the "Son of Dagon." The Mari texts also speak of an older deity named Dagan, a "Lord of Abundance" who—like El—is depicted as a retired deity, who complained on one occasion that he was no longer consulted on the conduct of a certain war.

Other members of the pantheon included the Moon God, whom the Canaanites called Yeraḥ, the Akkadians Sin, and the Sumerians Nannar; the Sun God, commonly called Shamash; and other deities whose identities leave no doubt that Mari was a bridge (geographically and chronologically) connecting the lands and the peoples of the eastern Mediterranean with the Mesopotamian sources.

Among the finds at Mari, as elsewhere in the lands of Sumer, there were dozens of statues of the people themselves: kings, nobles, priests, singers. They were invariably depicted with their hands clasped in prayer, their gaze frozen forever toward their gods. (Fig. 42)

Fig. 42

Who were these Gods of Heaven and Earth, divine yet human, always headed by a pantheon or inner circle of twelve deities?

We have entered the temples of the Greeks and the Aryans, the Hittites and the Hurrians, the Canaanites, the Egyptians, and the Amorites. We have followed paths that took us across continents and seas, and clues that carried us over several millennia.

And all the corridors of all the temples have led us to one source: *Sumer.*

4
·
SUMER: LAND OF THE GODS

THERE IS NO DOUBT that the "olden words," which for thousands of years constituted the language of higher learning and religious scriptures, was the language of Sumer. There is also no doubt that the "olden gods" were the gods of Sumer; records and tales and genealogies and histories of gods older than those pertaining to the gods of Sumer have not been found anywhere.

When these gods (in their original Sumerian forms or in the later Akkadian, Babylonian, or Assyrian) are named and counted, the list runs into the hundreds. But once they are classified, it is clear that they were not a hodgepodge of divinities. They were headed by a pantheon of Great Gods, governed by an Assembly of the Deities, and related to each other. Once the numerous lesser nieces, nephews, grandchildren, and the like are excluded, a much smaller and coherent group of deities emerges—each with a role to play, each with certain powers or responsibilities.

There were, the Sumerians believed, gods that were "of the heavens." Texts dealing with the time "before things were created" talk of such heavenly gods as Apsu, Tiamat, Anshar, Kishar. No claim is ever made that the gods of this category ever appeared upon Earth. As we look closer at these "gods," who existed before Earth was created, we shall realize that they were the celestial bodies that make up our solar system; and, as we shall show, the so-called Sumerian myths regarding these celestial beings are, in fact, precise and scientifically plausible cosmologic concepts regarding the creation of our solar system.

There were also lesser gods who were "of Earth." Their cult centers were mostly provincial towns; they were no more than local deities. At best, they were given charge of some limited operation—as, for example, the goddess NIN.KASHI ("lady-beer"), who supervised the preparation of beverages. Of them, no heroic tales were told. They possessed no awesome weapons, and the other gods did not shudder at their command. They remind one very much of the company of young gods that marched last in the procession depicted on the rocks of Hittite Yazilikaya.

Between the two groups there were the Gods of Heaven and Earth, the ones called "the ancient gods." They were the "olden gods" of the epic tales, and, in the Sumerian belief, they had come down to Earth from the heavens.

These were no mere local deities. They were national gods—indeed, international gods. Some of them were present and active upon Earth even before there were Men upon Earth. Indeed, the very existence of Man was deemed to have been the result of a deliberate creative enterprise on the part of these gods. They were powerful, capable of feats beyond mortal ability or comprehension. Yet these gods not only looked like humans but ate and drank like them and displayed virtually every human emotion of love and hate, loyalty and infidelity.

Although the roles and hierarchical standing of some of the principal deities shifted over the millennia, a number of them never lost their paramount position and their national and international veneration. As we take a close look at this central group, there emerges a picture of a dynasty of gods, a divine family, closely related yet bitterly divided.

•

The head of this family of Gods of Heaven and Earth was AN (or Anu in the Babylonian/Assyrian texts). He was the Great Father of the Gods, the King of the Gods. His realm was the expanse of the heavens, and his symbol was a star. In the Sumerian pictographic writing, the sign of a star also stood for An, for "heavens," and for "divine being," or "god" (descended of An). This fourfold meaning of the symbol remained through the ages, as the script moved from the Sumerian pictographic to the cuneiform Akkadian, to the stylized Babylonian and Assyrian. (Fig. 43)

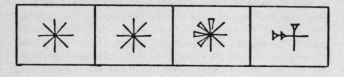

AN = Star = Heavens = "god"

Fig. 43

From the very earliest times until the cuneiform script faded away—from the fourth millennium B.C. almost to the time of Christ—this symbol preceded the names of the gods, indicating that the name written in the text was not of a mortal, but of a deity of heavenly origin.

Anu's abode, and the seat of his Kingship, was in the heavens. That was where the other Gods of Heaven and Earth went when they needed individual advice or favor, or where they met in assembly to settle disputes among themselves or to reach major decisions. Numerous texts describe Anu's palace (whose portals were guarded by a god of the Tree of Truth and a god of the Tree of Life), his throne, the manner in which other gods approached him, and how they sat in his presence.

The Sumerian texts could also recall instances when not only the other gods but even some chosen mortals were permitted to go up to Anu's abode, mostly with the object of escaping mortality. One such tale pertained to Adapa ("model of Man"). He was so perfect and so loyal to the god Ea, who had created him, that Ea arranged for him to be taken to Anu. Ea then described to Adapa what to expect.

> Adapa,
> thou art going before Anu, the King;
> The road to Heaven thou wilt take.
> When to Heaven thou hast ascended,
> and hast approached the gate of Anu,
> the "Bearer of Life" and the "Grower of Truth"
> at the gate of Anu will be standing.

Guided by his creator, Adapa "to Heaven went up . . . ascended to Heaven and approached the gate of Anu." But when he was offered the chance to become immortal, Adapa refused to eat the Bread of Life, thinking that the angry Anu offered him poisoned food. He was thus returned to Earth as an anointed priest but still a mortal.

The Sumerian claim that not only gods but also selected mortals could ascend to the Divine Abode in the heavens is echoed in the Old Testament tales of the ascents to the heavens by Enoch and the prophet Elijah.

Though Anu lived in a Heavenly Abode, the Sumerian texts reported instances when he came down to Earth— either at times of great crisis, or on ceremonial visits (when he was accompanied by his spouse ANTU), or (at least once) to make his great-granddaughter IN.ANNA his consort on Earth.

Since he did not permanently reside on Earth, there was apparently no need to grant him exclusivity over his own city or cult center; and the abode, or "high house," erected for him was located at Uruk (the biblical Erech), the domain of the goddess Inanna. The ruins of Uruk include to this day a huge man-made mound, where archaeologists have found evidence of the construction and reconstruction of a high temple—the temple of Anu; no less than eighteen strata or distinct phases were discovered there, indicating the existence of compelling reasons to maintain the temple at that sacred site.

The temple of Anu was called E.ANNA ("house of An"). But this simple name applied to a structure that, at least at some of its phases, was quite a sight to behold. It was, according to Sumerian texts, "the hallowed E-Anna, the pure sanctuary." Traditions maintained that the Great Gods themselves "had fashioned its parts." "Its cornice was like copper," "its great wall touching the clouds—a lofty dwelling place"; "it was the House whose charm was irresistible, whose allure was unending." And the texts also made clear the temple's purpose, for they called it "the House for descending from Heaven."

A tablet that belonged to an archive at Uruk enlightens us as to the pomp and pageantry that accompanied the arrival of Anu and his spouse on a "state visit." Because of damage to the tablet, we can read of the ceremonies only from some midpoint, when Anu and Antu were already seated in the temple's courtyard. The gods, "exactly in the same order as before," then formed a procession ahead of and behind the bearer of the scepter. The protocol then instructed:

> They shall then descend to the Exalted Court,
> and shall turn towards the god Anu.
> The Priest of Purification shall libate the Scepter,
> and the Scepter-bearer shall enter and be seated.
> The deities Papsukal, Nusku and Shala
> shall then be seated in the court of the god Anu.

Meanwhile, the goddesses, "The Divine Offspring of Anu, Uruk's Divine Daughters," bore a second object, whose name or purpose are unclear, to the E.NIR, "The House of the Golden Bed of the Goddess Antu." Then they returned in a procession to the courtyard, to the place where Antu was seated. While the evening meal was being prepared according to a strict ritual, a special priest smeared a mixture of "good oil" and wine on the door sockets of the sanctuary to which Anu and Antu were later to retire for the night—a thoughtful touch intended, it seems, to eliminate squeaking of the doors while the two deities slept.

While an "evening meal"—various drinks and appetizers—was being served, an astronomer-priest went up to the "topmost stage of the tower of the main temple" to observe the skies. He was to look out for the rising in a specific part of the sky of the planet named Great Anu of Heaven. Thereupon, he was to recite the compositions named "To the one who grows bright, the heavenly planet of the Lord Anu," and "The Creator's image has risen."

Once the planet had been sighted and the poems recited, Anu and Antu washed their hands with water out of a golden basin and the first part of the feast began. Then, the seven Great Gods also washed their hands from seven large golden trays and the second part of the feast began. The "rite of washing of the mouth" was then performed; the priests recited the hymn "The planet of Anu is Heaven's hero." Torches were lit, and the gods, priests, singers, and food-bearers arranged themselves in a procession, ac-

companying the two visitors to their sanctuary for the
night.

Four major deities were assigned to remain in the
courtyard and keep watch until daybreak. Others were
stationed at various designated gates. Meanwhile, the
whole country was to light up and celebrate the presence
of the two divine visitors. On a signal from the main
temple, the priests of all the other temples of Uruk were
"to use torches to start bonfires"; and the priests in other
cities, seeing the bonfires at Uruk, were to do likewise.
Then:

> The people of the Land shall light fires in their homes,
> and shall offer banquets to all the gods. . . .
> The guards of the cities shall light fires
> in the streets and in the squares.

The departure of the two Great Gods was also planned,
not only to the day but to the minute.

> On the seventeenth day,
> forty minutes after sunrise,
> the gate shall be opened before the gods Anu and
>   Antu,
> bringing to an end their overnight stay.

While the end of this tablet has broken off, another text
in all probability describes the departure: the morning
meal, the incantations, the handshakes ("grasping of the
hands") by the other gods. The Great Gods were then
carried to their point of departure on thronelike litters
carried on the shoulders of temple functionaries. An
Assyrian depiction of a procession of deities (though
from a much later time) probably gives us a good idea of
the manner in which Anu and Antu were carried during
their procession in Uruk. (Fig. 44)

Special incantations were recited when the procession
was passing through "the street of the gods"; other psalms
and hymns were sung as the procession neared "the holy
quay" and when it reached "the dike of the ship of Anu."
Good-byes were then said, and yet more incantations were
recited and sung "with hand-raising gestures."

Fig. 44

Then all the priests and temple functionaries who carried the gods, led by the great priest, offered a special "prayer of departure." "Great Anu, may Heaven and Earth bless you!" they intoned seven times. They prayed for the blessing of the seven celestial gods and invoked the gods that were in Heaven and the gods that were upon Earth. In conclusion, they bade farewell to Anu and Antu, thus:

> May the Gods of the Deep,
> and the Gods of the Divine Abode,
> bless you!
> May they bless you daily—
> every day of every month of every year!

Among the thousands upon thousands of depictions of the ancient gods that have been uncovered, none seems to depict Anu. Yet he peers at us from every statue and every portrait of every king that ever was, from antiquity to our very own days. For Anu was not only the Great King, King of the Gods, but also the one by whose grace others could be crowned as kings. By Sumerian tradition, rulership flowed from Anu; and the very term for "Kingship" was *Anutu* ("Anu-ship"). The insignia of Anu were the tiara (the divine headdress), the scepter (symbol of power), and the staff (symbolizing the guidance provided by the shepherd).

The shepherd's staff may now be found more in the hands of bishops than of kings. But the crown and scepter are still held by whatever kings Mankind has left on some thrones.

•

The second most powerful deity of the Sumerian pantheon was EN.LIL. His name meant "lord of the airspace" —the prototype and father of the later Storm Gods that were to head the pantheons of the ancient world.

He was Anu's eldest son, born at his father's Heavenly Abode. But at some point in the earliest times he descended to Earth, and was thus the principal God of Heaven *and* Earth. When the gods met in assembly at the Heavenly Abode, Enlil presided over the meetings alongside his father. When the gods met for assembly on Earth, they met at Enlil's court in the divine precinct of Nippur, the city dedicated to Enlil and the site of his main temple, the E.KUR ("house which is like a mountain").

Not only the Sumerians but the very gods of Sumer considered Enlil supreme. They called him Ruler of All the Lands, and made it clear that "in Heaven—he is the Prince; On Earth—he is the Chief." His "word [command] high above made the Heavens tremble, down below made the Earth quake":

> Enlil,
> Whose command is far reaching;
> Whose "word" is lofty and holy;
> Whose pronouncement is unchangeable;
> Who decrees destinies unto the distant future. . . .
> The Gods of Earth bow down willingly before him;
> The Heavenly gods who are on Earth
> humble themselves before him;
> They stand by faithfully, according to instructions.

Enlil, according to Sumerian beliefs, arrived on Earth well before Earth became settled and civilized. A "Hymn to Enlil, the All-Beneficent" recounts the many aspects of society and civilization that would not have existed had it not been for Enlil's instructions to "execute his orders, far and wide."

No cities would be built, no settlements founded;
No stalls would be built, no sheepfolds erected;
No king would be raised, no high priest born.

The Sumerian texts also stated that Enlil arrived on Earth before the "Black-Headed People"—the Sumerian nickname for Mankind—were created. During such pre-Mankind times, Enlil erected Nippur as his center, or "command post," at which Heaven and Earth were connected through some "bond." The Sumerian texts called this bond DUR.AN.KI ("bond heaven–earth") and used poetic language to describe Enlil's first actions on Earth:

Enlil,
When you marked off divine settlements on Earth,
Nippur you set up as your very own city.
The City of Earth, the lofty,
Your pure place whose water is sweet.
You founded the Dur-An-Ki
In the center of the four corners of the world.

In those early days, when gods alone inhabited Nippur and Man had not yet been created, Enlil met the goddess who was to become his wife. According to one version, Enlil saw his future bride while she was bathing in Nippur's stream—naked. It was love at first sight, but not necessarily with marriage in mind:

The shepherd Enlil, who decrees the fates,
The Bright-Eyed One, saw her.
The lord speaks to her of intercourse;
she is unwilling.
Enlil speaks to her of intercourse;
she is unwilling:
"My vagina is too small [she said],
It knows no copulation;
My lips are too little,
they know not kissing."

But Enlil did not take no for an answer. He disclosed to his chamberlain Nushku his burning desire for "the young maid," who was called SUD ("the nurse"), and who lived with her mother at E.RESH ("scented house"). Nushku suggested a boat ride and brought up a boat. Enlil persuaded Sud to go sailing with him. Once they were in the boat, he raped her.

The ancient tale then relates that though Enlil was chief of the gods they were so enraged that they seized him and banished him to the Lower World. "Enlil, immoral one!" they shouted at him. "Get thyself out of the city!" This version has it that Sud, pregnant with Enlil's child, followed him, and he married her. Another version has the repentant Enlil searching for the girl and sending his chamberlain to her mother to ask for the girl's hand. One way or another, Sud did become the wife of Enlil, and he bestowed on her the title NIN.LIL ("lady of the airspace").

But little did he and the gods who banished him know that it was not Enlil who had seduced Ninlil, but the other way around. The truth of the matter was that Ninlil bathed naked in the stream on her mother's instructions, with the hope that Enlil—who customarily took his walks by the stream—would notice Ninlil and wish to "forthwith embrace you, kiss you."

In spite of the manner in which the two fell for each other, Ninlil was held in the highest esteem once she was given by Enlil "the garment of ladyship." With one exception, which (we believe) had to do with dynastic succession, Enlil is never known to have had other indiscretions. A votive tablet found at Nippur shows Enlil and Ninlil being served food and beverage at their temple. The tablet was commissioned by Ur-Enlil, the "Domestic of Enlil." (Fig. 45)

Apart from being chief of the gods, Enlil was also deemed the supreme Lord of Sumer (sometimes simply called "The Land") and its "Black-Headed People." A Sumerian psalm spoke in veneration of this god:

Lord who knows the destiny of The Land,
    trustworthy in his calling;
Enlil who knows the destiny of Sumer,
    trustworthy in his calling;
Father Enlil,
    Lord of all the lands;

Fig. 45

Father Enlil,
    Lord of the Rightful Command;
Father Enlil,
    Shepherd of the Black-Headed Ones. . . .
From the Mountain of Sunrise
    to the Mountain of Sunset,
There is no other Lord in the land;
    you alone are King.

The Sumerians revered Enlil out of both fear and
gratitude. It was he who made sure that decrees by the
Assembly of the Gods were carried out against Mankind;
it was his "wind" that blew obliterating storms against
offending cities. It was he who, at the time of the Deluge,
sought the destruction of Mankind. But when at peace
with Mankind, he was a friendly god who bestowed favors;
according to the Sumerian text, the knowledge of farming,
together with the plow and the pickax, were granted to
Mankind by Enlil.

Enlil also selected the kings who were to rule over
Mankind, not as sovereigns but as servants of the god entrusted with the administration of divine laws of justice.
Accordingly, Sumerian, Akkadian, and Babylonian kings
opened their inscriptions of self-adoration by describing
how Enlil had called them to Kingship. These "calls"—

issued by Enlil on behalf of himself and his father Anu—granted legitimacy to the ruler and outlined his functions. Even Ḥammurabi, who acknowledged a god named Marduk as the national god of Babylon, prefaced his code of laws by stating that "Anu and Enlil named me to promote the welfare of the people . . . to cause justice to prevail in the land."

God of Heaven and Earth, Firstborn of Anu, Dispenser of Kingship, Chief Executive of the Assembly of the Gods, Father of Gods and Men, Granter of Agriculture, Lord of the Airspace—these were some of the attributes of Enlil that bespoke his greatness and powers. His "command was far reaching," his "pronouncements unchangeable"; he "decreed the destinies." He possessed the "bond heaven–earth," and from his "awesome city Nippur" he could "raise the beams that search the heart of all the lands"—"eyes that could scan all the lands."

Yet he was as human as any young man enticed by a naked beauty; subject to moral laws imposed by the community of the gods, transgressions of which were punishable by banishment; and not even immune to mortal complaints. At least in one known instance, a Sumerian king of Ur complained directly to the Assembly of the Gods that a series of troubles that had befallen Ur and her people could be traced back to the ill-fated fact that "Enlil did give the kingship to a worthless man . . . who is not of Sumerian seed."

As we go along, we shall see the central role that Enlil played in divine and mortal affairs on Earth, and how his several sons battled among themselves and with others for the divine succession, undoubtedly giving rise to the later tales of the battles of the gods.

•

The third Great God of Sumer was another son of Anu; he bore two names, E.A and EN.KI. Like his brother Enlil, he, too, was a God of Heaven and Earth, a deity originally of the heavens, who had come down to Earth.

His arrival on Earth is associated in Sumerian texts with a time when the waters of the Persian Gulf reached inland much farther than nowadays, turning the southern part of the country into marshlands. Ea (the name meant literally "house-water"), who was a master engineer, planned and supervised the construction of canals, the diking of the rivers, and the draining of the marshlands. He loved to go sailing on these waterways, and especially in the marshlands. The waters, as his name denoted, were indeed his home. He built his "great house" in the city he had founded at the edge of the marshlands, a city appropriately named ḤA.A.KI ("place of the water-fishes"); it was also known as E.RI.DU ("home of going afar").

Ea was "Lord of the Saltwaters," the seas and oceans. Sumerian texts speak repeatedly of a very early time when the three Great Gods divided the realms among them. "The seas they had given to Enki, the Prince of Earth," thereby giving Enki "the rulership of the Apsu" (the "Deep"). As Lord of the Seas, Ea built ships that sailed to far lands, and especially to places from which precious metals and semiprecious stones were brought to Sumer.

The earliest Sumerian cylinder seals depicted Ea as a deity surrounded by flowing streams that were sometimes shown to contain fish. The seals associated Ea, as shown here, with the Moon (indicated by its crescent), an association stemming perhaps from the fact that the Moon caused the tides of the seas. It was no doubt in reference to such an astral image that Ea was given the epithet NIN.IGI.KU ("lord bright-eye"). (Fig. 46)

Fig. 46

According to the Sumerian texts, including a truly amazing autobiography by Ea himself, he was born in the heavens and came down to Earth before there was any settlement or civilization upon Earth. "When I approached the land, there was much flooding," he stated. He then proceeded to describe the series of actions taken by him to make the land habitable: He filled the Tigris River with fresh, "life-giving waters"; he appointed a god to supervise the construction of canals, to make the Tigris and Euphrates navigable; and he unclogged the marshlands, filling them up with fish and making them a haven for birds of all kinds, and causing to grow there reeds that were a useful building material.

Turning from the seas and rivers to the dry land, Ea claimed that it was he who "directed the plow and the yoke . . . opened the holy furrows . . . built the stalls . . . erected sheepfolds." Continuing, the self-adulatory text (named by scholars "Enki and the World Order") credited the god with bringing to Earth the arts of brickmaking, construction of dwellings and cities, metallurgy, and so on.

Presenting the deity as Mankind's greatest benefactor, the god who brought about civilization, many texts also depicted him as Mankind's chief protagonist at the councils of the gods. Sumerian and Akkadian Deluge texts, on which the biblical account must have drawn, depict Ea as the god who—in defiance of the decision of the Assembly of the Gods—enabled a trusted follower (the Mesopotamian "Noah") to escape the disaster.

Indeed, the Sumerian and Akkadian texts, which (like the Old Testament) adhered to the belief that a god or the gods created Man through a conscious and deliberate act, attribute to Ea a key role: As the chief scientist of the gods, he outlined the method and the process by which Man was to be created. With such affinity to the "creation" or emergence of Man, no wonder that it was Ea who guided Adapa—the "model man" created by Ea's "wisdom" —to the abode of Anu in the heavens, in defiance of the gods' determination to withhold "eternal life" from Mankind.

Was Ea on the side of Man simply because he had a hand in his creation, or did he have other, more subjective motives? As we scan the record, we find that invariably Ea's defiance—in mortal and divine matters alike—was aimed mostly at frustrating decisions or plans emanating from Enlil.

The record is replete with indications of Ea's burning jealousy of his brother Enlil. Indeed, Ea's other (and perhaps first) name was EN.KI ("lord of Earth"), and the texts dealing with the division of the world among the three gods hint that it may have been simply by a drawing of lots that Ea lost mastery of Earth to his brother Enlil.

> The gods had clasped hands together,
> Had cast lots and had divided.
> Anu then went up to Heaven;
> To Enlil the Earth was made subject.
> The seas, enclosed as with a loop,
> They had given to Enki, the Prince of Earth.

As bitter as Ea/Enki may have been about the results of this drawing, he appears to have nurtured a much deeper resentment. The reason is given by Enki himself in his autobiography: It was he, not Enlil, who was firstborn, Enki claimed; it was then he, and not Enlil, who was entitled to be the heir apparent to Anu:

> "My father, the king of the universe,
> brought me forth in the universe. . . .
> I am the fecund seed,
> engendered by the Great Wild Bull;
> I am the first born son of Anu.
> I am the Great Brother of the gods. . . .
> I am he who has been born
> as the first son of the divine Anu."

Since the codes of laws by which men lived in the ancient Near East were given by the gods, it stands to reason that the social and family laws applying to men were copies of those applying to the gods. Court and family records found at such sites as Mari and Nuzi have confirmed that the biblical customs and laws by which the Hebrew patriarchs lived were the laws by which kings and noblemen were bound throughout the ancient Near East. The succession problems the patriarchs faced are therefore instructive.

Abraham, deprived of a child by the apparent barrenness of his wife Sarah, had a firstborn son by her maidservant. Yet this son (Ishmael) was excluded from the patriarchal succession as soon as Sarah herself bore Abraham a son, Isaac.

Isaac's wife Rebecca was pregnant with twins. The one who was technically firstborn was Esau—a reddish, hairy, and rugged fellow. Holding onto Esau's heel was the more refined Jacob, whom Rebecca cherished. When the aging and half-blind Isaac was about to proclaim his testament, Rebecca used a ruse to have the blessing of succession bestowed on Jacob rather than on Esau.

Finally, Jacob's succession problems resulted from the fact that though he served Laban for twenty years to get the hand of Rachel in marriage, Laban forced him to marry her older sister Leah first. It was Leah who bore Jacob his first son (Reuben), and he had more sons and a daughter by her and by two concubines. Yet when Rachel finally bore him *her* firstborn son (Joseph), Jacob preferred him over his brothers.

Against the background of such customs and succession laws, one can understand the conflicting claims between Enlil and Ea/Enki. Enlil, by all records the son of Anu and his official consort Antu, was the *legal* firstborn. But the anguished cry of Enki: "*I* am the fecund seed . . . I am the first born son of Anu," must have been a statement of fact. Was he then born to Anu, but by another goddess who was only a concubine? The tale of Isaac and Ishmael, or the story of the twins Esau and Jacob, may have had a prior parallel in the Heavenly Abode.

Though Enki appears to have accepted Enlil's succession prerogatives, some scholars see enough evidence to show a continuing power struggle between the two gods. Samuel N. Kramer has titled one of the ancient texts "Enki and His Inferiority Complex." As we shall see later on, several biblical tales—of Eve and the serpent in the Garden of Eden, or the tale of the Deluge—involve in their original Sumerian versions instances of defiance by Enki of his brother's edicts.

At some point, it seems, Enki decided that there was no sense to his struggle for the Divine Throne; and he put his efforts into making a son of his—rather than a son of Enlil—the third-generation successor. This he sought to achieve, at least at first, with the aid of his sister NIN.ḪUR.SAG ("lady of the mountainhead").

She, too, was a daughter of Anu, but evidently not by Antu, and therein lay another rule of succession. Scholars have wondered in years past why both Abraham and Isaac advertised the fact that their respective wives were also their sisters—a puzzling claim in view of the biblical prohibition against sexual relations with a sister. But as the legal documents were unearthed at Mari and Nuzi, it became clear that a man could marry a half-sister. Moreover, when all the children of all the wives were considered, the son born of such a wife—being fifty percent more of the "pure seed" than a son by an unrelated wife—was the legal heir whether or not he was the firstborn son. This, incidentally, led (in Mari and Nuzi) to the practice of adopting the preferred wife as a "sister" in order to make her son the unchallenged legal heir.

It was of such a half-sister, Ninḫursag, that Enki sought to have a son. She, too, was "of the heavens," having come to Earth in earliest times. Several texts state that when the gods were dividing Earth's domains among themselves, she was given the Land of Dilmun—"a pure place . . . a pure land . . . a place most bright." A text named by the scholars "Enki and Ninḫursag—a Paradise Myth" deals with Enki's trip to Dilmun for conjugal purposes. Ninḫursag, the text repeatedly stresses, "was alone"—unattached, a spinster. Though in later times she was depicted as an old matron, she must have been very attractive when she was younger, for the text informs us unabashedly that, when Enki neared her, the sight of her "caused his penis to water the dikes."

Instructing that they be left alone, Enki "poured the semen in the womb of Ninḫursag. She took the semen into the womb, the semen of Enki"; and then, "after the nine months of Womanhood . . . she gave birth at the bank of the waters." But the child was a daughter.

Having failed to obtain a male heir, Enki then proceeded to make love to his own daughter. "He embraced her, he kissed her; Enki poured the semen into the womb." But she, too, bore him a daughter. Enki then went after his granddaughter and made her pregnant, too; but once again the offspring was a female. Determined to stop these efforts, Ninhursag put a curse on him whereby Enki, having eaten some plants, became mortally sick. The other gods, however, forced Ninhursag to remove the curse.

While these events had great bearing on divine affairs, other tales pertaining to Enki and Ninhursag have great bearing on human affairs; for, according to the Sumerian texts, Man was created by Ninhursag following processes and formulas devised by Enki. She was the chief nurse, the one in charge of medical facilities; it was in that role that the goddess was called NIN.TI ("lady-life"). (Fig. 47)

Some scholars read in *Adapa* (the "model man" of Enki) the biblical *Adama*, or Adam. The double meaning of the Sumerian TI also raises biblical parallels. For *ti* could mean both "life" and "rib," so that Ninti's name meant both "lady of life" and "lady of the rib." The biblical Eve —whose name meant "life"—was created out of Adam's rib, so Eve, too, was in a way a "lady of life" and a "lady of the rib."

As giver of life to gods and Man alike, Ninhursag was spoken of as the Mother Goddess. She was nicknamed "Mammu"—the forerunner of our "mom" or "mamma"— and her symbol was the "cutter"—the tool used in antiquity by midwives to cut the umbilical cord after birth. (Fig. 48)

•

Enlil, Enki's brother and rival, did have the good fortune to achieve such a "rightful heir" by his sister Ninhursag. The youngest of the gods upon Earth who were born in the heavens, his name was NIN.UR.TA ("lord who completes the foundation"). He was "the heroic son of Enlil who went forth with net and rays of light" to battle for his father; "the avenging son . . . who launched bolts of light." (Fig. 49) His spouse BA.U was also a nurse or a doctor; her epithet was "lady who the dead brings back to life."

The ancient portraits of Ninurta showed him holding a unique weapon—no doubt the very one that could shoot "bolts of light." The ancient texts hailed him as a mighty

Fig. 47

Fig. 48

Fig. 49

hunter, a fighting god renowned for his martial abilities. But his greatest heroic fight was not in behalf of his father but for his own sake. It was a wide-ranging battle with an evil god named ZU ("wise"), and it involved no less a prize than the leadership of the gods on Earth; for Zu had illegally captured the insignia and objects Enlil had held as Chief of the Gods.

The texts describing these events are broken at the beginning, and the story becomes legible only from the point when Zu arrives at the E-Kur, the temple of Enlil. He is apparently known, and of some rank, for Enlil welcomes him, "entrusting to him the guarding of the entrance to his shrine." But the "evil Zu" was to repay trust with betrayal, for it was "the removal of the Enlilship"—the seizing of the divine powers—that "he conceived in his heart."

To do so, Zu had to take possession of certain objects, including the magical Tablet of Destinies. The wily Zu seized his opportunity when Enlil undressed and went into the pool for his daily swim, leaving his paraphernalia unattended.

> At the entrance of the sanctuary,
> which he had been viewing,
> Zu awaits the start of day.
> As Enlil was washing with pure water—
> his crown having been removed
> and deposited on the throne—
> Zu seized the Tablet of Destinies in his hands,
> took away the Enlilship.

As Zu fled in his MU (translated "name," but indicating a flying machine) to a faraway hideaway, the consequences of his bold act were beginning to take effect.

> Suspended were the Divine Formulas;
> Stillness spread all over; silence prevailed. . . .
> The Sanctuary's brilliance was taken off.

"Father Enlil was speechless." "The gods of the land gathered one by one at the news." The matter was so grave that even Anu was informed at his Heavenly Abode.

He reviewed the situation and concluded that Zu must be apprehended so that the "formulas" could be restored. Turning "to the gods, his children," Anu asked, "Which of the gods will smite Zu? His name shall be greatest of all!"

Several gods known for their valor were called in. But they all pointed out that having taken the Tablet of Destinies, Zu now possessed the same powers as Enlil, so that "he who opposes him becomes like clay." At this point, Ea had a great idea: Why not call upon Ninurta to take up the hopeless fight?

The assembled gods could not have missed Ea's ingenious mischief. Clearly, the chances of the succession falling to his own offspring stood to increase if Zu were defeated; likewise, he could benefit if Ninurta were killed in the process. To the amazement of the gods, Ninhursag (in this text called NIN.MAH—"great lady"), agreed. Turning to her son Ninurta, she explained to him that Zu robbed not only Enlil but Ninurta, too, of the Enlilship. "With shrieks of pain I gave birth," she shouted, and it was she who "made certain for my brother and for Anu" the continued "Kingship of Heaven." So that her pains not be in vain, she instructed Ninurta to go out and fight to win:

> Launch thy offensive . . . capture the fugitive Zu. . . .
> Let thy terrifying offensive rage against him. . . .
> Slit his throat! Vanquish Zu! . . .
> Let thy seven ill Winds go against him. . . .
> Cause the entire Whirlwind to attack him. . . .
> Let thy Radiance go against him. . . .
> Let thy Winds carry his Wings to a secret place. . . .
> Let sovereignty return to Ekur;
> Let the Divine Formulas return
> to the father who begot thee.

The various versions of the epic then provide thrilling descriptions of the battle that ensued. Ninurta shot "arrows" at Zu, but "the arrows could not approach Zu's body . . . while he bore the Tablet of Destinies of the gods in his hand." The launched "weapons were stopped in the midst" of their flight. As the inconclusive battle wore on, Ea advised Ninurta to add a *til-lum* to his weapons, and shoot it into the "pinions," or small cog-wheels, of Zu's "wings." Following this advice, and shouting "Wing to wing," Ninurta shot the *til-lum* at Zu's pinions. Thus hit, the pinions began to scatter, and the "wings" of Zu fell in a swirl. Zu was vanquished, and the Tablets of Destiny returned to Enlil.

•

Who was Zu? Was he, as some scholars hold, a "mythological bird"?

Evidently he could fly. But so can any man today who takes a plane, or any astronaut who goes up in a spaceship. Ninurta, too, could fly, as skillfully as Zu (and perhaps better). But he himself was not a bird of any kind, as his many depictions, by himself or with his consort BA.U (also called GU.LA), make abundantly clear. Rather, he did his flying with the aid of a remarkable "bird," which was kept at his sacred precinct (the GIR.SU) in the city of Lagash.

Nor was Zu a "bird"; apparently he had at his disposal a "bird" in which he could fly away into hiding. It was from within such "birds" that the sky battle took place between the two gods. And there can be no doubt regarding the nature of the weapon that finally smote Zu's "bird." Called TIL in Sumerian and *til-lum* in Assyrian, it was written pictorially thus: $\succ\!\!-\!\!-\!\!\rhd\!\!-$, and it must have meant then what *til* means nowadays in Hebrew: "missile."

Zu, then, was a god—one of the gods who had reason to scheme at usurpation of the Enlilship; a god whom Ninurta, as the legitimate successor, had every reason to fight.

Was he perhaps MAR.DUK ("son of the pure mound"), Enki's firstborn by his wife DAM.KI.NA, impatient to seize by a ruse what was not legally his?

There is reason to believe that, having failed to achieve a son by his sister and thus produce a legal contender for the Enlilship, Enki relied on his son Marduk. Indeed, when the ancient Near East was seized with great social and military upheavals at the beginning of the second millennium B.C., Marduk was elevated in Babylon to the status of national god of Sumer and Akkad. Marduk was proclaimed King of the Gods, replacing Enlil, and the other gods were required to pledge allegiance to him and to come to reside in Babylon, where their activities could easily be supervised. (Fig. 50)

Fig. 50

This usurpation of the Enlilship (long after the incident with Zu) was accompanied by an extensive Babylonian effort to forge the ancient texts. The most important texts were rewritten and altered so as to make Marduk appear as the Lord of Heavens, the Creator, the Benefactor, the Hero, instead of Anu or Enlil or even Ninurta. Among the texts altered was the "Tale of Zu"; and according to the Babylonian version it was Marduk (not Ninurta) who fought Zu. In this version, Marduk boasted: *"Maḥaṣti moḥ il Zu"* ("I have crushed the skull of the god Zu"). Clearly, then, Zu could not have been Marduk.

Nor would it stand to reason that Enki, "God of Sciences," would have coached Ninurta regarding the choice and use of the successful weapons against his own son Marduk. Enki, to judge by his behavior as well as by his urging Ninurta to "cut the throat of Zu," expected to gain from the fight, no matter who lost. The only logical conclusion is that Zu, too, was in some way a *legal* contender to the Enlilship.

This leaves only one god: Nanna, the firstborn of Enlil by his official consort Ninlil. For if Ninurta were eliminated, Nanna would be in the unobstructed line of succession.

Nanna (short for NAN.NAR—"bright one") has come down to us through the ages better known by his Akkadian (or "Semitic") name Sin.

As firstborn of Enlil, he was granted sovereignty over Sumer's best-known city-state, UR (*"The* City"). His temple there was called E.GISH.NU.GAL ("house of the seed of the throne"). From that abode, Nanna and his consort NIN.GAL ("great lady") conducted the affairs of the city and its people with great benevolence. The people of Ur reciprocated with great affection for their divine rulers, lovingly calling their god "Father Nanna" and other affectionate nicknames.

The prosperity of Ur was attributed by its people directly to Nanna. Shulgi, a ruler of Ur (by the god's grace) at the end of the third millennium B.C., described the "house" of Nanna as "a great stall filled with abundance," a "bountiful place of bread offerings," where sheep multiplied and oxen were slaughtered, a place of sweet music where the drum and timbrel sounded.

Under the administration of its god-protector Nanna, Ur became the granary of Sumer, the supplier of grains as well as of sheep and cattle to other temples elsewhere. A "Lamentation over the Destruction of Ur" informs us, in a negative way, of what Ur was like before its demise:

In the granaries of Nanna there was no grain.
The evening meals of the gods were suppressed;
in their great dining halls, wine and honey ended. . . .
In his temple's lofty oven, oxen and sheep are not
    prepared;

The hum has ceased at Nanna's great Place of
  Shackles:
that house where commands for the ox were
  shouted—
its silence is overwhelming. . . .
Its grinding mortar and pestle lie inert. . . .
The offering boats carried no offerings. . . .
Did not bring offering bread to Enlil in Nippur.
Ur's river is empty, no barge moves on it. . . .
No foot trods its banks; long grasses grow there.

Another lamentation, bewailing the "sheepfolds that
have been delivered to the wind," the abandoned stables,
the shepherds and herdsmen that were gone, is most unusual: It was not written by the people of Ur, but by the
god Nanna and his spouse Ningal themselves. These and
other lamentations over the fall of Ur disclose the trauma of some unusual event. The Sumerian texts inform
us that Nanna and Ningal left the city before its demise
became complete. It was a hasty departure, touchingly
described.

Nanna, who loved his city,
  departed from the city.
Sin, who loved Ur,
  no longer stayed in his House.
Ningal . . .
  fleeing her city through enemy territory,
hastily put on a garment,
  departed from her House.

The fall of Ur and the exile of its gods have been depicted in the lamentations as the results of a deliberate
decision by Anu and Enlil. It was to the two of them that
Nanna appealed to call off the punishment.

May Anu, the king of the gods,
  utter: "It is enough";
May Enlil, the king of the lands,
  decree a favorable fate!

Appealing directly to Enlil, "Sin brought his suffering

heart to his father; curtsied before Enlil, the father who begot him," and begged him:

> O my father who begot me,
> Until when will you look inimically
> upon my atonement?
> Until when? . . .
> On the oppressed heart that you have made
> flicker like a flame—
> please cast a friendly eye.

Nowhere do the lamentations disclose the *cause* of Anu's and Enlil's wrath. But if Nanna were Zu, the punishment would have justified his crime of usurpation. *Was* he Zu?

He certainly could have been Zu because Zu was in possession of some kind of flying machine—the "bird" in which he escaped and from which he fought Ninurta. Sumerian psalms spoke in adoration of his "Boat of Heaven."

> Father Nannar, Lord of Ur . . .
> Whose glory in the sacred Boat of Heaven is . . .
> Lord, firstborn son of Enlil.
> When in the Boat of Heaven thou ascendeth,
> Thou art glorious.
> Enlil hath adorned thy hand
> With a scepter everlasting
> When over Ur in the Sacred Boat thou mountest.

Before the discovery of new fragments of the Tale of Zu, it was assumed that because Nanna's other name, Sin, derived from SU.EN which was similar to ZU.EN, the two were identical. But we now know that Zu's full name was AN.ZU—yet another contender for the Enlilship.

Whatever the role of Nanna/Sin in those events, he was removed from direct contention for the Enlilship. Both Sumerian texts, as well as archaeological evidence, indicate that Sin and his spouse fled to Haran, the Hurrian city protected by several rivers and mountainous terrain. It is noteworthy that when Abraham's clan, led by his father Terah, left Ur, they also set their course to Haran, where they stayed for many years en route to the Promised Land.

Though Ur remained for all time a city dedicated to Nanna/Sin, Haran must have been his residence for a very long time, for it was made to resemble Ur—its temples, buildings, and streets—almost exactly. André Parrot (*Abraham et son temps*) sums up the similarities by saying that "there is every evidence that the cult of Harran was nothing but an exact replica of that of Ur."

When the temple of Sin at Haran—built and rebuilt over the millennia—was uncovered during excavations that lasted more than fifty years, the finds included two stelae (memorial stone pillars) on which a unique record was inscribed. It was a record dictated by Adadguppi, a high priestess of Sin, of how she prayed and planned for the return of Sin, for, at some unknown prior time,

> Sin, the king of all the gods,
> became angry with his city and his temple,
> and went up to Heaven.

That Sin, disgusted or despairing, just "packed up" and "went up to Heaven" is corroborated by other inscriptions. These tell us that the Assyrian king Ashurbanipal retrieved from certain enemies a sacred "cylinder seal of the costliest jasper" and "had it improved by drawing upon it a picture of Sin." He further inscribed upon the sacred stone "a eulogy of Sin, and hung it around the neck of the image of Sin." That stone seal of Sin must have been a relic of olden times, for it is further stated that "it is the one whose face had been damaged in those days, during the destruction wrought by the enemy."

The high priestess, who was born during the reign of Ashurbanipal, is assumed to have been of royal blood herself. In her appeals to Sin, she proposed a practical "deal": the restoration of his powers over his adversaries in return for helping her son Nabunaid become ruler of Sumer and Akkad. Historical records confirm that in the year 555 B.C. Nabunaid, then commander of the Babylonian armies, was named by his fellow officers to the throne. In this he was stated to have been directly helped by Sin. It was, the inscriptions by Nabunaid inform us, "on the first day of his appearance" that Sin, using "the weapon of Anu"—was able to "touch with a beam of light" the skies and crush the enemies down on Earth below.

Nabunaid kept his mother's promise to the god. He rebuilt Sin's temple E.HUL.HUL ("house of great joy") and declared Sin to be Supreme God. It was then that Sin was able to grasp in his hands "the power of the Anu-office, wield all the power of the Enlil-office, take over the power of the Ea-office—holding thus in his own hand all the Heavenly Powers." Thus defeating the usurper Marduk, even capturing the powers of Marduk's father Ea, Sin assumed the title of "Divine Crescent" and established his reputation as the so-called Moon God.

How could Sin, reported to have gone back to Heaven in disgust, have been able to perform such feats back on Earth?

Nabunaid, confirming that Sin had indeed "forgotten his angry command . . . and decided to return to the temple Ehulhul," claimed a miracle. A miracle "that has not happened to the Land since the days of old" had taken place: A deity "has come down from Heaven."

> This is the great miracle of Sin,
> That has not happened to the Land
> Since the days of old;
> That the people of the Land
> Have not seen, nor had written
> On clay tablets, to preserve forever:
> That Sin,
> Lord of all the gods and goddesses,
> Residing in Heaven,
> Has come down from Heaven.

Regrettably, no details are provided of the place and manner in which Sin landed back on Earth. But we do know that it was in the fields outside of Haran that Jacob, on his way from Canaan to find himself a bride in the "old country," saw "a ladder set up on the earth and its top reaching heavenward, and there were angels of the Lord ascending and descending by it."

●

At the same time that Nabunaid restored the powers and temples of Nanna/Sin, he also restored the temples and worship of Sin's twin children, IN.ANNA ("Anu's lady") and UTU ("the shining one").

The two were born to Sin by his official spouse Ningal, and were thus by birth members of the Divine Dynasty. Inanna was technically the firstborn, but her twin brother Utu was the firstborn *son,* and thus the legal dynastic heir. Unlike the rivalry that existed in the similar instance of Esau and Jacob, the two divine children grew up very close to each other. They shared experiences and adventures, came to each other's aid, and when Inanna had to choose a husband from one of two gods, she turned to her brother for advice.

Inanna and Utu were born in time immemorial, when only the gods inhabited Earth. Utu's city-domain Sippar was listed among the very first cities to have been established by the gods in Sumer. Nabunaid stated in an inscription that when he undertook to rebuild Utu's temple E.BABBARA ("shining house") in Sippar:

> I sought out its ancient foundation-platform,
> and I went down eighteen cubits into the soil.
> Utu, the Great Lord of Ebabbara . . .
> Showed me personally the foundation-platform
> of Naram-Sin, son of Sargon, which for 3,200 years
> no king preceding me had seen.

When civilization blossomed in Sumer, and Man joined the gods in the Land Between the Rivers, Utu became associated primarily with law and justice. Several early law codes, apart from invoking Anu and Enlil, were also presented as requiring acceptance and adherence because they were promulgated "in accordance with the true word of Utu." The Babylonian king Hammurabi inscribed his law code on a stela, at the top of which the king is depicted receiving the laws from the god. (Fig. 51)

Tablets uncovered at Sippar attest to its reputation in ancient times as a place of just and fair laws. Some texts depict Utu himself as sitting in judgment on gods and men alike; Sippar was, in fact, the seat of Sumer's "supreme court."

Fig. 51

The justice advocated by Utu is reminiscent of the Sermon on the Mount recorded in the New Testament. A "wisdom tablet" suggested the following behavior to please Utu:

> Unto your opponent do no evil;
> Your evildoer recompense with good.
> Unto your enemy, let justice be done. . . .
> Let not your heart be induced to do evil. . . .
> To the one begging for alms—
> give food to eat, give wine to drink. . . .
> Be helpful; do good.

Because he assured justice and prevented oppression—and perhaps for other reasons, too, as we shall see later on—Utu was considered the protector of travelers. Yet the most common and lasting epithets applied to Utu concerned his brilliance. From earliest times, he was called Babbar ("shining one"). He was "Utu, who sheds a wide light," the one who "lights up Heaven and Earth."

Hammurabi, in his inscription, called the god by his Akkadian name, Shamash, which in Semitic languages means "Sun." It has therefore been assumed by the scholars that Utu/Shamash was the Mesopotamian Sun God. We shall show, as we proceed, that while this god was assigned the Sun as his celestial counterpart, there was another aspect to the statements that he "shed a bright light" when he performed the special tasks assigned to him by his grandfather Enlil.

•

Just as the law codes and the court records are human testimonials to the actual presence among the ancient peoples of Mesopotamia of a deity named Utu/Shamash, so there exist endless inscriptions, texts, incantations, oracles, prayers, and depictions attesting to the physical presence and existence of the goddess Inanna, whose Akkadian name was Ishtar. A Mesopotamian king in the thirteenth century B.C. stated that he had rebuilt her temple in her brother's city of Sippar, on foundations that were eight hundred years old in his time. But in her central city, Uruk, tales of her went back to olden times.

Known to the Romans as Venus, to the Greeks as Aphrodite, to the Canaanites and the Hebrews as Astarte, to the Assyrians and Babylonians and Hittites and the other ancient peoples as Ishtar or Eshdar, to the Akkadians and the Sumerians as Inanna or Innin or Ninni, or by others of her many nicknames and epithets, she was at all times the Goddess of Warfare and the Goddess of Love, a fierce, beautiful female who, though only a great-granddaughter of Anu, carved for herself, by herself, a major place among the Great Gods of Heaven and Earth.

As a young goddess she was, apparently, assigned a domain in a far land east of Sumer, the Land of Aratta. It was there that "the lofty one, Inanna, queen of all the land," had her "house." But Inanna had greater ambitions. In the city of Uruk there stood the great temple of Anu, occupied only during his occasional state visits to Earth; and Inanna set her eyes on this seat of power.

Sumerian king lists state that the first nondivine ruler of Uruk was Meshkiaggasher, a son of the god Utu by a human mother. He was followed by his son Enmerkar, a great Sumerian king. Inanna, then, was the great-aunt of Enmerkar; and she found little difficulty in persuading him that she should really be the goddess of Uruk, rather than of the remote Aratta.

A long and fascinating text named "Enmerkar and the Lord of Aratta" describes how Enmerkar sent emissaries to Aratta, using every possible argument in a "war of nerves" to force Aratta to submit because "the lord Enmerkar who is the servant of Inanna made her queen of the House of Anu." The epic's unclear end hints at a happy ending: While Inanna moved to Uruk, she did not "abandon her House in Aratta." That she might have become a "commuting goddess" is not so improbable, for Inanna/Ishtar was known from other texts as an adventurous traveler.

Her occupation of Anu's temple in Uruk could not have taken place without his knowledge and consent; and the texts give us strong clues as to how such consent was obtained. Soon Inanna was known as "Anunitum," a nickname meaning "beloved of Anu." She was referred to in texts as "the holy mistress of Anu"; and it follows that Inanna shared not only Anu's temple but also his bed—whenever he came to Uruk, or on the reported occasions of her going up to his Heavenly Abode.

Having thus maneuvered herself into the position of goddess of Uruk and mistress of the temple of Anu, Ishtar proceeded to use trickery for enhancing Uruk's standing and her own powers. Farther down the Euphrates stood the ancient city of Eridu—Enki's center. Knowing of his great knowledge of all the arts and sciences of civilization, Inanna resolved to beg, borrow, or steal these secrets. Obviously intending to use her "personal charms" on Enki (her great-uncle), Inanna arranged to call on him alone. That fact was not unnoticed by Enki, who instructed his housemaster to prepare dinner for two.

Come my housemaster Isimud, hear my instructions;
a word I shall say to you, heed my words:
The maiden, all alone, has directed her step to the
  Abzu . . .

Have the maiden enter the Abzu of Eridu,
Give her to eat barley cakes with butter,
Pour for her cold water that freshens the heart,
Give her to drink beer. . . .

Happy and drunk, Enki was ready to do anything for
Inanna. She boldly asked for the divine formulas, which
were the basis of a high civilization. Enki granted her some
one hundred of them, including divine formulas pertaining
to supreme lordship, Kingship, priestly functions, weapons,
legal procedures, scribeship, woodworking, even the knowledge of musical instruments and of temple prostitution. By
the time Enki awoke and realized what he had done,
Inanna was already well on her way to Uruk. Enki ordered
after her his "awesome weapons," but to no avail, for
Inanna had sped to Uruk in her "Boat of Heaven."

Quite frequently, Ishtar was depicted as a naked
goddess; flaunting her beauty, she was sometimes even
depicted raising her skirts to reveal the lower parts of her
body. (Fig. 52)

Gilgamesh, a ruler of Uruk circa 2900 B.C. who was also
partly divine (having been born to a human father and a
goddess), reported how Inanna enticed him—even after
she already had an official spouse. Having washed himself
after a battle and put on "a fringe cloak, fastened with a
sash,"

Fig. 52

Glorious Ishtar raised an eye at his beauty.
"Come, Gilgamesh, be thou my lover!
Come, grant me your fruit.
Thou shall be my male mate, I will be thy female."

But Gilgamesh knew the score. "Which of thy lovers didst thou love forever?" he asked. "Which of thy shepherds pleased thee for all time?" Reciting a long list of her love affairs, he refused.

As time went on—as she assumed higher ranks in the pantheon, and with it the responsibility for affairs of state —Inanna/Ishtar began to display more martial qualities, and was often depicted as a Goddess of War, armed to the teeth. (Fig. 53)

The inscriptions left by Assyrian kings describe how they went to war for her and upon her command, how she directly advised when to wait and when to attack, how she sometimes marched at the head of the armies, and how, on at least one occasion, she granted a theophany and appeared before all the troops. In return for their loyalty, she promised the Assyrian kings long life and success. "From a Golden Chamber in the skies I will watch over thee," she assured them.

Was she turned into a bitter warrior because she, too, came upon hard times with the rise of Marduk to supremacy? In one of his inscriptions Nabunaid said: "Inanna of Uruk, the exalted princess who dwelt in a gold cella, who rode upon a chariot to which were harnessed seven lions— the inhabitants of Uruk changed her cult during the rule of king Erba-Marduk, removed her cella and unharnessed her team." Inanna, reported Nabunaid, "had therefore left the E-Anna angrily, and stayed hence in an unseemly place" (which he does not name). (Fig 54)

Seeking, perhaps, to combine love with power, the much-courted Inanna chose as her husband DU.MU.ZI, a younger son of Enki. Many ancient texts deal with the loves and quarrels of the two. Some are love songs of great beauty and vivid sexuality. Others tell how Ishtar—back from one of her journeys—found Dumuzi celebrating her absence. She arranged for his capture and disappearance into the Lower World—a domain ruled by her sister E.RESH.KI.GAL and her consort NER.GAL. Some of the

Fig. 53

Fig. 54

most celebrated Sumerian and Akkadian texts deal with the journey of Ishtar to the Lower World in search of her banished beloved.

•

Of the six known sons of Enki, three have been featured in Sumerian tales: the firstborn Marduk, who eventually usurped the supremacy; Nergal, who became ruler of the Lower World; and Dumuzi, who married Inanna/Ishtar.

Enlil, too, had three sons who played key roles in both divine and human affairs: Ninurta, who, having been born to Enlil by his sister Ninhursag, was the legal successor; Nanna/Sin, firstborn by Enlil's official spouse Ninlil; and a younger son by Ninlil named ISH.KUR ("mountainous," "far mountain land"), who was more frequently called Adad ("beloved").

As brother of Sin and uncle of Utu and Inanna, Adad appears to have felt more at home with them than at his own house. The Sumerian texts constantly grouped the four together. The ceremonies connected with the visit of Anu to Uruk also spoke of the four as a group. One text, describing the entrance to the court of Anu, states that the throne room was reached through "the gate of Sin, Shamash, Adad, and Ishtar." Another text, first published by V. K. Shileiko (Russian Academy of the History of Material Cultures) poetically described the four as retiring for the night together.

The greatest affinity seems to have existed between Adad and Ishtar, and the two were even depicted next to each other, as on this relief showing an Assyrian ruler being blessed by Adad (holding the ring and lightning) and by Ishtar, holding her bow. (The third deity is too nutilated to be identified.) (Fig. 55)

Was there more to this "affinity" than a platonic relationship, especially in view of Ishtar's "record"? It is noteworthy that in the biblical Song of Songs, the playful girl calls her lover *dod*—a word that means both "lover" and "uncle." Now, was Ishkur called Adad—a derivative from the Sumerian DA.DA—because he was the uncle who was the lover?

But Ishkur was not only a playboy; he was a mighty god, endowed by his father Enlil with the powers and prerogatives of a storm god. As such he was revered as the Hurrian/Hittite Teshub and the Urartian Teshubu ("wind blower"), the Amorite Ramanu ("thunderer"), the Canaanite Ragimu ("caster of hailstones"), the Indo-European Buriash ("light maker"), the Semitic Meir ("he who lights up" the skies). (Fig. 56)

A god list kept at the British Museum, as shown by Hans Schlobies (*Der Akkadische Wettergott in Mesopotamen*), clarifies that Ishkur was indeed the divine lord in lands far from Sumer and Akkad. As Sumerian texts reveal, this was no accident. Enlil, it seems, willfully dispatched his young son to become the "Resident Deity" in the mountain lands north and west of Mesopotamia.

Why did Enlil dispatch his youngest and beloved son away from Nippur?

Several Sumerian epic tales have been found about the arguments and even bloody struggles among the younger gods. Many cylinder seals depict scenes of god battling god (Fig. 57); it would seem that the original rivalry between Enki and Enlil was carried on and intensified between their sons, with brother sometimes turning against brother—a divine tale of Cain and Abel. Some of these battles were against a deity identified as Kur—in all probability, Ishkur/Adad. This may well explain why Enlil deemed it advisable to grant his younger son a far-off domain, to keep him out of the dangerous battles for the succession.

The position of the sons of Anu, Enlil, and Enki, and of their offspring, in the dynastic lineage emerges clearly through a unique Sumerian device: the allocation of *numerical rank* to certain gods. The discovery of this system also brings out the membership in the Great Circle of Gods of Heaven and Earth when Sumerian civilization blossomed. We shall find that this Supreme Pantheon was made up of *twelve* deities.

The first hint that a cryptographic number system was applied to the Great Gods came with the discovery that the names of the gods Sin, Shamash, and Ishtar were sometimes substituted in the texts by the numbers 30, 20, and 15, respectively. The highest unit of the Sumerian sexagesimal system—60—was assigned to Anu; Enlil "was" 50; Enki, 40; and Adad, 10. The number 10 and its six

Fig. 55

Fig. 56

Fig. 57

multiples within the prime number 60 were thus assigned to *male* deities, and it would appear plausible that the numbers ending with 5 were assigned to the female deities. From this, the following cryptographic table emerges:

Male	Female
60—Anu	55—Antu
50—Enlil	45—Ninlil
40—Ea/Enki	35—Ninki
30—Nanna/Sin	25—Ningal
20—Utu/Shamash	15—Inanna/Ishtar
10—Ishkur/Adad	5—Ninhursag
6 male deities	6 female deities

Ninurta, we should not be surprised to learn, was assigned the number 50, like his father. In other words, his dynastic rank was conveyed in a cryptographic message: If Enlil goes, you, Ninurta, step into his shoes; but until then, you are not one of the Twelve, for the rank of "50" is occupied.

Nor should we be surprised to learn that when Marduk usurped the Enlilship, he insisted that the gods bestow on him "the *fifty* names" to signify that the rank of "50" had become his.

There were many other gods in Sumer—children, grandchildren, nieces, and nephews of the Great Gods; there were also several hundred rank-and-file gods, called Anunnaki, who were assigned (one may say) "general duties." But only *twelve* made the Great Circle. They, their family relationships, and, above all, the line of dynastic succession can better be referred to if we show them in a chart:

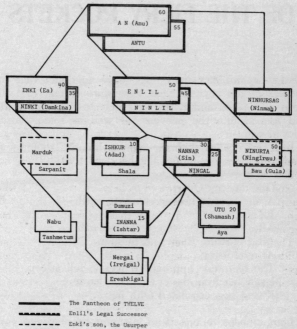

The Pantheon of TWELVE
Enlil's Legal Successor
Enki's son, the Usurper

60   The Rank Number of Succession

5
·
THE NEFILIM: PEOPLE OF THE FIERY ROCKETS

SUMERIAN AND AKKADIAN texts leave no doubt that the peoples of the ancient Near East were certain that the Gods of Heaven and Earth were able to rise from Earth and ascend into the heavens, as well as roam Earth's skies at will.

In a text dealing with the rape of Inanna/Ishtar by an unidentified person, he justifies his deed thus:

> One day my Queen,
> After crossing heaven, crossing earth—
> Inanna,
> After crossing heaven, crossing earth—
> After crossing Elam and Shubur,
> After crossing . . .
> The hierodule approached weary, fell asleep.
> I saw her from the edge of my garden;
> Kissed her, copulated with her.

Inanna, here described as roaming the heavens over many lands that lie far apart—feats possible only by *flying*—herself spoke on another occasion of her flying. In a text which S. Langdon (in *Revue d'Assyriologie et d'Archéologie Orientale*) named "A Classical Liturgy to Innini," the goddess laments her expulsion from her city. Acting on the instructions of Enlil, an emissary, who "brought to me the word of Heaven," entered her throne room, "his unwashed hands put on me," and, after other indignities,

> Me, from my temple,
> they caused to fly;
> A Queen am I whom, from my city,
> like a bird they caused to fly.

Such a capability, by Inanna as well as the other major gods, was often indicated by the ancient artists by depicting the gods—anthropomorphic in all other respects, as we have seen—with wings. The wings, as can be seen from numerous depictions, were not part of the body—not natural wings—but rather a decorative attachment to the god's clothing. (Fig. 58)

Inanna/Ishtar, whose far-flung travels are mentioned in many ancient texts, commuted between her initial distant domain in Aratta and her coveted abode in Uruk. She called upon Enki in Eridu and Enlil in Nippur, and visited

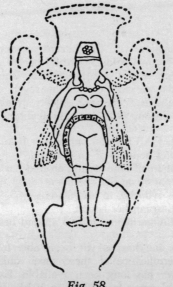

Fig. 58

her brother Utu at his headquarters in Sippar. But her most celebrated journey was to the Lower World, the domain of her sister Ereshkigal. The journey was the subject not only of epic tales but also of artistic depictions on cylinder seals—the latter showing the goddess with wings, to stress the fact that she flew over from Sumer to the Lower World. (Fig. 59)

Fig. 59

The texts dealing with this hazardous journey describe how Inanna very meticulously put on herself seven objects prior to the start of the voyage, and how she had to give them up as she passed through the seven gates leading to her sister's abode. Seven such objects are also mentioned in other texts dealing with Inanna's skyborne travels:

1. The SHU.GAR.RA she put on her head.
2. "Measuring pendants," on her ears.
3. Chains of small blue stones, around her neck.
4. Twin "stones," on her shoulders.
5. A golden cylinder, in her hands.
6. Straps, clasping her breast.
7. The PALA garment, clothed around her body.

Though no one has as yet been able to explain the nature and significance of these seven objects, we feel that the answer has long been available. Excavating the Assyrian capital Assur from 1903 to 1914, Walter Andrae and his colleagues found in the Temple of Ishtar a battered statue of the goddess showing her with various "contraptions" attached to her chest and back. In 1934 archaeologists excavating at Mari came upon a similar but intact statue buried in the ground. It was a life-size likeness of a beautiful woman. Her unusual headdress was adorned with a pair of horns, indicating that she was a goddess. Standing around the 4,000-year-old statue, the archaeologists were thrilled by her lifelike appearance (in a snapshot, one can hardly distinguish between the statue and the living men). They named her *The Goddess with a Vase* because she was holding a cylindrical object. (Fig. 60)

Unlike the flat carvings or bas-reliefs, this life-size, three-dimensional representation of the goddess reveals interesting features about her attire. On her head she wears not a milliner's chapeau but a special helmet; protruding from it on both sides and fitted over the ears are objects that remind one of a pilot's earphones. On her neck and upper chest the goddess wears a necklace of many small (and probably precious) stones; in her hands she holds a cylindrical object which appears too thick and heavy to be a vase for holding water.

Over a blouse of see-through material, two parallel straps run across her chest, leading back to and holding in place an unusual box of rectangular shape. The box is held tight against the back of the goddess's neck and is firmly attached to the helmet with a horizontal strap. Whatever the box held inside must have been heavy, for the contraption is further supported by two large shoulder pads. The weight of the box is increased by a hose that is connected to its base by a circular clasp. The complete package of instruments—for this is what they undoubtedly were—is held in place with the aid of the two sets of straps that crisscross the goddess's back and chest.

The parallel between the seven objects required by Inanna for her aerial journeys and the dress and objects worn by the statue from Mari (and probably also the mutilated one found at Ishtar's temple in Ashur) is easily proved. We see the "measuring pendants"—the earphones—on her ears; the rows or "chains" of small stones around her neck; the "twin stones"—the two shoulder pads—on her shoulders; the "golden cylinder" in her hands, and the

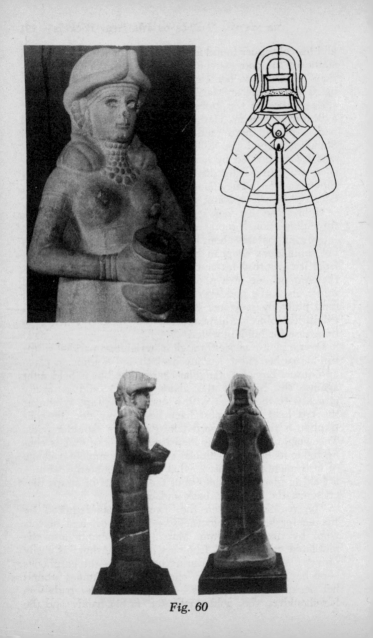

Fig. 60

clasping straps that crisscross her breast. She is indeed clothed in a "PALA garment" ("ruler's garment"), and on her head she wears the SHU.GAR.RA helmet—a term that literally means "that which makes go far into universe."

All this suggests to us that the attire of Inanna was that of an aeronaut or an astronaut.

The Old Testament called the "angels" of the Lord *malachim*—literally, "emissaries," who carried divine messages and carried out divine commands. As so many instances reveal, they were divine airmen: Jacob saw them going up a sky ladder, Hagar (Abraham's concubine) was addressed by them from the sky, and it was they who brought about the aerial destruction of Sodom and Gomorrah.

The biblical account of the events preceding the destruction of the two sinful cities illustrates the fact that these emissaries were, on the one hand, anthropomorphic in all respects, and, on the other hand, they could be identified as "angels" as soon as they were observed. We learn that their appearance was sudden. Abraham "raised his eyes and, lo and behold, there were three *men* standing by him." Bowing and calling them "My Lords," he pleaded with them, "Do not pass *over* thy servant," and prevailed on them to wash their feet, rest, and eat.

Having done as Abraham had requested, two of the angels (the third "man" turned out to be the Lord himself) then proceeded to Sodom. Lot, the nephew of Abraham, "was sitting at the gate of Sodom; and when he saw them he rose up to meet them and bowed to the ground, and said: If it pleases my Lords, pray come to the house of thy servant and wash your feet and sleep overnight." Then "he made for them a feast, and they ate." When the news of the arrival of the two spread in the town, "all the town's people, young and old, surrounded the house, and called out to Lot and said: Where are the *men* who came this night unto thee?"

How were these men—who ate, drank, slept, and washed their tired feet—nevertheless so instantly recognizable as angels of the Lord? The only plausible explanation is that what they wore—their helmets or uniforms—or what they carried—their weapons—made them immediately recognizable. That they carried distinctive weapons is certainly a possibility: The two "men" at Sodom, about to be lynched by the crowd, "smote the people at the entrance of the house with blindness . . . and they were unable to find the doorway." And another angel, this time appearing to Gideon, as he was chosen to be a Judge in Israel, gave him a divine sign by touching a rock with his baton, whereupon a fire jumped out of the rock.

The team headed by Andrae found yet another unusual depiction of Ishtar at her temple in Ashur. More a wall sculpture than the usual relief, it showed the goddess with a tight-fitting decorated helmet with the "earphones" extended as though they had their own flat antennas, and wearing very distinct goggles that were part of the helmet. (Fig. 61)

Needless to say, any man seeing a person—male or female—so clad, would at once realize that he is encountering a divine aeronaut.

Clay figurines found at Sumerian sites and believed to be some 5,500 years old may well be crude representations of such *malachim* holding wandlike weapons. In one instance the face is seen through a helmet's visor. In the other instance, the "emissary" wears the distinct divine conical headdress and a uniform studded with circular objects of unknown function. (Figs. 62, 63)

The eye slots or "goggles" of the figurines are a most interesting feature because the Near East in the fourth millennium B.C. was literally swamped with wafer-like figurines that depicted in a stylized manner the upper part of the deities, exaggerating their most prominent feature: a conical helmet with elliptical visors or goggles. (Fig. 64) A hoard of such figurines was found at Tell Brak, a prehistoric site on the Khabur River, the river on whose banks Ezekiel saw the divine chariot millennia later.

It is undoubtedly no mere coincidence that the Hittites, linked to Sumer and Akkad via the Khabur area, adopted as their written sign for "gods" the symbol ⟨ΦⲞ⟩, clearly borrowed from the "eye" figurines. It is also no wonder that this symbol or hieroglyph for "divine being," expressed in artistic styles, came to dominate the art not only of Asia Minor but also of the early Greeks during the Minoan and Mycenaean periods. (Fig. 65)

Fig. 61

Fig. 62

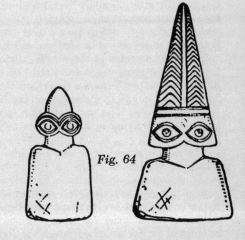

Fig. 63

Fig. 64

Fig. 65

The ancient texts indicate that the gods put on such special attire not only for their flights in Earth's skies but also when they ascended to the distant heavens. Speaking of her occasional visits to Anu at his Celestial Abode, Inanna herself explained that she could undertake such journeys because "Enlil himself fastened the divine ME-attire about my body." The text quoted Enlil as saying to her:

> You have lifted the ME,
> You have tied the ME to your hands,
> You have gathered the ME,
> You have attached the ME to your breast. . . .
> O Queen of all the ME, O radiant light
> Who with her hand grasps the seven ME.

An early Sumerian ruler invited by the gods to ascend to the heavens was named EN.ME.DUR.AN.KI, which literally meant "ruler whose *me* connect Heaven and Earth." An inscription by Nebuchadnezzar II, describing the reconstruction of a special pavilion for Marduk's "celestial chariot," states that it was part of the "fortified house of the seven *me* of Heaven and Earth."

The scholars refer to the *me* as "divine power objects." Literally, the term stems from the concept of "swimming in celestial waters." Inanna described them as parts of the "celestial garment" that she put on for her journeys in the Boat of Heaven. The *me* were thus parts of the special gear worn for flying in Earth's skies as well as into outer space.

The Greek legend of Icarus had him attempt to fly by attaching feathered wings to his body with wax. The evidence from the ancient Near East shows that though the gods may have been depicted with wings to indicate their flying capabilities—or perhaps sometimes put on winged uniforms as a mark of their airmanship—they never attempted to use attached wings for flying. Instead, they used vehicles for such travels.

The Old Testament informs us that the patriarch Jacob, spending the night in a field outside of Haran, saw "a ladder set up on Earth and its top reaching heavenwards," on which "angels of the Lord" were busily going up and down. The Lord himself stood at the top of the ladder. And the astounded Jacob "was fearful, and he said":

> Indeed, a God is present in this place,
> and I knew it not. . . .
> How awesome is this place!
> Indeed, this is none but the Lord's Abode
> and this is the Gateway to Heaven.

There are two interesting points in this tale. The first is that the divine beings going up and down at this "Gateway to Heaven" were using a mechanical facility—a "ladder." The second is that the sight took Jacob by complete surprise. The "Lord's Abode," the "ladder," and the "angels of the Lord" using it were not there when Jacob lay down to sleep in the field. Suddenly, there was the awesome "vision." And by morning the "Abode," the "ladder," and their occupants were gone.

We may conclude that the equipment used by the divine beings was some kind of craft that could appear over a place, hover for a while, and disappear from sight once again.

The Old Testament also reports that the prophet Elijah did not die on Earth, but "went up into Heaven by a Whirlwind." This was not a sudden and unexpected event: The ascent of Elijah to the heavens was prearranged. He was told to go to Beth-El ("the lord's house") on a specific day. Rumors had already spread among his disciples that he was about to be taken up to the heavens. When they queried his deputy whether the rumor was true, he confirmed that, indeed, "the Lord will take away the Master today." And then:

> There appeared a chariot of fire,
> and horses of fire. . . .
> And Elijah went up into Heaven
> by a Whirlwind.

Even more celebrated, and certainly better described, was the heavenly chariot seen by the prophet Ezekiel, who dwelt among the Judaean deportees on the banks of the Khabur River in northern Mesopotamia.

The Heavens were opened,
and I saw the appearances of the Lord.

What Ezekiel saw was a Manlike being, surrounded by brilliance and brightness, sitting on a throne that rested on a metal "firmament" within the chariot. The vehicle itself, which could move whichever way upon wheels-within-wheels and rise off the ground vertically, was described by the prophet as a glowing whirlwind.

And I saw
a Whirlwind coming from the north,
as a great cloud with flashes of fire
and brilliance all around it.
And within it, from within the fire,
there was a radiance like a glowing halo.

Some recent students of the biblical description (such as Josef F. Blumrich of the U.S. National Aeronautics and Space Administration) have concluded that the "chariot" seen by Ezekiel was a helicopter consisting of a cabin resting on four posts, each equipped with rotary wings— a "whirlwind" indeed.

About two millennia earlier, when the Sumerian ruler Gudea commemorated his building the temple for his god Ninurta, he wrote that there appeared to him "a man that shone like Heaven . . . by the helmet on his head, he was a god." When Ninurta and two divine companions appeared to Gudea, they were standing beside Ninurta's "divine black wind bird." As it turned out, the main purpose of the temple's construction was to provide a secure zone, an inner special enclosure within the temple grounds, for this "divine bird."

The construction of this enclosure, Gudea reported, required huge beams and massive stones imported from afar. Only when the "divine bird" was placed within the enclosure was the construction of the temple deemed completed. And, once in place, the "divine bird" "could lay hold on heaven" and was capable of "bringing together Heaven and Earth." The object was so important—"sacred" —that it was constantly protected by two "divine weapons," the "supreme hunter" and the "supreme killer"

—weapons that emitted beams of light and death-dealing rays.

The similarity of the biblical and Sumerian descriptions, both of the vehicles and the beings within them, is obvious. The description of the vehicles as "bird," "wind bird," and "whirlwind" that could rise heavenward while emitting a brilliance, leaves no doubt that they were some kind of flying machine.

Enigmatic murals uncovered at Tell Ghassul, a site east of the Dead Sea whose ancient name is unknown, may shed light on our subject. Dating to circa 3500 B.C., the murals depict a large eight-pointed "compass," the head of a helmeted person within a bell-shaped chamber, and two designs of mechanical craft that could well have been the "whirlwinds" of antiquity. (Fig. 66)

Fig. 66

The ancient texts also describe some vehicle used to lift aeronauts into the skies. Gudea stated that, as the "divine bird" rose to circle the lands, it "flashed upon the raised bricks." The protected enclosure was described as MU.NA.DA.TUR.TUR ("strong stone resting place of the MU"). Urukagina, who ruled in Lagash, said in regard to the "divine black wind bird": "The MU that lights up as a fire I made high and strong." Similarly, Lu-Utu, who ruled in Umma in the third millennium B.C., constructed a place for a *mu*, "which in a fire comes forth," for the god Utu, "in the appointed place within his temple."

The Babylonian king Nebuchadnezzar II, recording his rebuilding of Marduk's sacred precinct, said that within fortified walls made of burned brick and gleaming onyx marble:

> I raised the head of the boat ID.GE.UL
> the Chariot of Marduk's princeliness;
> The boat ZAG.MU.KU, whose approach is observed,
> the supreme traveler between Heaven and Earth,
> in the midst of the pavilion I enclosed,
> screening off its sides.

ID.GE.UL, the first epithet employed to describe this "supreme traveler," or "Chariot of Marduk," literally means "high to heaven, bright at night." ZAG.MU.KU, the second epithet describing the vehicle—clearly a "boat" nesting in a special pavilion—means "the bright MU which is for afar."

That a *mu*—an oval-topped, conical object—was indeed installed in the inner, sacred enclosure of the temples of the Great Gods of Heaven and Earth can, fortunately, be proved. An ancient coin found at Byblos (the biblical Gebal) on the Mediterranean coast of present-day Lebanon depicts the Great Temple of Ishtar. Though shown as it stood in the first millennium B.C., the requirement that temples be built and rebuilt upon the same site and in accordance with the original plan undoubtedly means that we see the basic elements of the original temple of Byblos, traced to millennia earlier.

The coin depicts a two-part temple. In front stands the main temple structure, imposing with its columned gateway. Behind it is an inner courtyard, or "sacred area," hidden and protected by a high, massive wall. It is clearly a raised area, for it can be reached only by ascending many stairs. (Fig. 67)

In the center of this sacred area stands a special platform, its crossbeam construction resembling that of the Eiffel Tower, as though built to withstand great weight. And on the platform stands the object of all this security and protection: an object that can only be a *mu*.

Fig. 67

Like most Sumerian syllabic words, *mu* had a primary meaning; in the case of *mu*, it was "that which rises straight." Its thirty-odd nuances encompassed the meanings "heights," "fire," "command," "a counted period," as well as (in later times) "that by which one is remembered." If we trace the written sign for *mu* from its Assyrian and Babylonian cuneiform stylizations to its original Sumerian pictographs, the following pictorial evidence emerges:

We clearly see a conical chamber, depicted by itself or with a narrow section attached to it. "From a golden chamber-in-the-sky I will watch over thee," Inanna promised to the Assyrian king. Was this *mu* the "heavenly chamber"?

A hymn to Inanna/Ishtar and her journeys in the Boat of Heaven clearly indicates that the *mu* was the vehicle in which the gods roamed the skies far and high:

Lady of Heaven:
She puts on the Garment of Heaven;
She valiantly ascends towards Heaven.
Over all the peopled lands
she flies in her MU.
Lady, who in her MU
to the heights of Heaven joyfully wings.
Over all the resting places
she flies in her MU.

There is evidence to show that the people of the eastern
Mediterranean had seen such a rocket-like object not only
in a temple enclosure but actually in flight. Hittite glyphs,
for example, showed—against a background of starry
heavens—cruising missiles, rockets mounted on launch
pads, and a god inside a radiating chamber. (Fig. 68)

Fig. 68

Professor H. Frankfort *(Cylinder Seals)*, demonstrating
how both the art of making the Mesopotamian cylinder
seals and the subjects depicted on them spread throughout
the ancient world, reproduces the design on a seal found
in Crete and dated to the thirteenth century B.C. The seal
design clearly depicts a rocket ship moving in the skies and
propelled by flames escaping from its rear. (Fig. 69)

The winged horses, the entwined animals, the winged
celestial globe, and the deity with horns protruding from
his headdress are all known Mesopotamian themes. It can

certainly be assumed that the fiery rocket shown on the Cretan seal was also an object familiar throughout the ancient Near East.

Indeed, a rocket with "wings" or fins—reachable by a "ladder"—can be seen on a tablet excavated at Gezer, a town in ancient Canaan, west of Jerusalem. The double imprint of the same seal also shows a rocket resting on the ground next to a palm tree. The celestial nature or destination of the objects is attested by symbols of the Sun, Moon, and zodiacal constellations that adorn the seal. (Fig. 70)

•

The Mesopotamian texts that refer to the inner enclosures of temples, or to the heavenly journeys of the gods, or even to instances where mortals ascended to the heavens, employ the Sumerian term *mu* or its Semitic derivatives *shu-mu* ("that which is a *mu*"), *sham*, or *shem*. Because the term also connoted "that by which one is remembered," the word has come to be taken as meaning "name." But the universal application of "name" to early texts that spoke of an object used in flying has obscured the true meaning of the ancient records.

Thus G. A. Barton *(The Royal Inscriptions of Sumer and Akkad)* established the unchallenged translation of Gudea's temple inscription—that "Its MU shall hug the lands from horizon to horizon"—as "Its *name* shall fill the lands." A hymn to Ishkur, extolling his "ray-emitting MU" that could attain the heights of Heaven, was likewise rendered: "Thy *name* is radiant, it reaches Heaven's zenith." Sensing, however, that *mu* or *shem* may mean an object and not "name," some scholars have treated the term as a suffix or grammatical phenomenon not requiring translation and have thereby avoided the issue altogether.

It is not too difficult to trace the etymology of the term, and the route by which the "sky chamber" assumed the meaning of "name." Sculptures have been found that show a god inside a rocket-shaped chamber, as in this object of extreme antiquity (now in the possession of the University Museum, Philadelphia) where the celestial nature of the chamber is attested by the twelve globes decorating it. (Fig. 71)

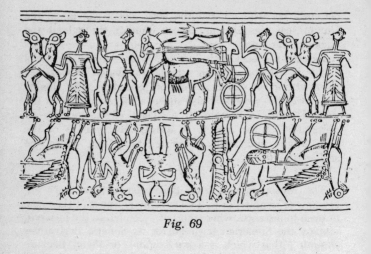

Fig. 69

Fig. 70

Many seals similarly depict a god (and sometimes two) within such oval "divine chambers"; in most instances, these gods within their sacred ovals were depicted as objects of veneration.

Wishing to worship their gods throughout the lands, and not only at the official "house" of each deity, the ancient peoples developed the custom of setting up imitations of the god within his divine "sky chamber." Stone pillars shaped to simulate the oval vehicle were erected at selected sites, and the image of the god was carved into the stone to indicate that he was within the object.

It was only a matter of time before kings and rulers—associating these pillars (called stelae) with the ability to ascend to the Heavenly Abode—began to carve their own images upon the stelae as a way of associating themselves with the Eternal Abode. If they could not escape a physical oblivion, it was important that at least their "name" be forever commemorated. (Fig. 72)

That the purpose of the commemorative stone pillars was to simulate a *fiery* skyship can further be gleaned from the term by which such stone stelae were known in antiquity. The Sumerians called them NA.RU ("stones that rise"). The Akkadians, Babylonians, and Assyrians called them *naru* ("objects that give off light"). The Amurru called them *nuras* ("fiery objects"—in Hebrew, *ner* still means a pillar that emits light, and thus today's "candle"). In the Indo-European tongues of the Hurrians and the Hittites, the stelae were called *hu-u-ashi* ("fire bird of stone").

Biblical references indicate familiarity with two types of commemorative monument, a *yad* and a *shem*. The prophet Isaiah conveyed to the suffering people of Judaea the Lord's promise of a better and safer future:

> And I will give them,
> In my House and within my walls,
> A *yad* and a *shem*.

Literally translated, this would amount to the Lord's promise to provide his people with a "hand" and a "name." Fortunately, however, from ancient monuments called *yad*'s that still stand in the Holy Land, we learn that they

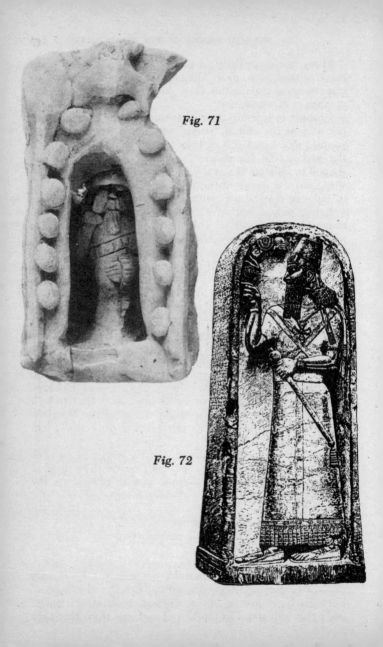

Fig. 71

Fig. 72

were distinguished by tops shaped like pyramidions. The *shem*, on the other hand, was a memorial with an *oval* top. Both, it seems evident, began as simulations of the "sky chamber," the gods' vehicle for ascending to the Eternal Abode. In ancient Egypt, in fact, the devout made pilgrimages to a special temple at Heliopolis to view and worship the *ben-ben*—a pyramidion-shaped object in which the gods had arrived on Earth in times immemorial. Egyptian pharaohs, on their deaths, were subjected to a ceremony of "opening of the mouth," in which they were supposed to be transported by a similar *yad* or a *shem* to the divine Abode of Eternal Life. (Fig. 73)

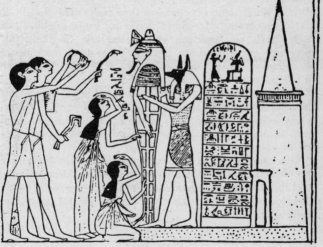

Fig. 73

The persistence of biblical translators to employ "name" wherever they encounter *shem* has ignored a farsighted study published more than a century ago by G. M. Redslob (in *Zeitschrift der Deutschen Morgenlandischen Gesellschaft*) in which he correctly pointed out that the term *shem* and the term *shamaim* ("heaven") stem from the root word *shamah*, meaning "that which is highward." When the Old Testament reports that King David "made a *shem*" to mark his victory over the Aramaeans, Redslob said, he did not "make a name" but set up a monument pointing skyward.

The realization that *mu* or *shem* in many Mesopotamian texts should be read not as "name" but as "sky vehicle" opens the way to the understanding of the true meaning of many ancient tales, including the biblical story of the Tower of Babel.

The Book of Genesis, in its eleventh chapter, reports on the attempt by humans to raise up a *shem*. The biblical account is given in concise (and precise) language that bespeaks historical fact. Yet generations of scholars and translators have sought to impart to the tale only an allegorical meaning because—as they understood it—it was a tale concerning Mankind's desire to "make a *name*" for itself. Such an approach voided the tale of its factual meaning; our conclusion regarding the true meaning of *shem* makes the tale as meaningful as it must have been to the people of antiquity themselves.

The biblical tale of the Tower of Babel deals with events that followed the repopulation of Earth after the Deluge, when some of the people "journeyed from the east, and they found a plain in the land of Shin'ar, and they settled there."

The Land of Shinar is, of course, the Land of Sumer, in the plain between the two rivers in southern Mesopotamia. And the people, already knowledgeable concerning the art of brickmaking and high-rise construction for an urban civilization, said:

"Let us build us a city,
and a tower whose top shall reach the heavens;
and let us make us a *shem*,
lest we be scattered upon the face of the Earth."

But this human scheme was not to God's liking.

And the Lord came down,
to see the city and the tower

which the Children of Adam had erected.
And he said: "Behold,
all are as one people with one language,
and this is just the beginning of their undertakings;
Now, anything which they shall scheme to do
shall no longer be impossible for them."

And the Lord said—to some colleagues whom the Old
Testament does not name:

"Come, let us go down,
and there confound their language;
So that they may not understand each other's speech."
And the Lord scattered them from there
upon the face of the whole Earth,
and they ceased to build the city.
Therefore was its name called Babel,
for there did the Lord mingle the Earth's tongue.

The traditional translation of *shem* as "name" has kept
the tale unintelligible for generations. Why did the ancient
residents of Babel—Babylonia—exert themselves to "make
a name," why was the "name" to be placed upon "a tower
whose top shall reach the heavens," and how could the
"making of a name" counteract the effects of Mankind's
scattering upon Earth?

If all that those people wanted was to make (as
scholars explain) a "reputation" for themselves, why did
this attempt upset the Lord so much? Why was the raising
of a "name" deemed by the Deity to be a feat after which
"anything which they shall scheme to do shall no longer
be impossible for them"? The traditional explanations
certainly are insufficient to clarify why the Lord found it
necessary to call upon other unnamed deities to go down
and put an end to this human attempt.

We believe that the answers to all these questions
become plausible—even obvious—once we read "skyborne
vehicle" rather than "name" for the word *shem*, which is
the term employed in the original Hebrew text of the Bible.
The story would then deal with the concern of Mankind
that, as the people spread upon Earth, they would lose
contact with one another. So they decided to build a "sky-

borne vehicle" and to erect a *launch tower* for such a vehicle so that they, too, could—like the goddess Ishtar, for example—fly in a *mu* "over all the peopled lands."

A portion of the Babylonian text known as the "Epic of Creation" relates that the first "Gateway of the Gods" was constructed in Babylon by the gods themselves. The Anunnaki, the rank-and-file gods, were ordered to

> Construct the Gateway of the Gods. . . .
> Let its brickwork be fashioned.
> Its *shem* shall be in the designated place.

For two years, the Anunnaki toiled—"applied the implement . . . molded bricks"—until "they raised high the top of Eshagila" ("house of Great Gods") and "built the stage tower as high as High Heaven."

It was thus some cheek on the part of Mankind to establish its own launch tower on a site originally used for the purpose by the gods, for the name of the place—Babili—literally meant "Gateway of the Gods."

Is there any other evidence to corroborate the biblical tale and our interpretation of it?

The Babylonian historian-priest Berossus, who in the third century B.C. compiled a history of Mankind, reported that the "first inhabitants of the land, glorying in their own strength . . . undertook to raise a tower whose 'top' should reach the sky." But the tower was overturned by the gods and heavy winds, "and the gods introduced a diversity of tongues among men, who till that time had all spoken the same language."

George Smith *(The Chaldean Account of Genesis)* found in the writings of the Greek historian Hestaeus a report that, in accordance with "olden traditions," the people who had escaped the Deluge came to Senaar in Babylonia but were driven away from there by a diversity of tongues. The historian Alexander Polyhistor (first century B.C.) wrote that all men formerly spoke the same language. Then some undertook to erect a large and lofty tower so that they might "climb up to heaven." But the chief god confounded their design by sending a whirlwind; each tribe was given a different language. "The city where it happened was Babylon."

There is little doubt by now that the biblical tales, as well as the reports of the Greek historians of 2,000 years ago and of their predecessor Berossus, all stem from earlier —Sumerian—origins. A. H. Sayce (*The Religion of the Babylonians*) reported reading on a fragmentary tablet in the British Museum "the Babylonian version of the building of the Tower of Babel." In all instances, the attempt to reach the heavens and the ensuing confusion of tongues are basic elements of the version. There are other Sumerian texts that record the deliberate confusion of Man's tongue by an irate god.

Mankind, presumably, did not possess at that time the technology required for such an aerospace project; the guidance and collaboration of a knowledgeable god was essential. Did such a god defy the others to help Mankind? A Sumerian seal depicts a confrontation between armed gods, apparently over the disputed construction by men of a stage tower. (Fig. 74)

A Sumerian stela now on view in Paris in the Louvre may well depict the incident reported in the Book of Genesis. It was put up circa 2300 B.C. by Naram-Sin, king of Akkad, and scholars have assumed that it depicts the king victorious over his enemies. But the large central figure is that of a deity and not of the human king, for the person is wearing a helmet adorned with horns—the identifying mark exclusive to the gods. Furthermore, this central figure does not appear to be the leader of the smaller-sized humans, but to be trampling upon them. These humans, in turn, do not seem to be engaged in any warlike activities, but to be marching toward, and standing in adoration of, the same large conical object on which the deity's attention is also focused. Armed with a bow and lance, the deity seems to view the object menacingly rather than with adoration. (Fig. 75)

The conical object is shown reaching toward three celestial bodies. If its size, shape, and purpose indicate that it was a *shem*, then the scene depicted an angry and fully armed god trampling upon people celebrating the raising of a *shem*.

Both the Mesopotamian texts and the biblical account impart the same moral: The flying machines were meant for the gods and not for Mankind.

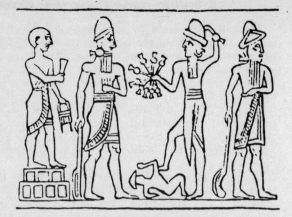

Fig. 74

Fig. 75

Men—assert both Mesopotamian and biblical texts—could ascend to the Heavenly Abode only upon the express wish of the gods. And therein lie more tales of ascents to the heavens and even of space flights.

•

The Old Testament records the ascent to the heavens of several mortal beings.

The first was Enoch, a pre-Diluvial patriarch whom God befriended and who "walked with the Lord." He was the seventh patriarch in the line of Adam and the great-grandfather of Noah, hero of the Deluge. The fifth chapter of the Book of Genesis lists the genealogies of all these patriarchs and the ages at which they died—except for Enoch, "who was gone, for the Lord had taken him." By implication and tradition, it was heavenward, to escape mortality on Earth, that God took Enoch. The other mortal was the prophet Elijah, who was lifted off Earth and taken heavenward in a "whirlwind."

A little-known reference to a third mortal who visited the Divine Abode and was endowed there with great wisdom is provided in the Old Testament, and it concerns the ruler of Tyre (a Phoenician center on the eastern Mediterranean coast). We read in Chapter 28 of the Book of Ezekiel that the Lord commanded the prophet to remind the king how, perfect and wise, he was enabled by the Deity to visit with the gods:

> Thou art molded by a plan,
> full of wisdom, perfect in beauty.
> Thou hast been in Eden, the garden of God;
> every precious stone was thy thicket. . . .
> Thou art an anointed Cherub, protected;
> and I have placed thee in the sacred mountain;
> as a god werest thou,
> moving within the Fiery Stones.

Predicting that the ruler of Tyre should die a death "of the uncircumcised" by the hand of strangers even if he called out to them "I am a Deity," the Lord then told Ezekiel the reason: After the king was taken to the Divine

Abode and given access to all wisdom and riches, his heart "grew haughty," he misused his wisdom, and he defiled the temples.

> Because thine heart is haughty, saying
> "A god am I;
> in the Abode of the Deity I sat,
> in the midst of the Waters";
> Though thou art a Man, not a god,
> thou set thy heart as that of a Deity.

The Sumerian texts also speak of several men who were privileged to ascend to the heavens. One was Adapa, the "model man" created by Ea. To him Ea "had given wisdom; eternal life he had not given him." As the years went by, Ea decided to avert Adapa's mortal end by providing him with a *shem* with which he was to reach the Heavenly Abode of Anu, there to partake of the Bread of Life and the Water of Life. When Adapa arrived at Anu's Celestial Abode, Anu demanded to know who had provided Adapa with a *shem* with which to reach the heavenly location.

There are several important clues to be found in both the biblical and the Mesopotamian tales of the rare ascents of mortals to the Abode of the Gods. Adapa, too, like the king of Tyre, was made of a perfect "mold." All had to reach and employ a *shem*—"fiery stone"—to reach the celestial "Eden." Some had gone up and returned to Earth; others, like the Mesopotamian hero of the Deluge, stayed there to enjoy the company of the gods. It was to find this Mesopotamian "Noah" and obtain from him the secret of the Tree of Life, that the Sumerian Gilgamesh set out.

The futile search by mortal Man for the Tree of Life is the subject of one of the longest, most powerful epic texts bequeathed to human culture by the Sumerian civilization. Named by modern scholars "The Epic of Gilgamesh," the moving tale concerns the ruler of Uruk who was born to a mortal father and a divine mother. As a result, Gilgamesh was considered to be "two-thirds of him god, one-third of him human," a circumstance that prompted him to seek escape from the death that was the fate of mortals. Tradition had informed him that one of his forefathers,

Utnapishtim—the hero of the Deluge—had escaped death, having been taken to the Heavenly Abode together with his spouse. Gilgamesh therefore decided to reach that place and obtain from his ancestor the secret of eternal life.

What prompted him to go was what he took to be an invitation from Anu. The verses read like a description of the sighting of the falling back to Earth of a spent rocket. Gilgamesh described it thus to his mother, the goddess NIN.SUN:

> My mother,
> During the night I felt joyful
> and I walked about among my nobles.
> The stars assembled in the Heavens.
> The handiwork of Anu descended toward me.
> I sought to lift it; it was too heavy.
> I sought to move it; move it I could not!
> The people of Uruk gathered about it,
> While the nobles kissed its legs.
> As I set my forehead, they gave me support.
> I raised it. I brought it to thee.

The interpretation of the incident by Gilgamesh's mother is mutilated in the text, and is thus unclear. But obviously Gilgamesh was encouraged by the sighting of the falling object—"the handiwork of Anu"—to embark on his adventure. In the introduction to the epic, the ancient reporter called Gilgamesh "the wise one, he who has experienced everything":

> Secret things he has seen,
> what is hidden to Man he knows;
> He even brought tidings
> of a time before the Deluge.
> He also took the distant journey,
> wearisome and under difficulties;
> He returned, and engraved all his toil
> upon a stone pillar.

The "distant journey" Gilgamesh undertook was, of course, his journey to the Abode of the Gods; he was accompanied by his comrade Enkidu. Their target was the

Land of Tilmun, for there Gilgamesh could raise a *shem* for himself. The current translations employ the expected "name" where the Sumerian *mu* or the Akkadian *shumu* appear in the ancient texts; we shall, however, employ *shem* instead so that the term's true meaning—a "skyborne vehicle"—will come through:

> The ruler Gilgamesh
> toward the Land of Tilmun set his mind.
> He says to his companion Enkidu:
> "O Enkidu . . .
> I would enter the Land, set up my *shem*.
> In the places where the *shem*'s were raised up
> I would raise my *shem*."

Unable to dissuade him, both the elders of Uruk and the gods whom Gilgamesh consulted advised him to first obtain the consent and assistance of Utu/Shamash. "If thou wouldst enter the Land—inform Utu," they cautioned him. "The Land, it is in Utu's charge," they stressed and re-stressed to him. Thus forewarned and advised, Gilgamesh appealed to Utu for permission:

> Let me enter the Land,
> Let me set up my *shem*.
> In the places where the *shem*'s are raised up,
> let me raise my *shem*. . . .
> Bring me to the landing place at. . . .
> Establish over me thy protection!

An unfortunate break in the tablet leaves us ignorant regarding the location of "the landing place." But, wherever it was, Gilgamesh and his companion finally reached its outskirts. It was a "restricted zone," protected by awesome guards. Weary and sleepy, the two friends decided to rest overnight before continuing.

No sooner had sleep overcome them than something shook them up and awoke them. "Didst thou arouse me?" Gilgamesh asked his comrade. "Am I awake?" he wondered, for he was witnessing unusual sights, so awesome that he wondered whether he was awake or dreaming. He told Enkidu:

> In my dream, my friend, the high ground toppled.
> It laid me low, trapped my feet. . . .
> The glare was overpowering!
> A man appeared;
> the fairest in the land was he.
> His grace . . .
> From under the toppled ground he pulled me out.
> He gave me water to drink; my heart quieted.

Who was this man, "the fairest in the land," who pulled Gilgamesh from under the landslide, gave him water, "quieted his heart"? And what was the "overpowering glare" that accompanied the unexplained landslide?

Unsure, troubled, Gilgamesh fell asleep again—but not for long.

> In the middle of the watch his sleep was ended.
> He started up, saying to his friend:
> "My friend, didst thou call me?
> Why am I awake?
> Didst thou not touch me?
> Why am I startled?
> Did not some god go by?
> Why is my flesh numb?"

Thus mysteriously reawakened, Gilgamesh wondered who had touched him. If it was not his comrade, was it "some *god*" who went by? Once more, Gilgamesh dozed off, only to be awakened a third time. He described the awesome occurrence to his friend.

> The vision that I saw was wholly awesome!
> The heavens shrieked, the earth boomed;
> Daylight failed, darkness came.
> Lightning flashed, a flame shot up.
> The clouds swelled, it rained death!
> Then the glow vanished; the fire went out.
> And all that had fallen had turned to ashes.

One needs little imagination to see in these few verses an ancient account of the witnessing of the launching of a rocket ship. First the tremendous thud as the rocket engines ignited ("the heavens shrieked"), accompanied by a marked shaking of the ground ("the earth boomed"). Clouds of smoke and dust enveloped the launching site ("daylight failed, darkness came"). Then the brilliance of the ignited engines showed through ("lightning flashed"); as the rocket ship began to climb skyward, "a flame shot up." The cloud of dust and debris "swelled" in all directions; then, as it began to fall down, "it rained death!" Now the rocket ship was high in the sky, streaking heavenward ("the glow vanished; the fire went out"). The rocket ship was gone from sight; and the debris "that had fallen had turned to ashes."

Awed by what he saw, yet as determined as ever to reach his destination, Gilgamesh once more appealed to Shamash for protection and support. Overcoming a "monstrous guard," he reached the mountain of Mashu, where one could see Shamash "rise up to the vault of Heaven."

He was now near his first objective—the "place where the *shem*'s are raised up." But the entrance to the site, apparently cut into the mountain, was guarded by fierce guards:

> Their terror is awesome, their glance is death.
> Their shimmering spotlight sweeps the mountains.
> They watch over Shamash,
> As he ascends and descends.

A seal depiction (Fig. 76) showing Gilgamesh *(second from left)* and his companion Enkidu *(far right)* may well depict the intercession of a god with one of the robot-like guards who could sweep the area with spotlights and emit

Fig. 76

death rays. The description brings to mind the statement in the Book of Genesis that God placed "the revolving sword" at the entrance to the Garden of Eden, to block its access to humans.

When Gilgamesh explained his partly divine origins, the purpose of his trip ("About death and life I wish to ask Utnapishtim") and the fact that he was on his way with the consent of Utu/Shamash, the guards allowed him to go ahead.

Proceeding "along the route of Shamash," Gilgamesh found himself in utter darkness; "seeing nothing ahead or behind," he cried out in fright. Traveling for many *beru* (a unit of time, distance, or the arc of the heavens), he was still engulfed by darkness. Finally, "it had grown bright when twelve *beru* he attained."

The damaged and blurred text then has Gilgamesh arriving at a magnificent garden where the fruits and trees were carved of semi-precious stones. It was there that Utnapishtim resided. Posing his problem to his ancestor, Gilgamesh encountered a disappointing answer: Man, Utnapishtim said, cannot escape his mortal fate. However, he offered Gilgamesh a way to postpone death, revealing to him the location of the Plant of Youth—"Man becomes young in old age," it was called. Triumphant, Gilgamesh obtained the plant. But, as fate would have it, he foolishly lost it on his way back, and returned to Uruk empty-handed.

Putting aside the literary and philosophic values of the epic tale, the story of Gilgamesh interests us here primarily for its "aerospace" aspects. The *shem* that Gilgamesh required in order to reach the Abode of the Gods was undoubtedly a rocket ship, the launching of one of which he had witnessed as he neared the "landing place." The rockets, it would seem, were located inside a mountain, and the area was a well-guarded, restricted zone.

No pictorial depiction of what Gilgamesh saw has so far come to light. But a drawing found in the tomb of an Egyptian governor of a far land shows a rockethead above-ground in a place where date trees grow. The shaft of the rocket is clearly stored *underground,* in a man-made silo constructed of tubular segments and decorated with leopard skins. (Fig. 77)

Fig. 77

Very much in the manner of modern draftsmen, the ancient artists showed a cross-section of the underground silo. We can see that the rocket contained a number of compartments. The lower one shows two men surrounded by curving tubes. Above them there are three circular panels. Comparing the size of the rockethead—the *ben-ben* —to the size of the two men inside the rocket, and the people above the ground, it is evident that the rockethead —equivalent to the Sumerian *mu*, the "celestial chamber" —could easily hold one or two operators or passengers.

TIL.MUN was the name of the land to which Gilgamesh set his course. The name literally meant "land of the missiles." It was the land where the *shem*'s were raised, a land under the authority of Utu/Shamash, a place where one could see this god "rise up to the vault of heavens."

And though the celestial counterpart of this member of the Pantheon of Twelve was the Sun, we suggest that his name did not mean "Sun" but was an epithet describing his functions and responsibilities. His Sumerian name Utu meant "he who brilliantly goes in." His derivate Akkadian name—Shem-Esh—was more explicit: *Esh* means "fire," and we now know what *shem* originally meant.

Utu/Shamash was "he of the fiery rocket ships." He was, we suggest, the commander of the spaceport of the gods.

•

The commanding role of Utu/Shamash in matters of travel to the Heavenly Abode of the Gods, and the functions performed by his subordinates in this connection, are brought out in even greater detail in yet another Sumerian tale of a heavenward journey by a mortal.

The Sumerian king lists inform us that the thirteenth ruler of Kish was Etana, "the one who to Heaven ascended." This brief statement needed no elaboration, for the tale of the mortal king who journeyed up to the highest heavens was well known throughout the ancient Near East, and was the subject of numerous seal depictions.

Etana, we are told, was designated by the gods to bring Mankind the security and prosperity that Kingship—an organized civilization—was intended to provide. But Etana, it seems, could not father a son who would continue the dynasty. The only known remedy was a certain Plant of Birth that Etana could obtain only by fetching it down from the heavens.

Like Gilgamesh at a later time, Etana turned to Shamash for permission and assistance. As the epic unfolds, it becomes clear that Etana was asking Shamash for a *shem!*

> O Lord, may it issue from thy mouth!
> Grant thou me the Plant of Birth!
> Show me the Plant of Birth!
> Remove my handicap!
> Produce for me a *shem!*

Flattered by prayer and fattened by sacrificial sheep, Shamash agreed to grant Etana's request to provide him with a *shem*. But instead of speaking of a *shem*. Shamash told Etana that an "eagle" would take him to the desired heavenly place.

Directing Etana to the pit where the Eagle had been placed, Shamash also informed the Eagle ahead of time of the intended mission. Exchanging cryptic messages with "Shamash, his lord," the Eagle was told: "A man I will send to thee; he will take thy hand . . . lead him hither . . . do whatever he says . . . do as I say."

Arriving at the mountain indicated to him by Shamash, "Etana saw the pit," and, inside it, "there the Eagle was." "At the command of valiant Shamash," the Eagle entered into communication with Etana. Once more, Etana explained his purpose and destination; whereupon the Eagle began to instruct Etana on the procedure for "raising the Eagle from its pit." The first two attempts failed, but on the third one the Eagle was properly raised. At daybreak, the Eagle announced to Etana: "My friend . . . up to the Heaven of Anu I will bear thee!" Instructing him how to hold on, the Eagle took off—and they were aloft, rising fast.

As though reported by a modern astronaut watching Earth recede as his rocket ship rises, the ancient storyteller describes how Earth appeared smaller and smaller to Etana:

> When he had borne him aloft one *beru*,
> the Eagle says to him, to Etana:
> "See, my friend, how the land appears!
> Peer at the sea at the sides of the Mountain House:
> The land has indeed become a mere hill,
> The wide sea is just like a tub."

Higher and higher the Eagle rose; smaller and smaller Earth appeared. When he had borne him aloft a second *beru*, the Eagle said:

> "My friend,
> Cast a glance at how the land appears!
> The land has turned into a furrow. . . .
> The wide sea is just like a bread-basket." . . .

When he had borne him aloft a third *beru,*
The Eagle says to him, to Etana:
"See, my friend, how the land appears!
The land has turned into a gardener's ditch!"

And then, as they continued to ascend, Earth was suddenly out of sight.

As I glanced around, the land had disappeared,
and upon the wide sea mine eyes could not feast.

According to one version of this tale, the Eagle and Etana did reach the Heaven of Anu. But another version states that Etana got cold feet when he could no longer see Earth, and ordered the Eagle to reverse course and "plunge down" to Earth.

Once again, we find a biblical parallel to such an unusual report of seeing Earth from a great distance above it. Exalting the Lord Yahweh, the prophet Isaiah said of him: "It is he who sitteth upon the circle of the Earth, and the inhabitants thereof are as insects."

The tale of Etana informs us that, seeking a *shem*, Etana had to communicate with an Eagle inside a pit. A seal depiction shows a winged, tall structure (a launch tower?) above which an eagle flies off. (Fig. 78)

What or who was the Eagle who took Etana to the distant heavens?

We cannot help associating the ancient text with the message beamed to Earth in July 1969 by Neil Armstrong, commander of the Apollo 11 spacecraft: "Houston! Tranquility Base here. The *Eagle* has landed!"

He was reporting the first landing by Man on the Moon. "Tranquility Base" was the site of the landing; *Eagle* was the name of the lunar module that separated from the spacecraft and took the two astronauts inside it to the Moon (and then back to their mother craft). When the lunar module first separated to start its own flight in Moon orbit, the astronauts told Mission Control in Houston: "The *Eagle* has wings."

But "Eagle" could also denote the astronauts who manned the spacecraft. On the Apollo 11 mission, "Eagle" was also the symbol of the astronauts themselves, worn as an emblem on their suits. Just as in the Etana tale, they, too, were "Eagles" who could fly, speak, and communicate (Fig. 79)

How would an ancient artist have depicted the pilots of the skyships of the gods? Would he have depicted them, by some chance, as eagles?

That is exactly what we have found. An Assyrian seal engraving from circa 1500 B.C. shows two "eagle-men" saluting a *shem!* (Fig. 80)

Numerous depictions of such "Eagles"—the scholars call them "bird-men"—have been found. Most depictions show them flanking the Tree of Life, as if to stress that they, in their *shem*'s, provided the link with the Heavenly Abode where the Bread of Life and Water of Life were to be found. Indeed, the usual depiction of the Eagles showed them holding in one hand the Fruit of Life and in the other the Water of Life, in full conformity with the tales of Adapa, Etana, and Gilgamesh. (Fig. 81)

The many depictions of the Eagles clearly show that they were not monstrous "bird-men," but anthropomorphic beings wearing costumes or uniforms that gave them the appearance of eagles.

The Hittite tale concerning the god Telepinu, who had vanished, reported that "the great gods and the lesser gods began to search for Telepinu" and "Shamash sent out a swift Eagle" to find him.

In the Book of Exodus, God is reported to have reminded the Children of Israel, "I have carried you upon the wings of Eagles, and have brought you unto me," confirming, it seems, that the way to reach the Divine Abode was upon the wings of Eagles—just as the tale of Etana relates. Numerous biblical verses, as a matter of fact, describe the Deity as a winged being. Boaz welcomed Ruth into the Judaean community as "coming under the wings" of the God Yahweh. The Psalmist sought security "under the shadow of thy wings" and described the descent of the Lord from the heavens. "He mounted a Cherub and went flying; He soared upon windy wings." Analyzing the similarities between the biblical El (employed as a title or generic term for the Deity) and the Canaanite El, S. Langdon (*Semitic Mythology*) showed that both were depicted, in text and on coins, as winged gods.

Fig. 78

Fig. 79

Fig. 80

Fig. 81

Fig. 82

The Mesopotamian texts invariably present Utu/Shamash as the god in charge of the landing place of the *shem*'s and of the Eagles. And like his subordinates he was sometimes shown wearing the full regalia of an Eagle's costume. (Fig. 82)

In such a capacity, he could grant to kings the privilege of "flying on the wings of birds" and of "rising from the lower heavens to the lofty ones." And when he was launched aloft in a fiery rocket, it was he "who stretched over unknown distances, for countless hours." Appropriately, "his net was the Earth, his trap the distant skies."

•

The Sumerian terminology for objects connected with celestial travel was not limited to the *me*'s that the gods put on or the *mu*'s that were their cone-shaped "chariots."

Sumerian texts describing Sippar relate that it had a central part, hidden and protected by mighty walls. Within those walls stood the Temple of Utu, "a house which is like a house of the Heavens." In an inner courtyard of the temple, also protected by high walls, stood "erected upwards, the mighty APIN" ("an object that plows through," according to the translators).

A drawing found at the temple mound of Anu at Uruk depicts such an object. We would have been hard put a few decades ago to guess what this object was; but it is a multistage space rocket at the top of which rests the conical *mu*, or command cabin. (Fig. 83)

Fig. 83

The evidence that the gods of Sumer possessed not just "flying chambers" for roaming Earth's skies but space-going multistage rocket ships also emerges from the examination of texts describing the sacred objects at Utu's temple at Sippar. We are told that witnesses at Sumer's supreme court were required to take the oath in an inner courtyard, standing by a gateway through which they could see and face three "divine objects." These were named "the golden sphere" (the crew's cabin?), the GIR, and the *alikmahrati* —a term that literally meant "advancer that makes vessel go," or what we would call a motor, an engine.

What emerges here is a reference to a three-part rocket ship, with the cabin or command module at the top end, the engines at the bottom end, and the *gir* in the center. The latter is a term that has been used extensively in connection with space flight. The guards Gilgamesh encountered at the entrance to the landing place of Shamash were called *gir*-men. In the temple of Ninurta, the sacred or most guarded inner area was called the GIR.SU ("where the *gir* is sprung up").

Gir, it is generally acknowledged, was a term used to describe a sharp-edged object. A close look at the pictorial sign for *gir* provides a better understanding of the term's "divine" nature; for what we see is a long, arrow-shaped object, divided into several parts or compartments:

That the *mu* could hover in Earth's skies on its own, or fly over Earth's lands when attached to a *gir*, or become the command module atop a multistage *apin* is testimony to the engineering ingenuity of the gods of Sumer, the Gods of Heaven and Earth.

A review of the Sumerian pictographs and ideograms leaves no doubt that whoever drew those signs was familiar with the shapes and purposes of rockets with tails of billowing fire, missile-like vehicles, and celestial "cabins."

KA.GIR ("rocket's mouth") showed a fin-equipped *gir*, or rocket, inside a shaftlike underground enclosure.

ESH ("Divine Abode"), the chamber or command module of a space vehicle.

ZIK ("ascend"), a command module taking off?

Finally, let us look at the pictographic sign for "gods" in Sumerian. The term was a two-syllable word: DIN.GIR. We have already seen what the symbol for GIR was: a two-stage rocket with fins. DIN, the first syllable, meant "righteous," "pure," "bright." Put together, then, DIN.GIR as "gods" or "divine beings" conveyed the meaning "the righteous ones of the bright, pointed objects" or, more explicitly, "the pure ones of the blazing rockets."

The pictographic sign for *din* was this:

easily bringing to mind a powerful jet engine spewing flames from the end part, and a front part that is puzzlingly open. But the puzzle turns to amazement if we "spell" *dingir* by combining the two pictographs. The tail of the finlike *gir* fits perfectly into the opening in the front of *din!* (Figs. 84, 85)

DIN   GIR

Fig. 84   *Fig. 85*

The astounding result is a picture of a rocket-propelled spaceship, with a landing craft docked into it perfectly—just as the lunar module was docked with the Apollo 11 spaceship! It is indeed a three-stage vehicle, with each part fitting neatly into the other: the thrust portion containing the engines, the midsection containing supplies and equipment, and the cylindrical "sky chamber" housing the people named *dingir*—the gods of antiquity, the astronauts of millennia ago.

Can there be any doubt that the ancient peoples, in calling their deities "Gods of Heaven and Earth," meant literally that they were people from elsewhere who had come to Earth from the heavens?

The evidence thus far submitted regarding the ancient gods and their vehicles should leave no further doubt that they were once indeed living beings of flesh and blood, people who literally came down to Earth from the heavens.

Even the ancient compilers of the Old Testament—who dedicated the Bible to a single God—found it necessary to acknowledge the presence upon Earth in early times of such divine beings.

The enigmatic section—a horror of translators and theologians alike—forms the beginning of Chapter 6 of Genesis. It is interposed between the review of the spread of Mankind through the generations following Adam and the story of the divine disenchantment with Mankind that preceded the Deluge. It states—unequivocally—that, at that time,

the sons of the gods
saw the daughters of man, that they were good;
and they took them for wives,
of all which they chose.

The implications of these verses, and the parallels to the Sumerian tales of gods and their sons and grandsons, and of semidivine offspring resulting from cohabitation between gods and mortals, mount further as we continue to read the biblical verses:

The Nefilim were upon the Earth,
in those days and thereafter too,
when the sons of the gods
cohabited with the daughters of the Adam,
and they bore children unto them.
They were the mighty ones of Eternity—
The People of the *shem*.

The above is not a traditional translation. For a long time, the expression "The Nefilim were upon the Earth" has been translated as "There were giants upon the earth"; but recent translators, recognizing the error, have simply resorted to leaving the Hebrew term *Nefilim* intact in the translation. The verse "The people of the *shem*," as one could expect, has been taken to mean "the people who have a name," and, thus, "the people of renown." But as we have already established, the term *shem* must be taken in its original meaning—a rocket, a rocket ship.

What, then, does the term *Nefilim* mean? Stemming from the Semitic root *NFL* ("to be cast down"), it means exactly what it says: It means *those who were cast down upon Earth!*

Contemporary theologians and biblical scholars have tended to avoid the troublesome verses, either by explaining them away allegorically or simply by ignoring them altogether. But Jewish writings of the time of the Second Temple did recognize in these verses the echoes of ancient traditions of "fallen angels." Some of the early scholarly works even mentioned the names of these divine beings "who fell from Heaven and were on Earth in those days": Sham-Hazzai ("*shem*'s lookout"), Uzza ("mighty") and Uzi-El ("God's might").

Malbim, a noted Jewish biblical commentator of the nineteenth century, recognized these ancient roots and explained that "in ancient times the rulers of countries were the sons of the deities who arrived upon the Earth from the Heavens, and ruled the Earth, and married wives from among the daughters of Man; and their offspring included heroes and mighty ones, princes and sovereigns." These stories, Malbim said, were of the pagan gods, "sons of the deities, who in earliest times fell down from the Heavens upon the Earth . . . that is why they called themselves 'Nefilim,' i.e. Those Who Fell Down."

Irrespective of the theological implications, the literal and original meaning of the verses cannot be escaped: The sons of the gods who came to Earth from the heavens were the Nefilim.

And the Nefilim were the People of the Shem—the People of the Rocket Ships. Henceforward, we shall call them by their biblical name.

6

THE TWELFTH PLANET

THE SUGGESTION that Earth was visited by intelligent beings from elsewhere postulates the existence of another celestial body upon which intelligent beings established a civilization more advanced than ours.

Speculation regarding the possibility of Earth visitation by intelligent beings from elsewhere has centered, in the past, on such planets as Mars or Venus as their place of origin. However, now that it is virtually certain that these two planetary neighbors of Earth have neither intelligent life nor an advanced civilization upon them, those who believe in such Earth visitations look to other galaxies and to distant stars as the home of such extraterrestrial astronauts.

The advantage of such suggestions is that while they cannot be proved, they cannot be disproved, either. The disadvantage is that these suggested "homes" are fantastically distant from Earth, requiring years upon years of travel at the speed of light. The authors of such suggestions therefore postulate one-way trips to Earth: a team of astronauts on a no-return mission, or perhaps on a spaceship lost and out of control, crash-landing upon Earth.

This is definitely *not* the Sumerian notion of the Heavenly Abode of the Gods.

The Sumerians accepted the existence of such a "Heavenly Abode," a "pure place," a "primeval abode." While Enlil, Enki, and Ninḥursag went to Earth and made their home upon it, their father Anu remained in the Heavenly Abode as its ruler. Not only occasional references in various texts but also detailed "god lists" actually named twenty-one divine couples of the dynasty that preceded Anu on the throne of the "pure place."

Anu himself reigned over a court of great splendor and extent. As Gilgamesh reported (and the Book of Ezekiel confirmed), it was a place with an artificial garden sculpted wholly of semiprecious stones. There Anu resided with his official consort Antu and six concubines, eighty offspring (of which fourteen were by Antu), one Prime Minister, three Commanders in charge of the *Mu*'s (rocket ships), two Commanders of the Weapons, two Great Masters of Written Knowledge, one Minister of the Purse, two Chief Justices, two "who with sound impress," and two Chief Scribes, with five Assistant Scribes.

Mesopotamian texts refer frequently to the magnificence of the abode of Anu and the gods and weapons that guarded its gateway. The tale of Adapa reports that the god Enki, having provided Adapa with a *shem*,

> Made him take the road to Heaven,
> and to Heaven he went up.
> When he had ascended to Heaven,
> he approached the Gate of Anu.
> Tammuz and Gizzida were standing guard
> at the Gate of Anu.

Guarded by the divine weapons SHAR.UR ("royal hunter") and SHAR.GAZ ("royal killer"), the throne room of Anu was the place of the Assembly of the Gods. On such occasions a strict protocol governed the order of entering and seating:

> Enlil enters the throne room of Anu,
> seats himself at the place of the right tiara,
> on the right of Anu.
> Ea enters [the throne room of Anu],
> seats himself at the place of the sacred tiara,
> on the left of Anu.

The Gods of Heaven and Earth of the ancient Near East not only originated in the heavens but could also return to the Heavenly Abode. Anu occasionally came down to Earth on state visits; Ishtar went up to Anu at least twice. Enlil's center in Nippur was equipped as a "bond heaven–earth." Shamash was in charge of the Eagles and the launching place of the rocket ships. Gilgamesh went up to the Place of Eternity and returned to Uruk; Adapa, too, made the trip and came back to tell about it; so did the biblical king of Tyre.

A number of Mesopotamian texts deal with the *Apkallu*, an Akkadian term stemming from the Sumerian AB.GAL ("great one who leads," or "master who points the way"). A study by Gustav Guterbock (*Die Historische Tradition und Ihre Literarische Gestaltung bei Babylonier und Hethiten*) ascertained that these were the "bird-men" depicted as the "Eagles" that we have already shown. The texts that spoke of their feats said of one that he "brought down Inanna from Heaven, to the E-Anna temple made her descend." This and other references indicate that these Apkallu were the pilots of the spaceships of the Nefilim.

Two-way travel was not only possible but actually contemplated to begin with, for we are told that, having decided to establish in Sumer the Gateway of the Gods (Babili), the leader of the gods explained:

> When to the Primeval Source
> for assembly you shall ascend,
> There shall be a restplace for the night
> to receive you all.
> When from the Heavens
> for assembly you shall descend,
> There shall be a restplace for the night
> to receive you all.

Realizing that such two-way travel between Earth and the Heavenly Abode was both contemplated and practiced, the people of Sumer did not exile their gods to distant galaxies. The Abode of the Gods, their legacy discloses, was within our own solar system.

We have seen Shamash in his official uniform as Commander of the Eagles. On each of his wrists he wears a watchlike object held in place by metal clasps. Other depictions of the Eagles reveal that all the important ones wore such objects. Whether they were merely decorative or served a useful purpose, we do not know. But all scholars are agreed that the objects represented rosettes—a circular cluster of "petals" radiating from a central point. (Fig. 86)

The rosette was the most common decorative temple symbol throughout the ancient lands, prevalent in Mesopotamia, western Asia, Anatolia, Cyprus, Crete, and Greece. It is the accepted view that the rosette as a temple symbol was an outgrowth or stylization of a celestial phenomenon—a sun encircled by its satellites. That the ancient astronauts wore this symbol on their wrists adds credence to this view.

An Assyrian depiction of the Gateway of Anu in the Heavenly Abode (Fig. 87) confirms ancient familiarity with a celestial system such as our Sun and its planets. The gateway is flanked by two Eagles—indicating that their services are needed to reach the Heavenly Abode. The Winged Globe—the supreme divine emblem—marks the gateway. It is flanked by the celestial symbols of the number seven and the crescent, representing (we believe) Anu flanked by Enlil and Enki.

Where are the celestial bodies represented by these symbols? Where is the Heavenly Abode? The ancient artist answers with yet another depiction, that of a large celestial deity extending its rays to eleven smaller celestial bodies encircling it. It is a representation of a Sun, orbited by eleven planets.

That this was not an isolated representation can be shown by reproducing other depictions on cylinder seals, like this one from the Berlin Museum of the Ancient Near East. (Fig 88)

When the central god or celestial body in the Berlin seal is enlarged (Fig. 89), we can see that it depicts a large, ray-emitting star surrounded by eleven heavenly bodies—planets. These, in turn, rest on a chain of twenty-four smaller globes. Is it only a coincidence that the number of all the "moons," or satellites, of the planets in our solar system (astronomers exclude those of ten miles or less in diameter) is also exactly twenty-four?

Now there is, of course, a catch to claiming that these depictions—of a Sun and *eleven* planets—represent *our* solar system, for our scholars tell us that the planetary system of which Earth is a part comprises the Sun, Earth

Fig. 86

Fig. 87

Fig. 88

Fig. 89

and Moon, Mercury, Venus, Mars, Jupiter, Saturn, Uranus, Neptune, and Pluto. This adds up to the Sun and only ten planets (when the Moon is counted as one).

But that is not what the Sumerians said. They claimed that our system was made up of the Sun and eleven planets (counting the Moon), and held steadfastly to the opinion that, in addition to the planets known to us today, there has been a *twelfth* member of the solar system—the home planet of the Nefilim.

We shall call it the *Twelfth* Planet.

•

Before we check the accuracy of the Sumerian information, let us review the history of our own knowledge of Earth and the heavens around it.

We know today that beyond the giant planets Jupiter and Saturn—at distances insignificant in terms of the universe, but immense in human terms—two more major planets (Uranus and Neptune) and a third, small one (Pluto) belong to our solar system. But such knowledge is quite recent. Uranus was discovered, through the use of improved telescopes, in 1781. After observing it for some fifty years, some astronomers reached the conclusion that its orbit revealed the influence of yet another planet. Guided by such mathematical calculations, the missing planet—named Neptune—was pinpointed by astronomers in 1846. Then, by the end of the nineteenth century, it became evident that Neptune itself was being subjected to unknown gravitational pull. Was there yet another planet in our solar system? The puzzle was solved in 1930 with the observation and location of Pluto.

Up to 1780, then, and for centuries before that, people believed there were *seven* members of our solar system: Sun, Moon, Mercury, Venus, Mars, Jupiter, Saturn. Earth was not counted as a planet because it was believed that these other celestial bodies circled Earth—the most important celestial body created by God, with God's most important creation, Man, upon it.

Our textbooks generally credit Nicolaus Copernicus with the discovery that Earth is only one of several planets in a heliocentric (Sun-centered) system. Fearing the wrath of the Christian church for challenging Earth's central position, Copernicus published his study *(De revolutionibus orbium coelestium)* only when on his deathbed, in 1543.

Spurred to reexamine centuries-old astronomical concepts primarily by the navigational needs of the Age of Discovery, and by the findings by Columbus (1492), Magellan (1520), and others that Earth was not flat but spherical, Copernicus depended on mathematical calculations and searched for the answers in ancient writings. One of the few churchmen who supported Copernicus, Cardinal Schonberg, wrote to him in 1536: "I have learned that you know not only the groundwork of the ancient mathematical doctrines, but that you have created a new theory . . . according to which the Earth is in motion and it is the Sun which occupies the fundamental and therefore the cardinal position."

The concepts then held were based on Greek and Roman traditions that Earth, which was flat, was "vaulted over" by the distant heavens, in which the stars were fixed. Against the star-studded heavens the planets (from the Greek word for "wanderer") moved around Earth. There were thus seven celestial bodies, from which the seven days of the week and their names originated: the Sun (Sunday), Moon (Monday), Mars *(mardi)*, Mercury *(mercredi)*, Jupiter *(jeudi)*, Venus *(vendredi)*, Saturn (Saturday). (Fig. 90)

These astronomical notions stemmed from the works and codifications of Ptolemy, an astronomer in the city of Alexandria, Egypt, in the second century A.D. His definite findings were that the Sun, Moon, and five planets moved in circles around Earth. Ptolemaic astronomy predominated for more than 1,300 years—until Copernicus put the Sun in the center.

While some have called Copernicus the "Father of Modern Astronomy," others view him more as a researcher and reconstructor of earlier ideas. The fact is that he pored over the writings of Greek astronomers who preceded Ptolemy, such as Hipparchus and Aristarchus of Samos. The latter suggested in the third century B.C. that the motions of the heavenly bodies could better be explained if the Sun—and not Earth—were assumed to be in the center. In fact, 2,000 years before Copernicus, Greek astronomers listed the planets in their correct order from

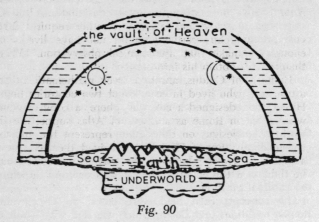

the vault of Heaven

Sea   Earth   Sea

UNDERWORLD

Fig. 90

the Sun, acknowledging thereby that the Sun, not Earth, was the solar system's focal point.

The heliocentric concept was only *re*discovered by Copernicus; and the interesting fact is that astronomers knew more in 500 b.c. than in a.d. 500 and 1500.

Indeed, scholars are now hard put to explain why first the later Greeks and then the Romans assumed that Earth was flat, rising above a layer of murky waters below which there lay Hades or "Hell," when some of the evidence left by Greek astronomers from earlier times indicates that they knew otherwise.

Hipparchus, who lived in Asia Minor in the second century b.c., discussed "the displacement of the sostitial and equinoctial sign," the phenomenon now called precession of the equinoxes. But the phenomenon can be explained only in terms of a "spherical astronomy," whereby Earth is surrounded by the other celestial bodies as a sphere within a spherical universe.

Did Hipparchus, then, know that Earth was a globe, and did he make his calculations in terms of a spherical astronomy? Equally important is yet another question. The phenomenon of the precession could be observed by relating the arrival of spring to the Sun's position (as seen from Earth) in a given zodiacal constellation. But the shift from one zodiacal house to another requires 2,160 years. Hipparchus certainly could not have lived long enough to make that astronomical observation. Where, then, did he obtain his information?

Eudoxus of Cnidus, another Greek mathematician and astronomer who lived in Asia Minor two centuries before Hipparchus, designed a celestial sphere, a copy of which was set up in Rome as a statue of Atlas supporting the world. The designs on the sphere represent the zodiacal constellations. But if Eudoxus conceived the heavens as a sphere, where in relation to the heavens was Earth? Did he think that the celestial globe rested on a *flat* Earth—a most awkward arrangement—or did he know of a spherical Earth, enveloped by a celestial sphere? (Fig. 91)

Fig. 91

The works of Eudoxus, lost in their originals, have come down to us thanks to the poems of Aratus, who in the third century B.C. "translated" the facts put forth by the astronomer into poetic language. In this poem (which must have been familiar to St. Paul, who quoted from it) the constellations are described in great detail, "drawn all around"; and their grouping and naming is ascribed to a very remote prior age. "Some men of yore a nomenclature thought of and devised, and appropriate forms found."

Who were the "men of yore" to whom Eudoxus attributed the designation of the constellations? Based on certain clues in the poem, modern astronomers believe that the Greek verses describe the heavens as they were observed in *Mesopotamia* circa 2200 B.C.

The fact that both Hipparchus and Eudoxus lived in Asia Minor raises the probability that they drew their knowledge from Hittite sources. Perhaps they even visited the Hittite capital and viewed the divine procession carved on the rocks there; for among the marching gods two bull-men hold up a globe—a sight that might well have inspired Eudoxus to sculpt Atlas and the celestial sphere. (Fig. 92)

Fig. 92

Were the earlier Greek astronomers, living in Asia Minor, better informed than their successors because they could draw on Mesopotamian sources?

Hipparchus, in fact, confirmed in his writings that his studies were based on knowledge accumulated and verified over many millennia. He named as his mentors "Babylonian astronomers of Erech, Borsippa, and Babylon." Geminus of Rhodes named the "Chaldeans" (the ancient

Babylonians) as the discoverers of the exact motions of the Moon. The historian Diodorus Siculus, writing in the first century B.C., confirmed the exactness of Mesopotamian astronomy; he stated that "the Chaldeans named the planets . . . in the center of their system was the Sun, the greatest light, of which the planets were 'offspring,' reflecting the Sun's position and shine."

The acknowledged source of Greek astronomical knowledge was, then, Chaldea; invariably, those earlier Chaldeans possessed greater and more accurate knowledge than the peoples that followed them. For generations, throughout the ancient world, the name "Chaldean" was synonymous with "stargazers," astronomers.

Abraham, who came out of "Ur of the Chaldeans," was told by God to gaze at the stars when the future Hebrew generations were discussed. Indeed, the Old Testament was replete with astronomical information. Joseph compared himself and his brothers to twelve celestial bodies, and the patriarch Jacob blessed his twelve descendants by associating them with the twelve constellations of the zodiac. The Psalms and the Book of Job refer repeatedly to celestial phenomena, the zodiacal constellations, and other star groups (such as the Pleiades). Knowledge of the zodiac, the scientific division of the heavens, and other astronomical information was thus prevalent in the ancient Near East well before the days of ancient Greece.

The scope of Mesopotamian astronomy on which the early Greek astronomers drew must have been vast, for even what archaeologists have found amounts to an avalanche of texts, inscriptions, seal impressions, reliefs, drawings, lists of celestial bodies, omens, calendars, tables of rising and setting times of the Sun and the planets, forecasts of eclipses.

Many such later texts were, to be sure, more astrological than astronomical in nature. The heavens and the movements of the heavenly bodies appeared to be a prime preoccupation of mighty kings, temple priests, and the people of the land in general; the purpose of the stargazing seemed to be to find in the heavens an answer to the course of affairs on Earth: war, peace, abundance, famine.

Compiling and analyzing hundreds of texts from the first millennium B.C., R. C. Thompson (*The Reports of the*

Magicians and Astrologers of Nineveh and Babylon) was able to show that these stargazers were concerned with the fortunes of the land, its people, and its ruler from a national point of view, and not with individual fortunes (as present-day "horoscopic" astrology is):

> When the Moon in its calculated time is not seen, there will be an invasion of a mighty city.

> When a comet reaches the path of the Sun, field-flow will be diminished; an uproar will happen twice.

> When Jupiter goes with Venus, the prayers of the land will reach the heart of the gods.

> If the Sun stands in the station of the Moon, the king of the land will be secure on the throne.

Even this astrology required comprehensive and accurate astronomical knowledge, without which no omens were possible. The Mesopotamians, possessing such knowledge, distinguished between the "fixed" stars and the planets that "wandered about" and knew that the Sun and the Moon were neither fixed stars nor ordinary planets. They were familiar with comets, meteors, and other celestial phenomena, and could calculate the relationships between the movements of the Sun, Moon, and Earth, and predict eclipses. They followed the motions of the celestial bodies and related them to Earth's orbit and rotation through the heliacal system—the system still in use today, which measures the rising and setting of stars and planets in Earth's skies relative to the Sun.

To keep track of the movements of the celestial bodies and their positions in the heavens relative to Earth and to one another, the Babylonians and Assyrians kept accurate ephemerides. These were tables that listed and predicted the future positions of the celestial bodies. Professor George Sarton (*Chaldean Astronomy of the Last Three Centuries* B.C.) found that they were computed by two methods: a later one used in Babylon, and an older one from Uruk. His unexpected finding was that the older, Uruk method was more sophisticated and more accurate than the later system. He accounted for this surprising situation by concluding that the erroneous astronomical notions of the Greeks and Romans resulted from a shift to a philosophy that explained the world in geometric terms, while the astronomer-priests of Chaldea followed the prescribed formulas and traditions of Sumer.

The unearthing of the Mesopotamian civilizations in the past one hundred years leaves no doubt that in the field of astronomy, as in so many others, the roots of our knowledge lie deep in Mesopotamia. In this field, too, we draw upon and continue the heritage of Sumer.

Sarton's conclusions have been reinforced by very comprehensive studies by Professor O. Neugebauer (*Astronomical Cuneiform Texts*), who was astonished to find that the ephemerides, precise as they were, were not based on observations by the Babylonian astronomers who prepared them. Instead, they were calculated "from some fixed arithmetical schemes . . . which were given and were not to be interfered with" by the astronomers who used them.

Such automatic adherence to "arithmetical schemes" was achieved with the aid of "procedure texts" that accompanied the ephemerides, which "gave the rules for computing ephemerides step by step" according to some "strict mathematical theory." Neugebauer concluded that the Babylonian astronomers were ignorant of the theories on which the ephemerides and their mathematical calculations were based. He also admitted that "the empirical and theoretical foundation" of these accurate tables, to a large extent, escapes modern scholars as well. Yet he is convinced that ancient astronomical theories "must have existed, because it is impossible to devise computational schemes of high complication without a very elaborate plan."

Professor Alfred Jeremias (*Handbuch der Altorientalischen Geistkultur*) concluded that the Mesopotamian astronomers were acquainted with the phenomenon of retrograde, the apparent erratic and snakelike course of the planets as seen from Earth, caused by the fact that Earth orbits the Sun either faster or slower than the other planets. The significance of such knowledge lies not only in the fact that retrograde is a phenomenon related to orbits around the Sun, but also in the fact that very long periods of observation were required to grasp and track it.

Where were these complicated theories developed, and who made the observations without which they could not have been developed? Neugebauer pointed out that "in the procedure texts, we meet a great number of technical terms of wholly unknown reading, if not unknown meaning." Someone, much earlier than the Babylonians, possessed astronomical and mathematical knowledge far superior to that of later culture in Babylon, Assyria, Egypt, Greece, and Rome.

The Babylonians and Assyrians devoted a substantial part of their astronomical efforts to keeping an accurate calendar. Like the Jewish calendar to this very day, it was a solar-lunar calendar, correlating ("intercalating") the solar year of just over 365 days with a lunar month of just under 30 days. While a calendar was important for business and other mundane needs, its accuracy was required primarily to determine the precise day and moment of the New Year, and other festivals and worship of the gods.

To measure and correlate the intricate movements of Sun, Earth, Moon, and planets, the Mesopotamian astronomer-priests relied on a complex spherical astronomy. Earth was taken to be a sphere with an equator and poles; the heavens, too, were divided by imaginary equatorial and polar lines. The passage of celestial bodies was related to the ecliptic, the projection of the plane of Earth's orbit around the Sun upon the celestial sphere; the equinoxes (the points and the times at which the Sun in its apparent annual movement north and south crosses the celestial equator); and the solstices (the time when the Sun during its apparent annual movement along the ecliptic is at its greatest declination north or south). All these are astronomical concepts used to this very day.

But the Babylonians and Assyrians did not invent the calendar or the ingenious methods for its calculation. Their calendars—as well as our own—originated in Sumer. There the scholars have found a calendar, in use from the very earliest times, that is the basis for *all* later calendars. The principal calendar and model was the calendar of Nippur, the seat and center of Enlil. Our present-day one is modeled on that Nippurian calendar.

The Sumerians considered the New Year to begin at

the exact moment when the Sun crossed the spring equinox. Professor Stephen Langdon *(Tablets from the Archives of Drehem)* found that records left by Dungi, a ruler of Ur circa 2400 B.C., show that the Nippurian calendar selected a certain celestial body by whose setting against the sunset it was possible to determine the exact moment of the New Year's arrival. This, he concluded, was done "perhaps 2,000 years before the era of Dungi"—that is, circa 4400 B.C.!

Can it really be that the Sumerians, without actual instruments, nevertheless had the sophisticated astronomical and mathematical know-how required by a spherical astronomy and geometry? Indeed they had, as their language shows.

They had a term—DUB—that meant (in astronomy) the 360-degree "circumference of the world," in relation to which they spoke of the curvature or arc of the heavens. For their astronomical and mathematical calculations they drew the AN.UR—an imagined "heavenly horizon" against which they could measure the rising and setting of celestial bodies. Perpendicular to this horizon they extended an imagined vertical line, the NU.BU.SAR.DA; with its aid they obtained the zenith point and called it the AN.PA. They traced the lines we call meridians, and called them "the graded yokes"; latitude lines were called "middle lines of heaven." The latitude line marking the summer solstice, for example, was called AN.BIL ("fiery point of the heavens").

The Akkadian, Hurrian, Hittite, and other literary masterpieces of the ancient Near East, being translations or versions of Sumerian originals, were replete with Sumerian loanwords pertaining to celestial bodies and phenomena. Babylonian and Assyrian scholars who drew up star lists or wrote down calculations of planetary movements often noted the Sumerian originals on the tablets that they were copying or translating. The 25,000 texts devoted to astronomy and astrology said to have been included in the Nineveh library of Ashurbanipal frequently bore acknowledgments of Sumerian origins.

A major astronomical series that the Babylonians called "The Day of the Lord" was declared by its scribes to have been copied from a Sumerian tablet written in the time of

Sargon of Akkad—in the third millennium B.C. A tablet dated to the third dynasty of Ur, also in the third millennium B.C., describes and lists a series of celestial bodies so clearly that modern scholars had little difficulty in recognizing the text as a classification of constellations, among them Ursa Major, Draco, Lyra, Cygnus and Cepheus, and Triangulum in the northern skies; Orion, Canis Major, Hydra, Corvus, and Centaurus in the southern skies; and the familiar zodiacal constellations in the central celestial band.

In ancient Mesopotamia the secrets of celestial knowledge were guarded, studied, and transmitted by astronomer-priests. It was thus perhaps fitting that three scholars who are credited with giving back to us this lost "Chaldean" science were Jesuit priests: Joseph Epping, Johann Strassman, and Franz X. Kugler. Kugler, in a masterwork (*Sternkunde und Sterndienst in Babel*), analyzed, deciphered, sorted out, and explained a vast number of texts and lists. In one instance, by mathematically "turning the skies backwards," he was able to show that a list of thirty-three celestial bodies in the Babylonian skies of 1800 B.C. was neatly arranged according to present-day groupings!

After much work deciding which are true groups and which are merely subgroups, the world's astronomical community agreed (in 1925) to divide the heavens as seen from Earth into *three* regions—northern, central, and southern—and group the stars therein into eighty-eight constellations. As it turned out, there was nothing new in this arrangement, for the Sumerians were the first to divide the heavens into three bands or "ways"—the northern "way" was named after Enlil, the southern after Ea, and the central band was the "Way of Anu"—and to assign to them various constellations. The present-day central band, the band of the twelve constellations of the zodiac, corresponds *exactly* to the Way of Anu, in which the Sumerians grouped the stars into twelve houses.

In antiquity, as today, the phenomenon was related to the concept of the zodiac. The great circle of Earth around the Sun was divided into twelve equal parts, of thirty degrees each. The stars seen in each of these segments, or "houses," were grouped together into a constellation, each of which was then named according to the shape the stars of the group seemed to form.

Because the constellations and their subdivisions, and even individual stars within the constellations, have reached Western civilization with names and descriptions borrowed heavily from Greek mythology, the Western world tended for nearly two millennia to credit the Greeks with this achievement. But it is now apparent that the early Greek astronomers merely adopted into their language and mythology a ready-made astronomy obtained from the Sumerians. We have already noted how Hipparchus, Eudoxus, and others obtained their knowledge. Even Thales, the earliest Greek astronomer of consequence, who is said to have predicted the total solar eclipse of May 28, 585 B.C., which stopped the war between the Lydians and the Medians, allowed that the sources of his knowledge were of pre-Semitic Mesopotamian origins, namely— Sumerian.

We have acquired the name "zodiac" from the Greek *zodiakos kyklos* ("animal circle") because the layout of the star groups was likened to the shape of a lion, fishes, and so on. But those imaginary shapes and names were actually originated by the Sumerians, who called the twelve zodiacal constellations UL.ḪE ("shiny herd"):

1. GU.AN.NA ("heavenly bull"), *Taurus*.
2. MASH.TAB.BA ("twins"), our *Gemini*.
3. DUB ("pincers," "tongs"), the Crab or *Cancer*.
4. UR.GULA ("lion"), which we call *Leo*.
5. AB.SIN ("her father was Sin"), the Maiden, *Virgo*.
6. ZI.BA.AN.NA ("heavenly fate"), the scales of *Libra*.
7. GIR.TAB ("which claws and cuts"), *Scorpio*.
8. PA.BIL ("defender"), the Archer, *Sagittarius*.
9. SUḪUR.MASH ("goat-fish"), *Capricorn*.
10. GU ("lord of the waters"), the Water Bearer, *Aquarius*.
11. SIM.MAḪ ("fishes"), *Pisces*.
12. KU.MAL ("field dweller"), the Ram, *Aries*.

The pictorial representations or signs of the zodiac, like their names, have remained virtually intact since their introduction in Sumer. (Fig. 93)

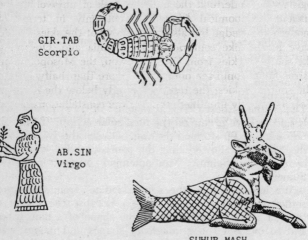

GIR.TAB
Scorpio

AB.SIN
Virgo

SUHUR.MASH
Capricorn

Fig. 93

Until the introduction of the telescope, European astronomers accepted the Ptolemaic recognition of only nineteen constellations in the northern skies. By 1925, when the current classification was agreed upon, twenty-eight constellations had been recognized in what the Sumerians called the Way of Enlil. We should no longer be surprised to find out that, unlike Ptolemy, the earlier Sumerians recognized, identified, grouped, named, and listed *all* the constellations of the northern skies!

Of the celestial bodies *in* the Way of Enlil, twelve were deemed to be *of* Enlil—paralleling the twelve zodiacal celestial bodies in the Way of Anu. Likewise, in the southern portion of the skies—the Way of Ea—twelve constellations were listed, not merely as present *in* the southern skies, but as *of* the god Ea. In addition to these twelve principal constellations of Ea, several others were listed for the southern skies—though not so many as are recognized today.

The Way of Ea posed serious problems to the Assyriologists who undertook the immense task of unraveling the ancient astronomical knowledge not only in terms of modern knowledge but also based on what the skies should have looked like centuries and millennia ago. Observing the southern skies from Ur or Babylon, the Mesopotamian astronomers could see only a little more than halfway into the southern skies; the rest was already below the horizon. Yet, if correctly identified, some of the constellations of the Way of Ea lay well beyond the horizon. But there was an even greater problem: If, as the scholars assumed, the Mesopotamians believed (as the Greeks did in later times) that Earth was a mass of dry land resting upon the chaotic darkness of a netherworld (the Greek Hades)—a flat disc over which the heavens arched in a semicircle—then there should have been no southern skies at all!

Restricted by the assumption that the Mesopotamians were beholden to a flat-Earth concept, modern scholars could not permit their conclusions to take them too much below the equatorial line dividing north and south. The evidence, however, shows that the three Sumerian "ways" encompassed the complete skies of a global, not flat, Earth.

In 1900 T. G. Pinches reported to the Royal Asiatic Society that he was able to reassemble and reconstruct a complete Mesopotamian astrolabe (literally, "taker of stars"). He showed it to be a circular disc, divided like a pie into twelve segments and three concentric rings, resulting in a field of thirty-six portions. The whole design had the appearance of a rosette of twelve "leaves," each of which had the name of a month written in it. Pinches marked them I to XII for convenience, starting with Nisannu, the first month of the Mesopotamian calendar. (Fig. 94)

Each of the thirty-six portions also contained a name with a small circle below it, signifying that it was the name of a celestial body. The names have since been found in many texts and "star lists" and are undoubtedly the names of constellations, stars, or planets.

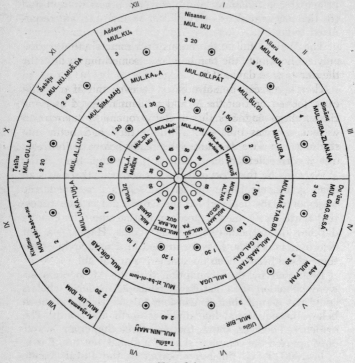

Fig. 94

Each of the thirty-six segments also had a number written below the name of the celestial body. In the innermost ring, the numbers ranged from 30 to 60; in the central ring, from 60 (written as "1") to 120 (this "2" in the sexagesimal system meant $2 \times 60 = 120$); and in the outermost ring, from 120 to 240. What did these numbers represent?

Writing nearly fifty years after the presentation by Pinches, the astronomer and Assyriologist O. Neugebauer (*A History of Ancient Astronomy: Problems and Methods*) could only say that "the whole text constitutes some kind of schematic celestial map . . . in each of the thirty-six fields we find the name of a constellation and simple numbers whose significance is not yet clear." A leading expert on the subject, B. L. Van der Waerden *(Babylonian Astronomy: The Thirty-Six Stars)*, reflecting on the apparent rise and fall of the numbers in some rhythm, could only suggest that "the numbers have something to do with the duration of daylight."

The puzzle can be solved, we believe, only if one discards the notion that the Mesopotamians believed in a flat Earth, and recognizes that their astronomical knowledge was as good as ours—not because they had better instruments than we do, but because their source of information was the Nefilim.

We suggest that the enigmatic numbers represent degrees of the celestial arc, with the North Pole as the starting point, and that the astrolabe was a planisphere, the representation of a sphere upon a flat surface.

While the numbers increase and decrease, those in the opposite segments for the Way of Enlil (such as Nisannu —50, Tashritu—40) add up to 90; all those for the Way of Anu add up to 180; and all those for the Way of Ea add up to 360 (such as Nisannu 200, Tashritu 160). These figures are too familiar to be misunderstood; they represent segments of a complete spherical circumference: a quarter of the way (90 degrees), halfway (180 degrees), or full circle (360 degrees).

The numbers given for the Way of Enlil are so paired as to show that this Sumerian segment of the northern skies stretched over 60 degrees from the North Pole, bordering on the Way of Anu at 30 degrees above the equator. The Way of Anu was equidistant on both sides of the equator, reaching to 30 degrees south below the equator. Then, farther south and farthest away from the North Pole, lay the Way of Ea—that part of Earth and of the celestial globe that lay between 30 degrees south and the South Pole. (Fig. 95)

The numbers in the Way of Ea segments add up to 180 degrees in Addaru (February–March) and Ululu (August–September). The only point that is 180 degrees away from the North Pole, whether you go south on the east or on the west, is the South Pole. And this can hold true only if one deals with a sphere.

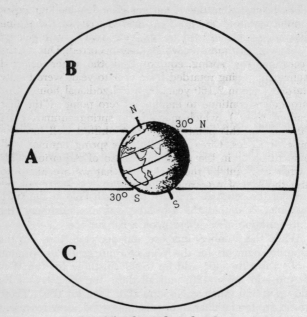

Fig. 95. The Celestial Sphere

A. The Way of Anu, the celestial band of the Sun, planets and the constellations of the Zodiac
B. The way of Enlil, the northern skies
C. The Way of Ea, the southern skies

Precession is the phenomenon caused by the wobble of Earth's north–south axis, causing the North Pole (the one pointing at the North Star) and the South Pole to trace a grand circle in the heavens. The apparent retardation of Earth against the starry constellations amounts to about fifty seconds of an arc in one year, or one degree in seventy-two years. The grand circle—the time it takes the Earth's North Pole to point again at the same North Star—therefore lasts 25,920 years (72 × 360), and that is what the astronomers call the Great Year or the Platonian Year (for apparently Plato, too, was aware of the phenomenon).

The rising and setting of various stars deemed significant in antiquity, and the precise determination of the spring equinox (which ushered in the New Year), were related to the zodiacal house in which they occurred. Due to the precession, the spring equinox and the other celestial phenomena, being retarded from year to year, were finally retarded once in 2,160 years by a full zodiacal house. Our astronomers continue to employ a "zero point" ("the first point of Aries"), which marked the spring equinox circa 900 B.C., but this point has by now shifted well into the house of Pisces. Circa A.D. 2100 the spring equinox will begin to occur in the preceding house of Aquarius. This is what is meant by those who say that we are about to enter the Age of Aquarius. (Fig. 96).

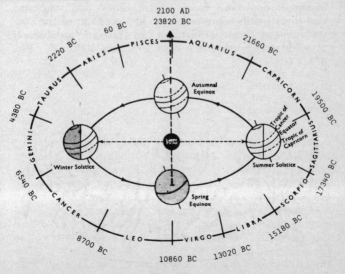

Fig. 96

Because the shift from one zodiacal house to another takes more than two millennia, scholars wondered how and where Hipparchus could have learned of the precession in the second century B.C. It is now clear that his source was Sumerian. Professor Langdon's findings reveal that the Nippurian calendar, established circa 4400 B.C., in the Age of Taurus, reflects knowledge of the precession and the shift of zodiacal houses that took place 2,160 years *earlier* than that. Professor Jeremias, who correlated Mesopotamian astronomical texts with Hittite astronomical texts, was also of the opinion that older astronomical tablets recorded the change from Taurus to Aries; and he concluded that the Mesopotamian astronomers predicted and anticipated the shift from Aries to Pisces.

Subscribing to these conclusions, Professor Willy Hartner *(The Earliest History of the Constellations in the Near East)* suggested that the Sumerians left behind plentiful pictorial evidence to that effect. When the spring equinox was in the zodiac of Taurus, the summer solstice occurred in the zodiac of Leo. Hartner drew attention to the recurrent motif of a bull-lion "combat" appearing in Sumerian depictions from earliest times, and suggested that these motifs represented the key positions of the constellations of Taurus (Bull) and Leo (Lion) to an observer at 30 degrees north (such as at Ur) circa 4000 B.C. (Fig. 97)

Fig. 97

Most scholars consider the Sumerian stress of Taurus as their first constellation as evidence not only of the antiquity of the zodiac—dating to circa 4000 B.C.—but also as testifying to the time when Sumerian civilization so suddenly began. Professor Jeremias (*The Old Testament in the Light of the Ancient East*) found evidence showing that the Sumerian zodiacal-chronological "point zero" stood precisely between the Bull and the Twins; from this and other data he concluded that the zodiac was devised in the Age of Gemini (the Twins)—that is, even *before* Sumerian civilization began. A Sumerian tablet in the Berlin Museum (VAT.7847) begins the list of zodiacal constellations with that of *Leo*—taking us back to circa 11,000 B.C., when Man had just begun to till the land.

Professor H. V. Hilprecht (*The Babylonian Expedition of the University of Pennsylvania*) went even farther. Studying thousands of tablets bearing mathematical tabulations, he concluded that "all the multiplication and division tables from the temple libraries of Nippur and Sippar, and from the library of Ashurbanipal [in Nineveh] are based upon [the number] 12960000." Analyzing this number and its significance, he concluded that it could be related only to the phenomenon of the precession, and that the Sumerians knew of the Great Year of 25,920 years.

This is indeed fantastic astronomical sophistication at an impossible time.

Just as it is evident that the Sumerian astronomers possessed knowledge that they could not possibly have acquired on their own, so is there evidence to show that a good deal of their knowledge was of no practical use to them.

This pertains not only to the very sophisticated astronomical methods that were used—who in ancient Sumer really needed to establish a celestial equator, for example? —but also to a variety of elaborate texts that dealt with the measurement of distances between stars.

One of these texts, known as AO.6478, lists the twenty-six major stars visible along the line we now call the Tropic of Cancer, and gives distances between them as measured in three different ways. The text first gives the distances between these stars by a unit called *mana shukultu* ("measured and weighed"). It is believed that this was an ingenious device that related the weight of escaping water to the passage of time. It made possible the determination of distances between two stars in terms of time.

The second column of distances was in terms of *degrees of the arc* of the skies. The full day (daylight and nighttime) was divided into twelve double hours. The arc of the heavens comprised a full circle of 360 degrees. Hence, one *beru* or "double hour" represented 30 degrees of the arc of the heavens. By this method, passage of time on Earth provided a measure of the distances in degrees between the named celestial bodies.

The third method of measurement was *beru ina shame* ("length in the skies"). F. Thureau-Dangin (*Distances entre Etoiles Fixes*) pointed out that while the first two methods were relative to other phenomena, this third method provided absolute measurements. A "celestial *beru,*" he and others believe, was equivalent to 10,692 of our present-day meters (11,693 yards). The "distance in the skies" between the twenty-six stars was calculated in the text as adding up to 655,200 "*beru* drawn in the skies."

The availability of three different methods of measuring distances between stars conveys the great importance attached to the matter. Yet, who among the men and women of Sumer needed such knowledge—and who among them could devise the methods and accurately use them? The only possible answer is: The Nefilim had the knowledge and the need for such accurate measurements.

Capable of space travel, arriving on Earth from another planet, roaming Earth's skies—they were the only ones who could, and did, possess at the dawn of Mankind's civilization the astronomical knowledge that required millennia to develop, the sophisticated methods and mathematics and concepts for an advanced astronomy, and the need to teach human scribes to copy and record meticulously table upon table of distances in the heavens, order of stars and groups of stars, heliacal risings and settings, a complex Sun-Moon-Earth calendar, and the rest of the remarkable knowledge of both Heaven and Earth.

Against this background, can it still be assumed that the Mesopotamian astronomers, guided by the Nefilim, were not aware of the planets beyond Saturn—that they did not know of Uranus, Neptune, and Pluto? Was their knowledge of Earth's own family, the solar system, less complete than that of distant stars, their order, and their distances?

Astronomical information from ancient times contained in hundreds of detailed texts lists celestial bodies, neatly arranged by their celestial order or by the gods or the months or the lands or the constellations with which they were associated. One such text, analyzed by Ernst F. Weidner (*Handbuch der Babylonischen Astronomie*), has come to be called "The Great Star List." It listed in five columns tens of celestial bodies as related to one another, to months, countries, and deities. Another text listed correctly the main stars in the zodiacal constellations. A text indexed as B.M.86378 arranged (in its unbroken part) seventy-one celestial bodies by their location in the heavens; and so on and on and on.

In efforts to make sense of this legion of texts, and in particular to identify correctly the planets of our solar system, a succession of scholars came up with confusing results. As we now know, their efforts were doomed to failure because they incorrectly assumed that the Sumerians and their successors were unaware that the solar system was heliocentric, that Earth was but another planet, and that there were more planets beyond Saturn.

Ignoring the possibility that some names in the star lists may have applied to Earth itself, and seeking to apply the great number of other names and epithets only to the five planets they believed were known to the Sumerians, scholars reached conflicting conclusions. Some scholars even suggested that the confusion was not theirs, but a Chaldean mix-up—for some unknown reason, they said, the Chaldeans had switched around the names of the five "known" planets.

The Sumerians referred to all celestial bodies (planets, stars, or constellations) as MUL ("who shine in the heights"). The Akkadian term *kakkab* was likewise applied by the Babylonians and Assyrians as a general term for any celestial body. This practice further frustrated the scholars seeking to unravel the ancient astronomical texts. But some *mul*'s that were termed LU.BAD clearly designated planets of our solar system.

Knowing that the Greek name for the planets was "wanderers," the scholars have read LU.BAD as "wandering sheep," deriving from LU ("those which are shepherded") and BAD ("high and afar"). But now that we have shown that the Sumerians were fully aware of the true nature of the solar system, the other meanings of the term *bad* ("the olden," "the foundation," "the one where death is") assume direct significance.

These are appropriate epithets for the Sun, and it follows that by *lubad* the Sumerians meant not mere "wandering sheep" but "sheep" shepherded by the Sun—the planets of our Sun.

The location and relation of the *lubad* to each other and to the Sun were described in many Mesopotamian astronomical texts. There were references to those planets that are "above" and those that are "below," and Kugler correctly guessed that the reference point was Earth itself.

But mostly the planets were spoken of in the framework of astronomical texts dealing with MUL.MUL—a term that kept the scholars guessing. In the absence of a better solution, most scholars have agreed that the term *mulmul* stood for the Pleiades, a cluster of stars in the zodiacal constellation of Taurus, and the one through which the axis of the spring equinox passed (as viewed from Babylon) circa 2200 B.C. Mesopotamian texts often indicated that the *mulmul* included seven LU.MASH (seven "wanderers that are familiar"), and the scholars assumed that these were the brightest members of the Pleiades, which can be seen with the naked eye. The fact that, depending on classification, the group has either six or nine such bright stars, and not seven, posed a problem; but it was brushed aside for lack of any better ideas as to the meaning of *mulmul*.

Franz Kugler (*Sternkunde und Sterndienst in Babel*), reluctantly accepted the Pleiades as the solution, but expressed his astonishment when he found it stated unambiguously in Mesopotamian texts that *mulmul* included not only "wanderers" (planets) but also the Sun and the Moon —making it impossible to retain the Pleiades idea. He also came upon texts that clearly stated that "*mulmul ul-shu 12*" ("*mulmul* is a band of twelve"), of which ten formed a distinct group.

We suggest that the term *mulmul* referred to the solar system, using the repetitive (MUL.MUL) to indicate the group as a whole, as "the celestial body comprising all celestial bodies."

Charles Virolleaud (*L'Astrologie Chaldéenne*), transliterated a Mesopotamian text (K.3558) that describes the members of the *mulmul* or *kakkabu/kakkabu* group. The text's last line is explicit:

Kakkabu/kakkabu.
The number of its celestial bodies is twelve.
The stations of its celestial bodies twelve.
The complete months of the Moon is twelve.

The texts leave no doubt: The *mulmul*—our solar system—was made up of *twelve* members. Perhaps this should not come as a surprise, for the Greek scholar Diodorus, explaining the three "ways" of the Chaldeans and the consequent listing of thirty-six celestial bodies, stated that "of those celestial gods, twelve hold chief authority; to each of these the Chaldeans assign a month and a sign of the zodiac."

Ernst Weidner (*Der Tierkreis und die Wege am Himmel*) reported that in addition to the Way of Anu and its twelve zodiac constellations, some texts also referred to the "way of the Sun," which was also made up of twelve celestial bodies: the Sun, the Moon, and ten others. Line 20 of the so-called TE-tablet stated: *"naphar 12 sheremesh ha.la sha kakkab.lu sha Sin u Shamash ina libbi ittiqu,"* which means, "all in all, 12 members where the Moon and Sun belong, where the planets orbit."

We can now grasp the significance of the number *twelve* in the ancient world. The Great Circle of Sumerian gods, and of all Olympian gods thereafter, comprised exactly twelve; younger gods could join this circle only if older gods retired. Likewise, a vacancy had to be filled to retain the divine number twelve. The principal celestial circle, the way of the Sun with its twelve members, set the pattern, according to which each other celestial band was divided into twelve segments or was allocated twelve principal celestial bodies. Accordingly, there were twelve months in a year, twelve double-hours in a day. Each division of Sumer was assigned twelve celestial bodies as a measure of good luck.

Many studies, such as the one by S. Langdon *(Babylonian Menologies and the Semitic Calendar)* show that the division of the year into twelve months was, from its very beginnings, related to the twelve Great Gods. Fritz Hommel *(Die Astronomie der alten Chaldäer)* and others after him have shown that the twelve months were closely connected with the twelve zodiacs and that both derived from twelve principal celestial bodies. Charles F. Jean *(Lexicologie Sumerienne)* reproduced a Sumerian list of twenty-four celestial bodies that paired twelve zodiacal constellations with twelve members of our solar system.

In a long text, identified by F. Thureau-Dangin *(Ritueles Accadiens)* as a temple program for the New Year Festival in Babylon, the evidence for the consecration of twelve as the central celestial phenomenon is persuasive. The great temple, the Esagila, had twelve gates. The powers of all the celestial gods were vested in Marduk by reciting twelve times the pronouncement "My Lord, is He not my Lord." The mercy of the god was then invoked twelve times, and that of his spouse twelve times. The total of twenty-four was then matched with the twelve zodiacal constellations and twelve members of the solar system.

A boundary stone carved with the symbols of the celestial bodies by a king of Susa depicts those twenty-four signs: the familiar twelve signs of the zodiac, and symbols that stand for the twelve members of the solar system. These were the twelve astral gods of Mesopotamia, as well as of the Hurrian, Hittite, Greek, and all other ancient pantheons. (Fig. 98)

Although our natural counting base is the number ten, the number twelve permeated all matters celestial and divine long after the Sumerians were gone. There were twelve Greek Titans, twelve Tribes of Israel, twelve parts to the magical breastplate of the Israelite High Priest. The power of this celestial twelve carried over to the twelve Apostles of Jesus, and even in our decimal system we count from one to twelve, and only after twelve do we return to "ten and three" (thirteen), "ten and four," and so on.

Where did this powerful, decisive number *twelve* stem from? From the heavens.

For the solar system—the *mulmul*—included, in addition to all the planets known to us, also the planet of Anu,

Fig. 98

the one whose symbol—a radiant celestial body—stood in the Sumerian writing for the god Anu and for "divine." "The *kakkab* of the Supreme Scepter is one of the sheep in *mulmul*," explained an astronomical text. And when Marduk usurped the supremacy and replaced Anu as the god associated with this planet, the Babylonians said: "The planet of Marduk within *mulmul* appears."

Teaching humanity the true nature of Earth and the heavens, the Nefilim informed the ancient astronomer-priests not only of the planets beyond Saturn but also of the existence of the most important planet, the one from which they came:

THE TWELFTH PLANET.

7

•

THE EPIC OF CREATION

ON MOST OF THE ANCIENT cylinder seals that have been
found, symbols that stand for certain celestial bodies,
members of our solar system, appear above the figures of
gods or humans.

An Akkadian seal from the third millennium B.C., now at
the Vorderasiatische Abteilung of the State Museum in
East Berlin (catalogued VA/243), departs from the usual
manner of depicting the celestial bodies. It does not show
them individually but rather as a group of eleven globes
encircling a large, rayed star. It is clearly a depiction of
the solar system as it was known to the Sumerians: a
system consisting of *twelve* celestial bodies. (Fig. 99)

We usually show our solar system schematically as a line
of planets stretching away from the Sun in ever-increasing
distances. But if we depicted the planets, not in a line,
but one after the other in a *circle* (the closest, Mercury,
first, then Venus, then Earth, and so on), the result would
look something like Fig. 100. (All drawings are schematic
and not to scale; planetary orbits in the drawings that
follow are circular rather than elliptical for ease of
presentation.)

If we now take a second look at an enlargement of the
solar system depicted on cylinder seal VA/243, we shall
see that the "dots" encircling the star are actually globes
whose sizes and order conform to that of the solar system
depicted in Fig. 100. The small Mercury is followed by a
larger Venus. Earth, the same size as Venus, is accompanied by the small Moon. Continuing in a counterclockwise direction, Mars is shown correctly as smaller than
Earth but larger than the Moon or Mercury. (Fig. 101)

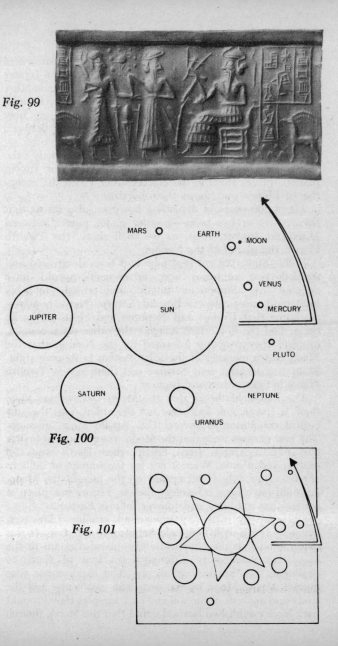

Fig. 99

Fig. 100

Fig. 101

The ancient depiction then shows a planet unknown to us—considerably larger than Earth, yet smaller than Jupiter and Saturn, which clearly follow it. Farther on, another pair perfectly matches our Uranus and Neptune. Finally, the smallish Pluto is also there, but not where we now place it (after Neptune); instead, it appears between Saturn and Uranus.

Treating the Moon as a proper celestial body, the Sumerian depiction fully accounts for all of our known planets, places them in the correct order (with the exception of Pluto), and shows them by size.

The 4,500-year-old depiction, however, also insists that there was—or has been—another major planet between Mars and Jupiter. It is, as we shall show, the Twelfth Planet, the planet of the Nefilim.

If this Sumerian celestial map had been discovered and studied two centuries ago, astronomers would have deemed the Sumerians totally uninformed, foolishly imagining more planets beyond Saturn. Now, however, we know that Uranus and Neptune and Pluto are really there. Did the Sumerians imagine the other discrepancies, or were they properly informed by the Nefilim that the Moon was a member of the solar system in its own right, Pluto was located near Saturn, and there was a Twelfth Planet between Mars and Jupiter?

The long-held theory that the Moon was nothing more than "a frozen golf ball" was not discarded until the successful conclusion of several U.S. Apollo Moon missions. The best guesses were that the Moon was a chunk of matter that had separated from Earth when Earth was still molten and plastic. Were it not for the impact of millions of meteorites, which left craters on the face of the Moon, it would have been a faceless, lifeless, history-less piece of matter that solidified and forever follows Earth.

Observations made by unmanned satellites, however, began to bring such long-held beliefs into question. It was determined that the chemical and mineral makeup of the Moon was sufficiently different from that of Earth to challenge the "breakaway" theory. The experiments conducted on the Moon by the American astronauts and the study and analysis of the soil and rock samples they brought back have established beyond doubt that the Moon, though presently barren, was once a "living planet." Like Earth it is layered, which means that it solidified from its own original molten stage. Like Earth it generated heat, but whereas Earth's heat comes from its radioactive materials, "cooked" inside Earth under tremendous pressure, the Moon's heat comes, apparently, from layers of radioactive materials lying very near the surface. These materials, however, are too heavy to have floated up. What, then, deposited them near the Moon's surface?

The Moon's gravity field appears to be erratic, as though huge chunks of heavy matter (such as iron) had not evenly sunk to its core but were scattered about. By what process or force, we might ask? There is evidence that the ancient rocks of the Moon were magnetized. There is also evidence that the magnetic fields were changed or reversed. Was it by some unknown internal process, or by an undetermined outside influence?

The Apollo 16 astronauts found on the Moon rocks (called breccias) that result from the shattering of solid rock and its rewelding together by extreme and sudden heat. When and how were these rocks shattered, then re-fused? Other surface materials on the Moon are rich in rare radioactive potassium and phosphorus, materials that on Earth are deep down inside.

Putting such findings together, scientists are now certain that the Moon and Earth, formed of roughly the same elements at about the same time, evolved as separate celestial bodies. In the opinion of the scientists of the U.S. National Aeronautics and Space Administration (NASA), the Moon evolved "normally" for its first 500 million years. Then, they said (as reported in *The New York Times*),

> The most cataclysmic period came 4 billion years ago, when celestial bodies the size of large cities and small countries came crashing into the Moon and formed its huge basins and towering mountains.
>
> The huge amounts of radioactive materials left by the collisions began heating the rock beneath the surface, melting massive amounts of it and forcing seas of lava through cracks in the surface.
>
> Apollo 15 found a rockslide in the crater Tsiolovsky six times greater than any rockslide on Earth. Apollo

16 discovered that the collision that created the Sea of Nectar deposited debris as much as 1,000 miles away.

Apollo 17 landed near a scarp eight times higher than any on Earth, meaning it was formed by a moonquake eight times more violent than any earthquake in history.

The convulsions following that cosmic event continued for some 800 million years, so that the Moon's makeup and surface finally took on their frozen shape some 3.2 billion years ago.

The Sumerians, then, were right to depict the Moon as a celestial body in its own right. And, as we shall soon see, they also left us a text that explains and describes the cosmic catastrophe to which the NASA experts refer.

The planet Pluto has been called "the enigma." While the orbits around the Sun of the other planets deviate only somewhat from a perfect circle, the deviation ("eccentricity") of Pluto is such that it has the most extended and elliptical orbit around the Sun. While the other planets orbit the Sun more or less within the same plane, Pluto is out of kilter by a whopping seventeen degrees. Because of these two unusual features of its orbit, Pluto is the only planet that cuts across the orbit of another planet, Neptune.

In size, Pluto is indeed in the "satellite" class: Its diameter, 3,600 miles, is not much greater than that of Triton, a satellite of Neptune, or Titan, one of the ten satellites of Saturn. Because of its unusual characteristics, it has been suggested that this "misfit" might have started its celestial life as a satellite that somehow escaped its master and went into orbit around the Sun on its own.

This, as we shall soon see, is indeed what happened—according to the Sumerian texts.

And now we reach the climax of our search for answers to primeval celestial events: the existence of the Twelfth Planet. Astonishing as it may sound, our astronomers have been looking for evidence that indeed such a planet once existed between Mars and Jupiter.

Toward the end of the eighteenth century, even before Neptune had been discovered, several astronomers demonstrated that "the planets were placed at certain distances from the Sun according to some definite law." The suggestion, which came to be known as Bode's Law, convinced astronomers that a planet ought to revolve in a place where hitherto no planet had been known to exist—that is, between the orbits of Mars and Jupiter.

Spurred by these mathematical calculations, astronomers began to scan the skies in the indicated zone for the "missing planet." On the first day of the nineteenth century, the Italian astronomer Giuseppe Piazzi discovered at the exact indicated distance a very small planet (485 miles across), which he named Ceres. By 1804 the number of asteroids ("small planets") found there rose to four; to date, nearly 3,000 asteroids have been counted orbiting the Sun in what is now called the asteroid belt. Beyond any doubt, this is the debris of a planet that had shattered to pieces. Russian astronomers have named it Phayton ("chariot").

While astronomers are certain that such a planet existed, they are unable to explain its disappearance. Did the planet self-explode? But then its pieces would have flown off in all directions and not stayed in a single belt. If a collision shattered the missing planet, where is the celestial body responsible for the collision? Did it also shatter? But the debris circling the Sun, when added up, is insufficient to account for even one whole planet, to say nothing of two. Also, if the asteroids comprise the debris of two planets, they should have retained the axial revolution of two planets. But all the asteroids have a single axial rotation, indicating they come from a single celestial body. How then was the missing planet shattered, and what shattered it?

The answers to these puzzles have been handed down to us from antiquity.

•

About a century ago the decipherment of the texts found in Mesopotamia unexpectedly grew into a realization that there—in Mesopotamia—texts existed that not only paralleled but also *preceded* portions of the Holy Scriptures. *Die Kielschriften und das alte Testament* by Eberhard Schräder in 1872 started an avalanche of books, articles, lectures, and debates that lasted half a century.

Was there a link, at some early time, between Babylon and the Bible? The headlines provocatively affirmed, or denounced: BABEL UND BIBEL.

Among the texts uncovered by Henry Layard in the ruins of the library of Ashurbanipal in Nineveh, there was one that told a tale of Creation not unlike the one in the Book of Genesis. The broken tablets, first pieced together and published by George Smith in 1876 (*The Chaldean Genesis*), conclusively established that there indeed existed an Akkadian text, written in the Old Babylonian dialect, that related how a certain deity created Heaven and Earth and all upon Earth, including Man.

A vast literature now exists that compares the Mesopotamian text with the biblical narrative. The Babylonian deity's work was done, if not in six "days," then over the span of six tablets. Parallel to the biblical God's seventh day of rest and enjoyment of his handiwork, the Mesopotamian epic devotes a seventh tablet to the exaltation of the Babylonian deity and his achievements. Appropriately, L. W. King named his authoritative text on the subject *The Seven Tablets of Creation.*

Now called "The Creation Epic," the text was known in antiquity by its opening words, *Enuma Elish* ("When in the heights"). The biblical tale of Creation begins with the creation of Heaven and Earth; the Mesopotamian tale is a true cosmogony, dealing with prior events and taking us to the beginning of time:

Enuma elish la nabu shamamu
   When in the heights Heaven had not been named
Shaplitu ammatum shuma la zakrat
   And below, firm ground [Earth] had not been called

It was then, the epic tells us, that two primeval celestial bodies gave birth to a series of celestial "gods." As the number of celestial beings increased, they made great noise and commotion, disturbing the Primeval Father. His faithful messenger urged him to take strong measures to discipline the young gods, but they ganged up on him and robbed him of his creative powers. The Primeval Mother sought to take revenge. The god who led the revolt against the Primeval Father had a new suggestion: Let his young son be invited to join the Assembly of the Gods and be given supremacy so that he might go to fight singlehanded the "monster" their mother turned out to be.

Granted supremacy, the young god—Marduk, according to the Babylonian version—proceeded to face the monster, and, after a fierce battle, vanquished her and split her in two. Of one part of her he made Heaven, and of the other, Earth.

He then proclaimed a fixed order in the heavens, assigning to each celestial god a permanent position. On Earth he produced the mountains and seas and rivers, established the seasons and vegetation, and created Man. In duplication of the Heavenly Abode, Babylon and its towering temple were built on Earth. Gods and mortals were given assignments, commandments, and rituals to be followed. The gods then proclaimed Marduk the supreme deity, and bestowed on him the "fifty names"—the prerogatives and numerical rank of the Enlilship.

As more tablets and fragments were found and translated, it became evident that the text was not a simple literary work: It was the most hallowed historical-religious epic of Babylon, read as part of the New Year rituals. Intended to propagate the supremacy of Marduk, the Babylonian version made him the hero of the tale of Creation. This, however, was not always so. There is enough evidence to show that the Babylonian version of the epic was a masterful religious-political forgery of earlier Sumerian versions, in which Anu, Enlil, and Ninurta were the heroes.

No matter, however, what the actors in this celestial and divine drama were called, the tale is certainly as ancient as Sumerian civilization. Most scholars see it as a philosophic work—the earliest version of the eternal struggle between good and evil—or as an allegorical tale of nature's winter and summer, sunrise and sunset, death and resurrection.

But why not take the epic at face value, as nothing more nor less than the statement of cosmologic facts as known to the Sumerians, as told them by the Nefilim? Using such a bold and novel approach, we find that the "Epic of Creation" perfectly explains the events that probably took place in our solar system.

The stage on which the celestial drama of *Enuma Elish*

unfolds is the primeval universe. The celestial actors are the ones who create as well as the ones being created. Act I:

> When in the heights Heaven had not been named,
> And below, Earth had not been called;
> Naught, but primordial APSU, their Begetter,
> MUMMU, and TIAMAT—she who bore them all;
> Their waters were mingled together.
>
> No reed had yet formed, no marshland had appeared.
> None of the gods had yet been brought into being,
> None bore a name, their destinies were undetermined;
> Then it was that gods were formed in their midst.

With a few strokes of the reed stylus upon the first clay tablet—in nine short lines—the ancient poet-chronicler manages to seat us in front row center, and boldly and dramatically raise the curtain on the most majestic show ever: the Creation of our solar system.

In the expanse of space, the "gods"—the planets—are yet to appear, to be named, to have their "destinies"—their orbits—fixed. Only three bodies exist: "primordial AP.SU" ("one who exists from the beginning"); MUM.MU ("one who was born"); and TIAMAT ("maiden of life"). The "waters" of Apsu and Tiamat were mingled, and the text makes it clear that it does not mean the waters in which reeds grow, but rather the primordial waters, the basic life-giving elements of the universe.

Apsu, then, is the Sun, "one who exists from the beginning."

Nearest him is Mummu. The epic's narrative makes clear later on that Mummu was the trusted aide and emissary of Apsu: a good description of Mercury, the small planet rapidly running around his giant master. Indeed, this was the concept the ancient Greeks and Romans had of the god-planet Mercury: the fast messenger of the gods.

Farther away was Tiamat. She was the "monster" that Marduk later shattered—the "missing planet." But in primordial times she was the very first Virgin Mother of the first Divine Trinity. The space between her and Apsu was not void; it was filled with the primordial elements of Apsu and Tiamat. These "waters" "commingled," and a pair of celestial gods—planets—were formed in the space between Apsu and Tiamat.

> Their waters were mingled together. . . .
> Gods were formed in their midst:
> Gor LAHMU and god LAHAMU were brought forth;
> By name they were called.

Etymologically, the names of these two planets stem from the root *LHM* ("to make war"). The ancients bequeathed to us the tradition that Mars was the God of War and Venus the Goddess of both Love and War. LAHMU and LAHAMU are indeed male and female names, respectively; and the identity of the two gods of the epic and the planets Mars and Venus is thus affirmed both etymologically and mythologically. It is also affirmed astronomically: As the "missing planet," Tiamat was located beyond Mars. Mars and Venus are indeed located in the space between the Sun (Apsu) and "Tiamat." We can illustrate this by following the Sumerian celestial map. (Figs. 102, 103)

The process of the formation of the solar system then went on. Lahmu and Lahamu—Mars and Venus—were brought forth, but even

> Before they had grown in age
> And in stature to an appointed size—
> God ANSHAR and god KISHAR were formed,
> Surpassing them [in size].
> As lengthened the days and multiplied the years,
> God ANU became their son—of his ancestors a rival.
> Then Anshar's first-born, Anu,
> As his equal and in his image begot NUDIMMUD.

With a terseness matched only by the narrative's precision, Act I of the epic of Creation has been swiftly played out before our very eyes. We are informed that Mars and Venus were to grow only to a limited size; but even before their formation was complete, another pair of planets was formed. The two were majestic planets, as evidenced by their names—AN.SHAR ("prince, foremost of the heavens") and KI.SHAR ("foremost of the firm lands"). They overtook in size the first pair, "surpassing them" in stature. The description, epithets, and location of this second pair easily identify them as Saturn and Jupiter. (Fig. 104).

Some time then passed ("multiplied the years"), and a third pair of planets was brought forth. First came ANU, smaller than Anshar and Kishar ("their son"), but larger than the first planets ("of his ancestors a rival" in size). Then Anu, in turn, begot a twin planet, "his equal and in his image." The Babylonian version names the planet NUDIMMUD, an epithet of Ea/Enki. Once again, the descriptions of the sizes and locations fit the next known pair of planets in our solar system, Uranus and Neptune.

There was yet another planet to be accounted for among these outer planets, the one we call Pluto. The "Epic of Creation" has already referred to Anu as "Anshar's first-born," implying that there was yet another planetary god "born" to Anshar/Saturn. The epic catches up with this celestial deity later on, when it relates how Anshar sent out his emissary GAGA on various missions to the other planets. Gaga appears in function and stature equal to Apsu's emissary Mummu; this brings to mind the many similarities between Mercury and Pluto. Gaga, then, was Pluto; but the Sumerians placed Pluto on their celestial map not beyond Neptune, but next to Saturn, whose "emissary," or satellite, it was. (Fig. 105)

As Act I of the "Epic of Creation" came to an end, there was a solar system made up of the Sun and nine planets:

SUN—*Apsu*, "one who existed from the beginning."
MERCURY—*Mummu*, counselor and emissary of Apsu.
VENUS—*Lahamu*, "lady of battles."
MARS—*Lahmu*, "deity of war."
??—*Tiamat*, "maiden who gave life."
JUPITER—*Kishar*, "foremost of firm lands."
SATURN—*Anshar*, "foremost of the heavens."
PLUTO—*Gaga*, counselor and emissary of Anshar.
URANUS—*Anu*, "he of the heavens."
NEPTUNE—*Nudimmud (Ea)*, "artful creator."

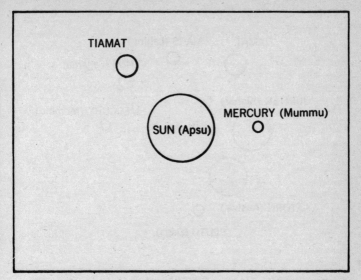

Fig. 102. I. In the Beginning: Sun, Mercury, "Tiamat."

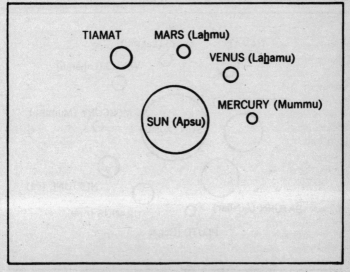

*Fig. 103. II. The Inner Planets—
the "gods in the midst"—come forth.*

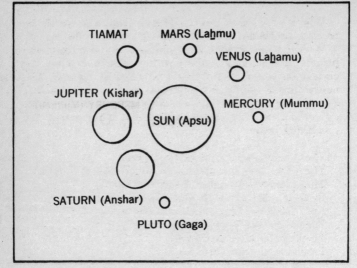

*Fig. 104. III. The SHAR's—the giant planets—
are created, together with their "emissary."*

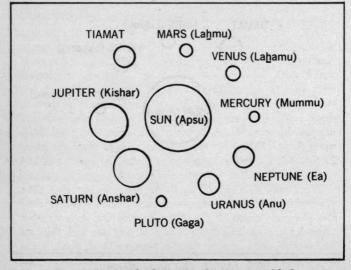

*Fig. 105. IV. The last two planets are added—
equal, in each other's image.*

Where were Earth and the Moon? They were yet to be created, products of the forthcoming cosmic collision.

With the end of the majestic drama of the birth of the planets, the authors of the Creation epic now raise the curtain on Act II, on a drama of celestial turmoil. The newly created family of planets was far from being stable. The planets were gravitating toward each other; they were converging on Tiamat, disturbing and endangering the primordial bodies.

> The divine brothers banded together;
> They disturbed Tiamat as they surged back and forth.
> They were troubling the "belly" of Tiamat
> By their antics in the dwellings of heaven.
> Apsu could not lessen their clamor;
> Tiamat was speechless at their ways.
> Their doings were loathsome. . . .
> Troublesome were their ways.

We have here obvious references to erratic orbits. The new planets "surged back and forth"; they got too close to each other ("banded together"); they interfered with Tiamat's orbit; they got too close to her "belly"; their "ways" were troublesome. Though it was Tiamat that was principally endangered, Apsu, too, found the planets' ways "loathsome." He announced his intention to "destroy, wreck their ways." He huddled with Mummu, conferred with him in secret. But "whatever they had plotted between them" was overheard by the gods, and the plot to destroy them left them speechless. The only one who did not lose his wits was Ea. He devised a ploy to "pour sleep upon Apsu." When the other celestial gods liked the plan, Ea "drew a faithful map of the universe" and cast a divine spell upon the primeval waters of the solar system.

What was this "spell" or force exerted by "Ea" (the planet Neptune)—then the outermost planet—as it orbited the Sun and circled all the other planets? Did its own orbit around the Sun affect the Sun's magnetism and thus its radioactive outpourings? Or did Neptune itself emit, upon its creation, some vast radiations of energy? Whatever the effects were, the epic likened them to a "pouring of sleep"

—a calming effect—upon Apsu (the Sun). Even "Mummu, the Counsellor, was powerless to stir."

As in the biblical tale of Samson and Delilah, the hero—overcome by sleep—could easily be robbed of his powers. Ea moved quickly to rob Apsu of his creative role. Quenching, it seems, the immense outpourings of primeval matter from the Sun, Ea/Neptune "pulled off Apsu's tiara, removed his cloak of aura." Apsu was "vanquished." Mummu could no longer roam about. He was "bound and left behind," a lifeless planet by his master's side.

By depriving the Sun of its creativity—stopping the process of emitting more energy and matter to form additional planets—the gods brought temporary peace to the solar system. The victory was further signified by changing the meaning and location of the Apsu. This epithet was henceforth to be applied to the "Abode of Ea." Any additional planets could henceforth come only from the new Apsu—from "the Deep"—the far reaches of space that the outermost planet faced.

How long was it before the celestial peace was broken once more? The epic does not say. But it does continue, with little pause, and raises the curtain on Act III:

In the Chamber of Fates, the place of Destinies,
A god was engendered, most able and wisest of gods;
In the heart of the Deep was MARDUK created.

A new celestial "god"—a new planet—now joins the cast. He was formed in the Deep, far out in space, in a zone where orbital motion—a planet's "destiny"—had been imparted to him. He was attracted to the solar system by the outermost planet: "He who begot him was Ea" (Neptune). The new planet was a sight to behold:

Alluring was his figure, sparkling the lift of his eyes;
Lordly was his gait, commanding as of olden times. . . .
Greatly exalted was he above the gods, exceeding
     throughout. . . .
He was the loftiest of the gods, surpassing was his height;
His members were enormous, he was exceedingly tall.

Appearing from outer space, Marduk was still a newborn planet, belching fire and emitting radiation. "When he moved his lips, fire blazed forth."

As Marduk neared the other planets, "they heaped upon him their awesome flashes," and he shone brightly, "clothed with the halo of ten gods." His approach thus stirred up electrical and other emissions from the other members of the solar system. And a single word here confirms our decipherment of the Creation epic: *Ten* celestial bodies awaited him—the Sun and only nine other planets.

The epic's narrative now takes us along Marduk's speeding course. He first passes by the planet that "begot" him, that pulled him into the solar system, the planet Ea/Neptune. As Marduk nears Neptune, the latter's gravitational pull on the newcomer grows in intensity. It rounds out Marduk's path, "making it good for its purpose."

Marduk must still have been in a very plastic stage at that time. As he passed by Ea/Neptune, the gravitational pull caused the side of Marduk to bulge, as though he had "a second head." No part of Marduk, however, was torn off at this passage; but as Marduk reached the vicinity of Anu/Uranus, chunks of matter began to tear away from him, resulting in the formation of four satellites of Marduk. "Anu brought forth and fashioned the four sides, consigned their power to the leader of the host." Called "winds," the four were thrust into a fast orbit around Marduk, "swirling as a whirlwind."

The order of passage—first by Neptune, then by Uranus—indicates that Marduk was coming into the solar system not in the system's orbital direction (counterclockwise) but from the opposite direction, moving clockwise. Moving on, the oncoming planet was soon seized by the immense gravitational and magnetic forces of the giant Anshar/Saturn, then Kishar/Jupiter. His path was bent even more inward—into the center of the solar system, toward Tiamat. (Fig. 106)

The approach of Marduk soon began to disturb Tiamat and the inner planets (Mars, Venus, Mercury). "He produced streams, disturbed Tiamat; the gods were not at rest, carried as in a storm."

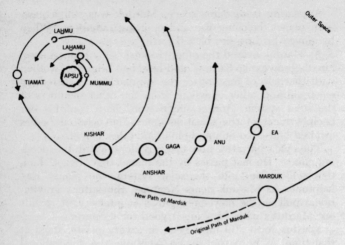

Fig. 106

Though the lines of the ancient text were partially damaged here, we can still read that the nearing planet "diluted their vitals . . . pinched their eyes." Tiamat herself "paced about distraught"—her orbit, evidently, disturbed.

The gravitational pull of the large approaching planet soon began to tear away parts of Tiamat. From her midst there emerged eleven "monsters," a "growling, raging" throng of satellites who "separated themselves" from her body and "marched at the side of Tiamat." Preparing herself to face the onrushing Marduk, Tiamat "crowned them with halos," giving them the appearance of "gods" (planets).

Of particular importance to the epic and to Mesopotamian cosmogony was Tiamat's chief satellite, who was named KINGU, "the first-born among the gods who formed her assembly."

> She exalted Kingu,
> In their midst she made him great. . . .
> The high command of the battle
> She entrusted into his hand.

Subjected to conflicting gravitational pulls, this large satellite of Tiamat began to shift toward Marduk. It was this granting to Kingu of a Tablet of Destinies—a planetary path of his own—that especially upset the outer planets. Who had granted Tiamat the right to bring forth new planets? Ea asked. He took the problem to Anshar, the giant Saturn.

All that Tiamat had plotted, to him he repeated:
". . . she has set up an Assembly and is furious with rage . . .
she has added matchless weapons, has borne
    monster-gods . . .
withal eleven of this kind she has brought forth;
from among the gods who formed her Assembly,
she has elevated Kingu, her first-born, made him chief . . .
she has given him a Tablet of Destinies, fastened it
on his breast."

Turning to Ea, Anshar asked him whether he could go to slay Kingu. The reply is lost due to a break in the tablets; but apparently Ea did not satisfy Anshar, for the continuing narrative has Anshar turning to Anu (Uranus) to find out whether he would "go and stand up to Tiamat." But Anu "was unable to face her and turned back."

In the agitated heavens, a confrontation builds; one god after another steps aside. Will no one do battle with the raging Tiamat?

Marduk, having passed Neptune and Uranus, is now nearing Anshar (Saturn) and his extended rings. This gives Anshar an idea: "He who is potent shall be our Avenger; he who is keen in battle: Marduk, the Hero!" Coming within reach of Saturn's rings ("he kissed the lips of Anshar"), Marduk answers:

"If I, indeed, as your Avenger
Am to vanquish Tiamat, save your lives—
Convene an Assembly to proclaim my Destiny supreme!"

The condition was audacious but simple: Marduk and his "destiny"—his orbit around the Sun—were to be supreme among all the celestial gods. It was then that

Gaga, Anshar/Saturn's satellite—and the future Pluto—was loosened from his course:

> Anshar opened his mouth,
> To Gaga, his Counsellor, a word he addressed. . . .
> "Be on thy way, Gaga,
> take the stand before the gods,
> and that which I shall tell thee
> repeat thou unto them."

Passing by the other god/planets, Gaga urged them to "fix your decrees for Marduk." The decision was as anticipated: The gods were only too eager to have someone else go to settle the score for them. "Marduk is king!" they shouted, and urged him to lose no more time: "Go and cut off the life of Tiamat!"

The curtain now rises on Act IV, the celestial battle.

The gods have decreed Marduk's "destiny"; their combined gravitational pull has now determined Marduk's orbital path so that he can go but one way—toward a "battle," a collision with Tiamat.

As befits a warrior, Marduk armed himself with a variety of weapons. He filled his body with a "blazing flame"; "he constructed a bow . . . attached thereto an arrow . . . in front of him he set the lightning"; and "he then made a net to enfold Tiamat therein." These are common names for what could only have been celestial phenomena—the discharge of electrical bolts as the two planets converged, the gravitational pull (a "net") of one upon the other.

But Marduk's chief weapons were his satellites, the four "winds" with which Uranus had provided him when Marduk passed by that planet: South Wind, North Wind, East Wind, West Wind. Passing now by the giants, Saturn and Jupiter, and subjected to their tremendous gravitational pull, Marduk "brought forth" three more satellites—Evil Wind, Whirlwind, and Matchless Wind.

Using his satellites as a "storm chariot," he "sent forth the winds that he had brought forth, the seven of them." The adversaries were ready for battle.

> The Lord went forth, followed his course;
> Towards the raging Tiamat he set his face. . . .

The Lord approached to scan the innerside of Tiamat—
The scheme of Kingu, her consort, to perceive.

But as the planets drew nearer each other, Marduk's
course became erratic:

As he looks on, his course becomes upset,
His direction is distracted, his doings are confused.

Even Marduk's satellites began to veer off course:

When the gods, his helpers,
Who were marching at his side,
Saw the valiant Kingu, blurred became their vision.

Were the combatants to miss each other after all?
But the die was cast, the courses irrevocably set on
collision. "Tiamat emitted a roar" . . . "the Lord raised
the flooding storm, his mighty weapon." As Marduk came
ever closer, Tiamat's "fury" grew; "the roots of her legs
shook back and forth." She commenced to cast "spells"
against Marduk—the same kind of celestial waves Ea had
earlier used against Apsu and Mummu. But Marduk kept
coming at her.

Tiamat and Marduk, the wisest of the gods,
Advanced against one another;
They pressed on to single combat,
They approached for battle.

The epic now turns to the description of the celestial
battle, in the aftermath of which Heaven and Earth were
created.

The Lord spread out his net to enfold her;
The Evil Wind, the rearmost, he unleashed at her face.
As she opened her mouth, Tiamat, to devour him—
He drove in the Evil Wind so that she close not her lips.
The fierce storm Winds then charged her belly;
Her body became distended; her mouth had opened wide.
He shot there through an arrow, it tore her belly;
It cut through her insides, tore into her womb.
Having thus subdued her, her life-breath he extinguished.

Here, then, (Fig. 107) is a most original theory explaining the celestial puzzles still confronting us. An unstable solar system, made up of the Sun and nine planets, was invaded by a large, comet-like planet from outer space. It first encountered Neptune; as it passed by Uranus, the giant Saturn, and Jupiter, its course was profoundly bent inward toward the solar system's center, and it brought forth seven satellites. It was unalterably set on a collision course with Tiamat, the next planet in line.

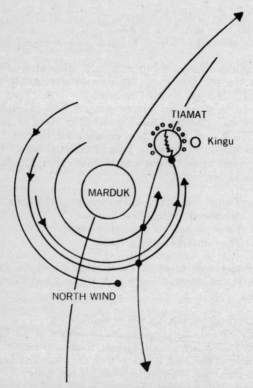

Fig. 107. The Celestial Battle

A. Marduk's "winds" colliding with Tiamat and her "host" (led by Kingu).

But the two planets did *not* collide, a fact of cardinal astronomical importance: It was the satellites of Marduk that smashed into Tiamat, and not Marduk himself. They "distended" Tiamat's body, made in her a wide cleavage. Through these fissures in Tiamat, Marduk shot an "arrow," a "divine lightning," an immense bolt of electricity that jumped as a spark from the energy-charged Marduk, the planet that was "filled with brilliance." Finding its way into Tiamat's innards, it "extinguished her life-breath"— neutralized Tiamat's own electric and magnetic forces and fields, and "extinguished" them.

The first encounter between Marduk and Tiamat left her fissured and lifeless; but her final fate was still to be determined by future encounters between the two. Kingu, leader of Tiamat's satellites, was also to be dealt with separately. But the fate of the other ten, smaller satellites of Tiamat was determined at once.

> After he had slain Tiamat, the leader,
> Her band was shattered, her host broken up.
> The gods, her helpers who marched at her side,
> Trembling with fear,
> Turned their backs about so as to save
>   and preserve their lives.

Can we identify this "shattered . . . broken" host that trembled and "turned their backs about"—reversed their direction?

By doing so we offer an explanation to yet another puzzle of our solar system—the phenomenon of the comets. Tiny globes of matter, they are often referred to as the solar system's "rebellious members," for they appear to obey none of the normal rules of the road. The orbits of the planets around the Sun are (with the exception of Pluto) almost circular; the orbits of the comets are elongated, and in most instances very much so—to the extent that some of them disappear from our view for hundreds or thousands of years. The planets (with the exception of Pluto) orbit the Sun in the same general plane; the comets' orbits lie in many diverse planes. Most significant, while all the planets known to us circle the

Sun in the same counterclockwise direction, many comets move in the reverse direction.

Astronomers are unable to say what force, what event created the comets and threw them into their unusual orbits. Our answer: Marduk. Sweeping in the reverse direction, in an orbital plane of his own, he shattered, broke the host of Tiamat into smaller comets and affected them by his gravitational pull, his so-called net:

Thrown into the net, they found themselves ensnared. . . .
The whole band of demons that had marched on her side
He cast into fetters, their hands he bound. . . .
Tightly encircled, they could not escape.

After the battle was over, Marduk took away from Kingu the Tablet of Destinies (Kingu's independent orbit) and attached it to his own (Marduk's) breast: his course was bent into permanent solar orbit. From that time on, Marduk was bound always to return to the scene of the celestial battle.

Having "vanquished" Tiamat, Marduk sailed on in the heavens, out into space, around the Sun, and back to retrace his passage by the outer planets: Ea/Neptune, "whose desire Marduk achieved," Anshar/Saturn, "whose triumph Marduk established." Then his new orbital path returned Marduk to the scene of his triumph, "to strengthen his hold on the vanquished gods," Tiamat and Kingu.

As the curtain is about to rise on Act V, it will be here —and only here, though this has not hitherto been realized —that the biblical tale of Genesis joins the Mesopotamian "Epic of Creation"; for it is only at this point that the tale of the Creation of Earth and Heaven really began.

Completing his first-ever orbit around the Sun, Marduk "then returned to Tiamat, whom he had subdued."

The Lord paused to view her lifeless body.
To divide the monster he then artfully planned.
Then, as a mussel, he split her into two parts.

Marduk himself now hit the defeated planet, splitting Tiamat in two, severing her "skull," or upper part. Then another of Marduk's satellites, the one called North Wind, crashed into the separated half. The heavy blow carried this part—destined to become Earth—to an orbit where no planet had been orbiting before:

> The Lord trod upon Tiamat's hinder part;
> With his weapon the connected skull he cut loose;
> He severed the channels of her blood;
> And caused the North Wind to bear it
> To places that have been unknown.

Earth had been created!

The lower part had another fate: on the second orbit, Marduk himself hit it, smashing it to pieces (Fig. 108):

> The [other] half of her he set up as a screen for the skies:
> Locking them together, as watchmen he stationed them. . . .
> He bent Tiamat's tail to form the Great Band as a bracelet.

The pieces of this broken half were hammered to become a "bracelet" in the heavens, acting as a screen between the inner planets and the outer planets. They were stretched out into a "great band." The asteroid belt had been created.

Astronomers and physicists recognize the existence of great differences between the inner, or "terrestrial," planets (Mercury, Venus, Earth and its Moon, and Mars) and the outer planets (Jupiter and beyond), two groups separated by the asteroid belt. We now find, in the Sumerian epic, ancient recognition of these phenomena.

Moreover, we are offered—for the first time—a coherent cosmogonic-scientific explanation of the celestial events that led to the disappearance of the "missing planet" and the resultant creation of the asteroid belt (plus the comets) and of Earth. After several of his satellites and his electric bolts split Tiamat in two, another satellite of Marduk shunted her upper half to a new orbit as our planet Earth; then Marduk, on his second orbit, smashed the lower half to pieces and stretched them in a great celestial band.

Every puzzle that we have mentioned is answered by the "Epic of Creation" as we have deciphered it. Moreover, we also have the answer to the question of why Earth's continents are concentrated on one side of it and a deep

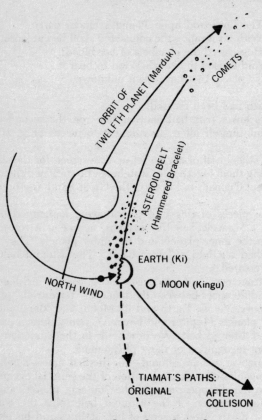

Fig. 108. The Celestial Battle

B. Tiamat has been split: its shattered half is the Heaven
—the Asteroid Belt; the other half, Earth, is thrust to a
new orbit by Marduk's satellite "North Wind." Tiamat's
chief satellite, Kingu, becomes Earth's Moon; her other
satellites now make up the comets.

cavity (the Pacific Ocean's bed) exists on the opposite side. The constant reference to the "waters" of Tiamat is also illuminating. She was called the Watery Monster, and it stands to reason that Earth, as part of Tiamat, was equally endowed with these waters. Indeed, some modern scholars describe Earth as "Planet Ocean"—for it is the only one of the solar system's known planets that is blessed with such life-giving waters.

New as these cosmologic theories may sound, they were accepted fact to the prophets and sages whose words fill the Old Testament. The prophet Isaiah recalled "the primeval days" when the might of the Lord "carved the Haughty One, made spin the watery monster, dried up the waters of *Tehom-Raba*." Calling the Lord Yahweh "my primeval king," the Psalmist rendered in a few verses the cosmogony of the epic of Creation. "By thy might, the waters thou didst disperse; the leader of the watery monsters thou didst break up." Job recalled how this celestial Lord also smote "the assistants of the Haughty One"; and with impressive astronomical sophistication exalted the Lord who:

The hammered canopy stretched out in the place of *Tehom*,
The Earth suspended in the void. . . .
His powers the waters did arrest,
His energy the Haughty One did cleave;
His Wind the Hammered Bracelet measured out;
His hand the twisting dragon did extinguish.

Biblical scholars now recognize that the Hebrew *Tehom* ("watery deep") stems from Tiamat; that *Tehom-Raba* means "great Tiamat," and that the biblical understanding of primeval events is based upon the Sumerian cosmologic epics. It should also be clear that first and foremost among these parallels are the opening verses of the Book of Genesis, describing how the Wind of the Lord hovered over the waters of *Tehom*, and how the lightning of the Lord (Marduk in the Babylonian version) lit the darkness of space as it hit and split Tiamat, creating Earth and the *Rakia* (literally, "the hammered bracelet"). This celestial band (hitherto translated as "firmament") is called "the Heaven."

The Book of Genesis (1:8) explicitly states that it is this "hammered out bracelet" that the Lord had named "heaven" (*shamaim*). The Akkadian texts also called this celestial zone "the hammered bracelet" (*rakkis*), and describe how Marduk stretched out Tiamat's lower part until he brought it end to end, fastened into a permanent great circle. The Sumerian sources leave no doubt that the specific "heaven," as distinct from the general concept of heavens and space, was the asteroid belt.

Our Earth and the asteroid belt are the "Heaven and Earth" of both Mesopotamian and biblical references, created when Tiamat was dismembered by the celestial Lord.

After Marduk's North Wind had pushed Earth to its new celestial location, Earth obtained its own orbit around the Sun (resulting in our seasons) and received its axial spin (giving us day and night). The Mesopotamian texts claim that one of Marduk's tasks after he created Earth was, indeed, to have "allotted [to Earth] the days of the Sun and established the precincts of day and night." The biblical concepts are identical:

> And God said:
> "Let there be Lights in the hammered Heaven,
> to divide between the Day and the Night;
> and let them be celestial signs
> and for Seasons and for Days and for Years."

Modern scholars believe that after Earth became a planet it was a hot ball of belching volcanoes, filling the skies with mists and clouds. As temperatures began to cool, the vapors turned to water, separating the face of Earth into dry land and oceans.

The fifth tablet of *Enuma Elish*, though badly mutilated, imparts exactly the same scientific information. Describing the gushing lava as Tiamat's "spittle," the Creation epic correctly places this phenomenon before the formation of the atmosphere, the oceans of Earth, and the continents. After the "cloud waters were gathered," the oceans began to form, and the "foundations" of Earth—its continents—were raised. As "the making of cold"—a cooling off—took place, rain and mist appeared. Meanwhile, the "spittle" continued to pour forth, "laying in layers," shaping Earth's topography.

Once again, the biblical parallel is clear:

And God said:
"Let the waters under the skies be gathered together, unto one place, and let dry land appear."
And it was so.

Earth, with oceans, continents, and an atmosphere, was now ready for the formation of mountains, rivers, springs, valleys. Attributing all Creation to the Lord Marduk, *Enuma Elish* continued the narration:

Putting Tiamat's head [Earth] into position,
He raised the mountains thereon.
He opened springs, the torrents to draw off.
Through her eyes he released the Tigris and Euphrates.
From her teats he formed the lofty mountains,
Drilled springs for wells, the water to carry off.

In perfect accord with modern findings, both the Book of Genesis and *Enuma Elish* and other related Mesopotamian texts place the beginning of life upon Earth in the waters, followed by the "living creatures that swarm" and "fowl that fly." Not until then did "living creatures after their kind: cattle and creeping things and beasts" appear upon Earth, culminating with the appearance of Man— the final act of Creation.

•

As part of the new celestial order upon Earth, Marduk "made the divine Moon appear . . . designated him to mark the night, define the days every month."

Who was this celestial god? The text calls him SHESH.KI ("celestial god who protects Earth"). There is no mention earlier in the epic of a planet by this name; yet there he is, "within *her* heavenly pressure [gravitational field]." And who is meant by "her": Tiamat or Earth?

The roles of, and references to, Tiamat and Earth appear to be interchangeable. Earth is Tiamat reincarnated. The Moon is called Earth's "protector"; that is exactly what Tiamat called Kingu, her chief satellite.

The Creation epic specifically excludes Kingu from the "host" of Tiamat that were shattered and scattered and put into reverse motion around the Sun as comets. After Marduk completed his own first orbit and returned to the scene of the battle, he decreed Kingu's separate fate:

> And Kingu, who had become chief among them,
> He made shrink;
> As god DUG.GA.E he counted him.
> He took from him the Tablet of Destinies,
> Not rightfully his.

Marduk, then, did not destroy Kingu. He punished him by taking away his independent orbit, which Tiamat had granted him as he grew in size. Shrunk to a smaller size, Kingu remained a "god"—a planetary member of our solar system. Without an orbit he could only become a satellite again. As Tiamat's upper part was thrown into a new orbit (as the new planet Earth), we suggest, Kingu was pulled along. Our Moon, we suggest, is Kingu, Tiamat's former satellite.

Transformed into a celestial *duggae*, Kingu had been stripped of his "vital" elements—atmosphere, waters, radioactive matter; he shrank in size and became "a mass of lifeless clay." These Sumerian terms fittingly describe our lifeless Moon, its recently discovered history, and the fate that befell this satellite that started out as KIN.GU ("great emissary") and ended up as DUG.GA.E ("pot of lead").

L. W. King (*The Seven Tablets of Creation*) reported the existence of three fragments of an astronomical-mythological tablet that presented another version of Marduk's battle with Tiamat, which included verses that dealt with the manner in which Marduk dispatched Kingu. "Kingu, her spouse, with a weapon not of war he cut away . . . the Tablets of Destiny from Kingu he took in his hand." A further attempt, by B. Landesberger (in 1923, in the *Archiv für Keilschriftforschung*), to edit and fully translate the text, demonstrated the interchangeability of the names Kingu/Ensu/Moon.

Such texts not only confirm our conclusion that Tiamat's main satellite became our Moon; they also explain NASA's findings regarding a huge collision "when celestial bodies the size of large cities came crashing into the Moon." Both the NASA findings and the text discovered by L. W. King describe the Moon as the "planet that was laid waste."

Cylinder seals have been found that depict the celestial battle, showing Marduk fighting a fierce female deity. One such depiction shows Marduk shooting his lightning at Tiamat, with Kingu, clearly identified as the Moon, trying to protect Tiamat, his creator. (Fig. 109)

Fig. 109

This pictorial evidence that Earth's Moon and Kingu were the same satellite is further enhanced by the etymological fact that the name of the god SIN, in later times associated with the Moon, derived from SU.EN ("lord of wasteland").

Having disposed of Tiamat and Kingu, Marduk once again "crossed the heavens and surveyed the regions." This time his attention was focused on "the dwelling of Nudimmud" (Neptune), to fix a final "destiny" for Gaga, the erstwhile satellite of Anshar/Saturn who was made an "emissary" to the other planets.

The epic informs us that as one of his final acts in the heavens, Marduk assigned this celestial god "to a hidden place," a hitherto unknown orbit facing "the deep" (outer space), and entrusted to him the "counsellorship of the Watery Deep." In line with his new position, the planet was renamed US.MI ("one who shows the way"), the outermost planet, our Pluto.

According to the Creation epic, Marduk had at one point boasted, "The ways of the celestial gods I will artfully alter . . . into two groups shall they be divided."

Indeed he did. He eliminated from the heavens the Sun's first partner-in-Creation, Tiamat. He brought Earth into being, thrusting it into a new orbit nearer the Sun. He hammered a "bracelet" in the heavens—the asteroid belt that does separate the group of inner planets from the group of outer planets. He turned most of Tiamat's satellites into comets; her chief satellite, Kingu, he put into orbit around Earth to become the Moon. And he shifted a satellite of Saturn, Gaga, to become the planet Pluto, imparting to it some of Marduk's own orbital characteristics (such as a different orbital plane).

The puzzles of our solar system—the oceanic cavities upon Earth, the devastation upon the Moon, the reverse orbits of the comets, the enigmatic phenomena of Pluto— all are perfectly answered by the Mesopotamian Creation epic, as deciphered by us.

Having thus "constructed the stations" for the planets, Marduk took for himself "Station Nibiru," and "crossed the heavens and surveyed" the *new* solar system. It was now made up of twelve celestial bodies, with twelve Great Gods as their counterparts. (Fig. 110)

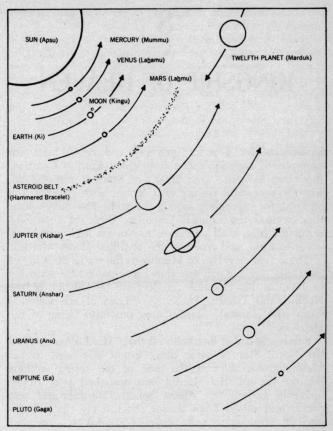

Fig. 110

8
·
KINGSHIP OF HEAVEN

STUDIES OF THE "EPIC OF CREATION" and parallel texts (for
example, S. Langdon's *The Babylonian Epic of Creation)*
show that sometime after 2000 B.C., Marduk, son of Enki,
was the successful winner of a contest with Ninurta, son
of Enlil, for supremacy among the gods. The Babylonians
then revised the original Sumerian "Epic of Creation,"
expunged from it all references to Ninurta and most references to Enlil, and renamed the invading planet Marduk.

The actual elevation of Marduk to the status of "King of
the Gods" upon Earth was thus accompanied by assigning
to him, as his celestial counterpart, the planet of the
Nefilim, the Twelfth Planet. As "Lord of the Celestial
Gods [the planets]" Marduk was thus also "King of the
Heavens."

Some scholars at first believed that "Marduk" was either
the North Star or some other bright star seen in the
Mesopotamian skies at the time of the spring equinox
because the celestial Marduk was described as a "bright
heavenly body." But Albert Schott *(Marduk und sein
Stern)* and others have shown conclusively that all the
ancient astronomical texts spoke of Marduk as a member
of the solar system.

Since other epithets described Marduk as the "Great
Heavenly Body" and the "One Who Illumines," the theory
was advanced that Marduk was a Babylonian Sun God,
parallel to the Egyptian god Ra, whom the scholars also
considered a Sun God. Texts describing Marduk as he "who
scans the heights of the distant heavens . . . wearing a
halo whose brilliance is awe-inspiring" appeared to support
this theory. But the same text continued to say that "he

surveys the lands like Shamash [the Sun]." If Marduk was in some respects *akin* to the Sun, he could not, of course, *be* the Sun.

If Marduk was not the Sun, which one of the planets was he? The ancient astronomical texts failed to fit any one planet. Basing their theories on certain epithets (such as Son of the Sun), some scholars pointed at Saturn. The description of Marduk as a reddish planet made Mars, too, a candidate. But the texts placed Marduk in *markas shame* ("in the center of Heaven"), and this convinced most scholars that the proper identification should be Jupiter, which is located in the center of the line of planets:

Jupiter

Mercury Venus Earth Mars   Saturn Uranus Neptune Pluto

This theory suffers from a contradiction. The same scholars who put it forward were the ones who held the view that the Chaldeans were unaware of the planets beyond Saturn. These scholars list Earth as a planet, while contending that the Chaldeans thought of Earth as a flat center of the planetary system. And they omit the Moon, which the Mesopotamians most definitely counted among the "celestial gods." The equating of the Twelfth Planet with Jupiter simply does not work out.

The "Epic of Creation" clearly states that Marduk was an invader from outside the solar system, passing by the outer planets (including Saturn and Jupiter) before colliding with Tiamat. The Sumerians called the planet NIBIRU, the "planet of crossing," and the Babylonian version of the epic retained the following astronomical information:

Planet NIBIRU:
The Crossroads of Heaven and Earth he shall occupy.
Above and below, they shall not go across;
They must await him.

Planet NIBIRU:
Planet which is brilliant in the heavens.
He holds the central position;
To him they shall pay homage.

Planet NIBIRU:
It is he who without tiring
The midst of Tiamat keeps crossing.
Let "CROSSING" be his name—
The one who occupies the midst.

These lines provide the additional and conclusive information that in dividing the other planets into two equal groups, the Twelfth Planet in "the midst of Tiamat keeps crossing": Its orbit takes it again and again to the site of the celestial battle, where Tiamat used to be.

We find that astronomical texts that dealt in a highly sophisticated manner with the planetary periods, as well as lists of planets in their celestial order, also suggested that Marduk appeared somewhere between Jupiter and Mars. Since the Sumerians did know of all the planets, the appearance of the Twelfth Planet in "the central position" confirms our conclusions:

Marduk

Mercury Venus Moon	Jupiter Saturn Uranus
Earth Mars	Neptune Pluto

If Marduk's orbit takes it to where Tiamat once was, relatively near us (between Mars and Jupiter), why have we not yet seen this planet, which is supposedly large and bright?

The Mesopotamian texts spoke of Marduk as reaching unknown regions of the skies and the far reaches of the universe. "He scans the hidden knowledge . . . he sees all the quarters of the universe." He was described as the "monitor" of all the planets, one whose orbit enables him to encircle all the others. "He keeps hold on their bands [orbits]," makes a "hoop" around them. His orbit was "loftier" and "grander" than that of any other planet. It thus occurred to Franz Kugler (*Sternkunde und Sterndienst in Babylon*) that Marduk was a fast-moving celestial body, orbiting in a great elliptical path just like a comet.

Such an elliptical path, focused on the Sun as a center of gravity, has an apogee—the point farthest from the Sun, where the return flight begins—and a perigee—the point nearest the Sun, where the return to outer space begins. We find that two such "bases" are indeed associated with Marduk in the Mesopotamian texts. The Sumerian texts described the planet as going from AN.UR ("Heaven's base") to E.NUN ("lordly abode"). The Creation epic said of Marduk:

> He crossed the Heaven and surveyed the regions. . . .
> The structure of the Deep the Lord then measured.
> *E-Shara* he established as his outstanding abode;
> *E-Shara* as a great abode in the Heaven he established.

One "abode" was thus "outstanding"—far in the deep regions of space. The other was established in the "Heaven," within the asteroid belt, between Mars and Jupiter. (Fig. 111)

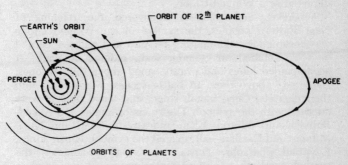

Fig. 111

Following the teachings of their Sumerian forefather, Abraham of Ur, the ancient Hebrews also associated their supreme deity with the supreme planet. Like the Mesopotamian texts, many books of the Old Testament describe the "Lord" as having his abode in the "heights of Heaven," where he "beheld the foremost planets as they were arisen"; a celestial Lord who, unseen, "in the heavens moves about in a circle." The Book of Job, having described the celestial collision, contains these significant verses telling us where the lordly planet had gone:

> Upon the Deep he marked out an orbit;
> Where light and darkness [merge]
> Is his farthest limit.

No less explicitly, the Psalms outlined the planet's majestic course:

> The Heavens bespeak the glory of the Lord;
> The Hammered Bracelet proclaims his handiwork. . . .
> He comes forth as a groom from the canopy;
> Like an athlete he rejoices to run the course.
> From the end of heavens he emanates,
> And his circuit is to their end.

Recognized as a great traveler in the heavens, soaring to immense heights at its apogee and then "coming down, bowing unto the Heaven" at its perigee, the planet was depicted as a Winged Globe.

Wherever archaeologists uncovered the remains of Near Eastern peoples, the symbol of the Winged Globe was conspicuous, dominating temples and palaces, carved on rocks, etched on cylinder seals, painted on walls. It accompanied kings and priests, stood above their thrones, "hovered" above them in battle scenes, was etched into their chariots. Clay, metal, stone, and wood objects were adorned with the symbol. The rulers of Sumer and Akkad, Babylon and Assyria, Elam and Urartu, Mari and Nuzi, Mitanni and Canaan—all revered the symbol. Hittite kings, Egyptian pharaohs, Persian *shar*'s—all proclaimed the symbol (and what it stood for) supreme. It remained so for millennia. (Fig. 112)

Central to the religious beliefs and astronomy of the ancient world was the conviction that the Twelfth Planet, the "Planet of the Gods," remained within the solar system and that its grand orbit returned it periodically to Earth's vicinity. The pictographic sign for the Twelfth Planet, the "Planet of Crossing," was a cross. This cuneiform sign, ⏗, which also meant "Anu" and "divine," evolved in the Semitic languages to the letter *tav*, which meant "the sign."

Fig. 112

Indeed, all the peoples of the ancient world considered the periodic nearing of the Twelfth Planet as a sign of upheavals, great changes, and new eras. The Mesopotamian texts spoke of the planet's periodic appearance as an anticipated, predictable, and observable event:

> The great planet:
> At his appearance, dark red.
> The Heaven he divides in half
> and stands as Nibiru.

Many of the texts dealing with the planet's arrival were omen texts prophesying the effect the event would have upon Earth and Mankind. R. Campbell Thompson (*Reports of the Magicians and Astronomers of Nineveh and Babylon*) reproduced several such texts, which trace the progress of the planet as it "ringed the station of Jupiter" and arrived at the point of crossing, Nibiru:

> When from the station of Jupiter
> the Planet passes towards the west,
> there will be a time of dwelling in security.
> Kindly peace will descend on the land.
> When from the station of Jupiter
> the Planet increases in brilliance
> and in the Zodiac of Cancer will become Nibiru,
> Akkad will overflow with plenty,
> the king of Akkad will grow powerful.
> When Nibiru culminates. . . .
> The lands will dwell securely,
> Hostile kings will be at peace,
> The gods will receive prayers and hear supplications.

The nearing planet, however, was expected to cause rains and flooding, as its strong gravitational effects have been known to do:

> When the Planet of the Throne of Heaven
> will grow brighter,
> there will be floods and rains. . . .
> When Nibiru attains its perigee,
> the gods will give peace;

troubles will be cleared up,
complications will be unravelled.
Rains and floods will come.

Like the Mesopotamian savants, the Hebrew prophets considered the time of the planet's approaching Earth and becoming visible to Mankind as ushering in a new era. The similarities between the Mesopotamian omens of peace and prosperity that would accompany the Planet of the Throne of Heaven, and the biblical prophesies of the peace and justice that would settle upon Earth after the Day of the Lord, can best be expressed in the words of Isaiah:

And it shall come to pass at the End of Days:
. . . the Lord shall judge among the nations
and shall rebuke many peoples.
They shall beat their swords into ploughshares
and their spears into pruning hooks;
nation shall not lift up sword against nation.

In contrast with the blessings of the new era following the Day of the Lord, the day itself was described by the Old Testament as a time of rains, inundations, and earthquakes. If we think of the biblical passages as referring, like their Mesopotamian counterparts, to the passage in Earth's vicinity of a large planet with a strong gravitational pull, the words of Isaiah can be plainly understood:

Like the noise of a multitude in the mountains,
a tumultuous noise like of a great many people,
of kingdoms of nations gathered together;
it is the Lord of Hosts,
commanding a Host to battle.
From a far away land they come,
from the end-point of the Heaven
do the Lord and his Weapons of wrath
come to destroy the whole Earth. . . .
Therefore will I agitate the Heaven
and Earth shall be shaken out of its place
when the Lord of Hosts shall be crossing,
the day of his burning wrath.

While on Earth "mountains shall melt . . . valleys shall be cleft," Earth's axial spin would also be affected. The prophet Amos explicitly predicted:

It shall come to pass on that Day,
sayeth the Lord God,
that I will cause the Sun to go down at noon
and I will darken the Earth in the midst of daytime.

Announcing, "Behold, the Day of the Lord is come!" the prophet Zechariah informed the people that this phenomenon of an arrest in Earth's spin around its own axis would last only one day:

And it shall come to pass on that Day
there shall be no light—uncommonly shall it freeze.
And there shall be one day, known to the Lord,
which shall be neither day nor night,
when at eve-time there shall be light.

On the Day of the Lord, the prophet Joel said, "the Sun and Moon shall be darkened, the stars shall withdraw their radiance"; "the Sun shall be turned into darkness, and the Moon shall be as red blood."

Mesopotamian texts exalted the planet's radiance and suggested that it could be seen even at daytime: "visible at sunrise, disappearing from view at sunset." A cylinder seal, found at Nippur, depicts a group of plowmen looking up with awe as the Twelfth Planet (depicted with its cross symbol) is visible in the skies. (Fig. 113)

Fig. 113

The ancient peoples not only expected the periodic arrival of the Twelfth Planet but also charted its advancing course.

Various biblical passages—especially in Isaiah, Amos, and Job—relate the movement of the celestial Lord to various constellations. "Alone he stretches out the heavens and treads upon the highest Deep; he arrives at the Great Bear, Orion and Sirius, and the constellations of the south." Or, "He smiles his face upon Taurus and Aries; from Taurus to Sagittarius he shall go." These verses describe a planet that not only spans the highest heavens but also comes in from the *south* and moves in a *clockwise* direction—just as we have deduced from the Mesopotamian data. Quite explicitly, the prophet Habakkuk stated: "The Lord from the south shall come . . . his glory shall fill the Earth . . . and Venus shall be as light, its rays of the Lord given."

Among the many Mesopotamian texts that dealt with the subject, one is quite clear:

Planet of the god Marduk:
Upon its appearance: Mercury.
Rising thirty degrees of the celestial arc: Jupiter.
When standing in the place of the celestial battle:
Nibiru.

As the accompanying schematic chart illustrates, the above texts do not simply call the Twelfth Planet by different names (as scholars have assumed). They deal rather with the movements of the planet and the three crucial points at which its appearance can be observed and charted from Earth. (Fig. 114)

The first opportunity to observe the Twelfth Planet as its orbit brings it back to Earth's vicinity, then, was when it aligned with Mercury (point A)—by our calculations, at an angle of 30 degrees to the imaginary celestial axis of Sun–Earth–perigee. Coming closer to Earth and thus appearing to "rise" farther in Earth's skies (another 30 degrees, to be exact), the planet crossed the orbit of Jupiter at point B. Finally, arriving at the place where the celestial battle had taken place, the perigee, or the Place of the Crossing, the planet is Nibiru, point C. Drawing an

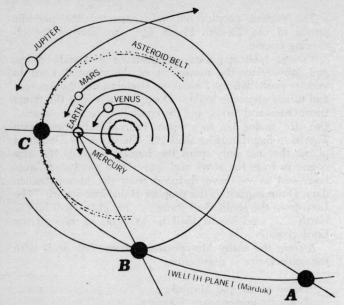

Fig. 114. The Re-appearance of the Twelfth Planet.

imaginary axis between Sun, Earth and the perigee of Marduk's orbit, observers on Earth first saw Marduk aligned with Mercury, at a 30° angle (point A). Progressing another 30°, Marduk crossed the orbital path of Jupiter at point B.

Then, at its perigee (point C) Marduk reached The Crossing: back at the site of the Celestial Battle, it was closest to Earth, and began its orbit back to distant space.

The anticipation of the Day of the Lord in the ancient Mesopotamian and Hebrew writings (which were echoed in the New Testament's expectations of the coming of the Kingship of Heaven) was thus based on the actual experiences of Earth's people: their witnessing the periodic return of the Planet of Kingship to Earth's vicinity.

The planet's periodic appearance and disappearance from Earth's view confirms the assumption of its permanence in solar orbit. In this it acts like many comets. Some of the known comets—like Halley's comet, which nears Earth every seventy-five years—disappeared from view for such long times that astronomers were hard-pressed to realize that they were seeing the same comet. Other comets have been seen only once in human memory, and are assumed to have orbital periods running into thousands of years. The comet Kohoutek, for example, first discovered in March 1973, came within 75,000,000 miles of Earth in January 1974, and disappeared behind the Sun soon thereafter. Astronomers calculate it will reappear anywhere from 7,500 to 75,000 years in the future.

Human familiarity with the Twelfth Planet's periodic appearances and disappearances from view suggests that its orbital period is shorter than that calculated for Kohoutek. If so, why are our astronomers not aware of the existence of this planet? The fact is that even an orbit half as long as the lower figure for Kohoutek would take the Twelfth Planet about six times farther away from us than Pluto—a distance at which such a planet would not be visible from Earth, since it would barely (if at all) reflect the Sun's light toward Earth. In fact, the known planets beyond Saturn were first discovered not visually but mathematically. The orbits of known planets, astronomers found, were apparently being affected by other celestial bodies.

This may also be the way in which astronomers will "discover" the Twelfth Planet. There has already been speculation that a "Planet X" exists, which, though unseen, may be "sensed" through its effects on the orbits of certain comets. In 1972, Joseph L. Brady of the Lawrence Livermore Laboratory of the University of California discovered that discrepancies in the orbit of Halley's comet could be caused by a planet the size of Jupiter orbiting the Sun every 1,800 years. At its estimated distance of 6,000,000,-000 miles, its presence could be detected only mathematically.

While such an orbital period cannot be ruled out, the Mesopotamian and biblical sources present strong evidence that the orbital period of the Twelfth Planet is 3,600 years. The number 3,600 was written in Sumerian as a large circle. The epithet for the planet—*shar* ("supreme ruler")

—also meant "a perfect circle," a "completed cycle." It also meant the number 3,600. And the identity of the three terms—planet/orbit/3,600—could not be a mere coincidence.

Berossus, the Babylonian priest-astronomer-scholar, spoke of ten rulers who reigned upon Earth before the Deluge. Summarizing the writings of Berossus, Alexander Polyhistor wrote: "In the second book was the history of the ten kings of the Chaldeans, and the periods of each reign, which consisted collectively of an hundred and twenty *shar*'s, or four hundred and thirty-two thousand years; reaching to the time of the Deluge."

Abydenus, a disciple of Aristotle, also quoted Berossus in terms of ten pre-Diluvial rulers whose total reign numbered 120 *shar*'s. He made clear that these rulers and their cities were located in ancient Mesopotamia:

It is said that the first king of the land was Alorus. . . . He reigned ten *shar*'s.

Now, a *shar* is esteemed to be three thousand six hundred years. . . .

After him Alaprus reigned three *shar*'s; to him succeeded Amillarus from the city of panti-Biblon, who reigned thirteen *shar*'s. . . .

After him Ammenon reigned twelve *shar*'s; he was of the city of panti-Biblon. Then Megalurus of the same place, eighteen *shar*'s.

Then Daos, the Shepherd, governed for the space of ten *shar*'s. . . .

There were afterwards other Rulers, and the last of all Sisithrus; so that in the whole, the number amounted to ten kings, and the term of their reigns to an hundred and twenty *shar*'s.

Apollodorus of Athens also reported on the prehistorical disclosures of Berossus in similar terms: Ten rulers reigned a total of 120 *shar*'s (432,000 years), and the reign of each one of them was also measured in the 3,600-year *shar* units.

With the advent of Sumerology, the "olden texts" to which Berossus referred were found and deciphered; these were Sumerian king lists, which apparently laid down the tradition of ten pre-Diluvial rulers who ruled Earth from the time when "Kingship was lowered from Heaven" until the "Deluge swept over the Earth."

One Sumerian king list, known as text W-B/144, records the divine reigns in five settled places or "cities." In the first city, Eridu, there were two rulers. The text prefixes both names with the title-syllable "A," meaning "progenitor."

> When kingship was lowered from Heaven,
> kingship was first in Eridu.
> In Eridu,
> A.LU.LIM became king; he ruled 28,800 years.
> A.LAL.GAR ruled 36,000 years.
> Two kings ruled it 64,800 years.

Kingship then transferred to other seats of government, where the rulers were called *en*, or "lord" (and in one instance by the divine title *dingir*).

> I drop Eridu;
> its kingship was carried to Bad-Tibira.
> In Bad-Tibira,
> EN.MEN.LU.AN.NA ruled 43,200 years;
> EN.MEN.GAL.AN.NA ruled 28,800 years.
> Divine DU.MU.ZI, Shepherd, ruled 36,000 years.
> Three kings ruled it for 108,000 years.

The list then names the cities that followed, Larak and Sippar, and their divine rulers; and last, the city of Shuruppak, where a human of divine parentage was king. The striking fact about the fantastic lengths of these rules is that, without exception, they are multiples of 3,600:

Alulim	— 8	× 3,600	=	28,800
Alalgar	—10	× 3,600	=	36,000
Enmenluanna	—12	× 3,600	=	43,200
Enmengalanna	— 8	× 3,600	=	28,800
Dumuzi	—10	× 3,600	=	36,000
Ensipazianna	— 8	× 3,600	=	28,800
Enmenduranna	— 6	× 3,600	=	21,600
Ubartutu	— 5	× 3,600	=	18,000

Another Sumerian text (W-B/62) added Larsa and its two divine rulers to the king list, and the reign periods it gives are also perfect multiples of the 3,600-year *shar*. With the aid of other texts, the conclusion is that there were indeed ten rulers in Sumer before the Deluge; each rule lasted so many *shar*'s; and altogether their reign lasted 120 *shar*'s—as reported by Berossus.

The conclusion that suggests itself is that these *shar*'s of rulership were related to the orbital period *shar* (3,600 years) of the planet "Shar," the "Planet of Kingship"; that Alulim reigned during eight orbits of the Twelfth Planet, Alalgar during ten orbits, and so on.

If these pre-Diluvial rulers were, as we suggest, Nefilim who came to Earth from the Twelfth Planet, then it should not be surprising that their periods of "reign" on Earth should be related to the orbital period of the Twelfth Planet. The periods of such tenure or Kingship would last from the time of a landing to the time of a takeoff; as one commander arrived from the Twelfth Planet, the other's time came up. Since the landings and takeoffs must have been related to the Twelfth Planet's approach to Earth, the command tenures could only have been measured in these orbital periods, of *shar*'s.

One may ask, of course, whether any one of the Nefilim, having landed on Earth, could remain in command here for the purported 28,800 or 36,000 years. No wonder scholars speak of the length of these reigns as "legendary."

But what is a year? Our "year" is simply the time it takes Earth to complete one orbit around the Sun. Because life developed on Earth when it was already orbiting the Sun, life on Earth is patterned by this length of orbit. (Even a more minor orbit time, like that of the Moon, or the day–night cycle is powerful enough to affect almost all life on Earth.) We live so many years because our biological clocks are geared to so many Earth orbits around the Sun.

There can be little doubt that life on another planet would be "timed" to the cycles of that planet. If the trajectory of the Twelfth Planet around the Sun were so extended that one orbit was completed in the same time it takes Earth to complete 100 orbits, then one year of the Nefilim would equal 100 of our years. If their orbit took

1,000 times longer than ours, then 1,000 Earth years would equal only one Nefilim year.

And what if, as we believe, their orbit around the sun lasted 3,600 Earth years? Then 3,600 of our years would amount to only one year in their calendar, and also only one year in their lifetime. The tenures of Kingship reported by the Sumerians and Berossus would thus be neither "legendary" nor fantastic: They would have lasted five or eight or ten Nefilim years.

We have noted, in earlier chapters, that Mankind's march to civilization—through the intervention of the Nefilim—passed through three stages, which were separated by periods of 3,600 years: the Mesolithic period (circa 11,000 B.C.), the pottery phase (circa 7400 B.C.), and the sudden Sumerian civilization (circa 3800 B.C.). It is not unlikely, then, that the Nefilim periodically reviewed (and resolved to continue) Mankind's progress, since they could meet in assembly each time the Twelfth Planet neared Earth.

Many scholars (for example, Heinrich Zimmern in *The Babylonian and Hebrew Genesis*) have pointed out that the Old Testament also carried traditions of pre-Diluvial chieftains, or forefathers, and that the line from Adam to Noah (the hero of the Deluge) listen ten such rulers. Putting the situation prior to the Deluge in perspective, the Book of Genesis (Chapter 6) described the divine disenchantment with Mankind. "And it repented the Lord that he had made Man on Earth . . . and the Lord said: I will destroy Man whom I had created."

And the Lord said:
My spirit shall not shield Man forever;
having erred, he is but flesh.
And his days were one hundred and twenty years.

Generations of scholars have read the verse "And his days shall be a hundred and twenty years" as God's granting a life span of 120 years to Man. But this just does not make sense. If the text dealt with God's intent to destroy Mankind, why would he in the same breath offer Man long life? And we find that no sooner had the Deluge subsided than Noah lived far longer than the supposed limit of 120 years, as did his descendants Shem (600), Arpakhshad (438), Shelah (433), and so on.

In seeking to apply the span of 120 years to Man, the scholars ignore the fact that the biblical language employs not the future tense—"His days *shall be*"—but the past tense—"And his days *were* one hundred and twenty years." The obvious question, then, is: *Whose* time span is referred to here?

Our conclusion is that the count of 120 years was meant to apply to the Deity.

Setting a momentous event in its proper time perspective is a common feature of the Sumerian and Babylonian epic texts. The "Epic of Creation" opens with the words *Enuma elish* ("when on high"). The story of the encounter of the god Enlil and the goddess Ninlil is placed at the time "*when* man had not yet been created," and so on.

The language and purpose of Chapter 6 of Genesis were geared to the same purpose—to put the momentous events of the great Flood in their proper time perspective. The very first word of the very first verse of Chapter 6 is *when:*

> When the Earthlings
> began to increase in number
> upon the face of the Earth,
> and daughters were born unto them.

This, the narrative continues, was the time when

> The sons of the gods
> saw the daughters of the Earthling
> that they were compatible;
> and they took unto themselves
> wives of whichever they chose.

It was the time when

> The Nefilim were upon the land
> in those days, and thereafter too;
> when the sons of the gods
> cohabited with the Earthling's daughters
> and they conceived.
> They were the Mighty Ones who are of Olam,
> the People of the *Shem*.

It was then, in those days, at that time that Man was about to be wiped off the face of the Earth by the Flood.

When exactly was that?

Verse 3 tells us unequivocally: when his, the Deity's count, was 120 years. One hundred twenty "years," not of Man and not of Earth, but as counted by the mighty ones, the "People of the Rockets," the Nefilim. And their year was the *shar*—3,600 Earth years.

This interpretation not only clarifies the perplexing verses of Genesis 6, it also shows how the verses match the Sumerian information: 120 *shar's*, 432,000 Earth years, had passed between the Nefilim's first landing on Earth and the Deluge.

Based on our estimates of when the Deluge occurred, we place the first landing of the Nefilim on Earth circa 450,000 years ago.

•

Before we turn to the ancient records regarding the voyages of the Nefilim to Earth and their settlement on Earth, two basic questions need to be answered: Could beings obviously not much different from us evolve on another planet? Could such beings have had the capability, half a million years ago, for interplanetary travel?

The first question touches upon a more fundamental question: Is there life as we know it anywhere besides the planet Earth? Scientists now know that there are innumerable galaxies like ours, containing countless stars like our Sun, with astronomical numbers of planets providing every imaginable combination of temperature and atmosphere and chemicals, offering billions of chances for Life.

They have also found that our own interplanetary space is not void. For example, there are water molecules in space, the remnants of what are believed to have been clouds of ice crystals that apparently envelop stars in their early stages of development. This discovery lends support to persistent Mesopotamian references to the waters of the Sun, which mingled with the waters of Tiamat.

The basic molecules of living matter have also been found "floating" in interplanetary space, and the belief that life can exist only within certain atmospheres or temperature ranges has also been shattered. Furthermore, the notion that the only source of energy and heat available to living organisms is the Sun's emissions has been discarded. Thus, the spacecraft *Pioneer 10* discovered that Jupiter, though much farther away from the Sun than Earth, was so hot that it must have its own sources of energy and heat.

A planet with an abundance of radioactive elements in its depths would not only generate its own heat; it would also experience substantial volcanic activity. Such volcanic activity provides an atmosphere. If the planet is large enough to exert a strong gravitational pull, it will keep its atmosphere almost indefinitely. Such an atmosphere, in turn, creates a hothouse effect: it shields the planet from the cold of outer space, and keeps the planet's own heat from dissipating into space—much as clothing keeps us warm by not letting the body's heat dissipate. With this in mind, the ancient texts' descriptions of the Twelfth Planet as "clothed with a halo" assume more than poetic significance. It was always referred to as a radiant planet—"most radiant of the gods he is"—and depictions of it showed it as a ray-emitting body. The Twelfth Planet could generate its own heat and retain the heat because of its atmospheric mantle. (Fig. 115)

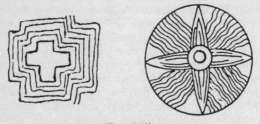

Fig. 115

Scientists have also come to the unexpected conclusion that not only could life have evolved upon the outer planets (Jupiter, Saturn, Uranus, Neptune) but it probably *did* evolve there. These planets are made up of the lighter elements of the solar system, have a composition more akin to that of the universe in general, and offer a profusion of hydrogen, helium, methane, ammonia, and probably neon and water vapor in their atmospheres—all the elements required for the production of organic molecules.

For life as we know it to develop, water is essential. The Mesopotamian texts left no doubt that the Twelfth Planet was a watery planet. In the "Epic of Creation," the planet's list of fifty names included a group exalting its watery aspects. Based on the epithet A.SAR ("watery king"), "who established water levels," the names described the planet as A.SAR.U ("lofty, bright watery king"), A.SAR.U.LU.DU ("lofty, bright watery king whose deep is plentiful"), and so on.

The Sumerians had no doubt that the Twelfth Planet was a verdant planet of life; indeed, they called it NAM.TIL.LA.KU, "the god who maintains life." He was also "bestower of cultivation," "creator of grain and herbs who causes vegetation to sprout . . . who opened the wells, apportioning waters of abundance"—the "irrigator of Heaven and Earth."

Life, scientists have concluded, evolved not upon the terrestrial planets, with their heavy chemical components, but in the outer fringes of the solar system. From these fringes of the solar system, the Twelfth Planet came into our midst, a reddish, glowing planet, generating and radiating its own heat, providing from its own atmosphere the ingredients needed for the chemistry of life.

If a puzzle exists, it is the appearance of life on Earth. Earth was formed some 4,500,000,000 years ago, and scientists believe that the simpler forms of life were already present on Earth within a few hundred million years thereafter. This is simply much too soon for comfort. There are also several indications that the oldest and simplest forms of life, more than 3,000,000,000 years old, had molecules of a biological, not a nonbiological, origin. Stated differently, this means that the life that was on Earth so soon after Earth was born was itself a descendant of some previous life form, and *not* the result of the combination of lifeless chemicals and gases.

What all this suggests to the baffled scientists is that life, which could not easily evolve on Earth, did not, in fact, evolve on Earth. Writing in the scientific magazine *Icarus* (September 1973), Nobel Prize winner Francis Crick and

Dr. Leslie Orgel advanced the theory that "life on Earth may have sprung from tiny organisms from a distant planet."

They launched their studies out of the known uneasiness among scientists over current theories of the origins of life on Earth. Why is there only *one* genetic code for all terrestrial life? If life started in a primeval "soup," as most biologists believe, organisms with a variety of genetic codes should have developed. Also, why does the element molybdenum play a key role in enzymatic reactions that are essential to life, when molybdenum is a very rare element? Why are elements that are more abundant on Earth, such as chromium or nickel, so unimportant in biochemical reactions?

The bizarre theory offered by Crick and Orgel was not only that all life on Earth may have sprung from an organism from another planet but that such "seeding" was *deliberate*—that intelligent beings from another planet launched the "seed of life" from their planet to Earth in a spaceship, for the express purpose of starting the life chain on Earth.

Without benefit of the data provided by this book, these two eminent scientists came close to the real fact. There was no premeditated "seeding"; instead, there was a celestial collision. A life-bearing planet, the Twelfth Planet and its satellites, collided with Tiamat and split it in two, "creating" Earth of its half.

During that collision the life-bearing soil and air of the Twelfth Planet "seeded" Earth, giving it the biological and complex early forms of life for whose early appearance there is no other explanation.

If life on the Twelfth Planet started even 1 percent sooner than on Earth, then it began there some 45,000,000 years earlier. Even by this minute margin, beings as developed as Man would already have been living upon the Twelfth Planet when the first small mammals had just begun to appear on Earth.

Given this earlier start for life on the Twelfth Planet, it was possible for its people to be capable of space travel a mere 500,000 years ago.

9
·
LANDING ON
PLANET EARTH

WE HAVE SET FOOT only on the Moon, and have probed only the planets closest to us with unmanned craft. Beyond our relatively close neighbors, both interplanetary and outer space are still outside the reach of even small scanning craft. But the Nefilim's own planet, with its vast orbit, has served as a traveling observatory, taking them through the orbits of all the outer planets and enabling them to observe at first hand most of the solar system.

No wonder, then, that when they landed on Earth, a good deal of the knowledge they brought with them concerned astronomy and celestial mathematics. The Nefilim, "Gods of Heaven" upon Earth, taught Man to look up unto the heavens—just as Yahweh urged Abraham to do.

No wonder, too, that even the earliest and crudest sculptures and drawings bore celestial symbols of constellations and planets; and that when the gods were to be represented or invoked, their celestial symbols were used as a graphic shorthand. By invoking the celestial ("divine") symbols, Man was no longer alone; the symbols connected Earthlings with the Nefilim, Earth with Heaven, Mankind with the universe.

Some of the symbols, we believe, also convey information that could be related only to space travel to Earth.

Ancient sources provide a profusion of texts and lists dealing with the celestial bodies and their associations with the various deities. The ancient habit of assigning several epithet names to both the celestial bodies and the deities has made identification difficult. Even in the case of established identifications, such as Venus/Ishtar, the picture is confused by the changes in the pantheon. Thus, in earlier times Venus was associated with Ninhursag.

Somewhat greater clarity has been obtained by scholars, such as E. D. Van Buren (*Symbols of the Gods in Mesopotamian Art*), who assembled and sorted out the more than eighty symbols—of gods and celestial bodies—that can be found on cylinder seals, sculptures, stelae, reliefs, murals, and (in great detail and clarity) on boundary stones (*kudurru* in Akkadian). When the classification of the symbols is made, it becomes evident that apart from standing for some of the better-known southern or northern constellations (such as the Sea Serpent for the constellation Hydra), they represented either the *twelve* constellations of the zodiac (for example, the Crab for Scorpio), or the *twelve* Gods of Heaven and Earth, or the *twelve* members of the solar system. The *kudurru* set up by Melishipak, king of Susa (see p. 185), shows the twelve symbols of the zodiac and the symbols of the twelve astral gods.

A stela erected by the Assyrian king Esarhaddon shows the ruler holding the Cup of Life while facing the twelve chief Gods of Heaven and Earth. We see four gods atop animals, of whom Ishtar on the lion and Adad holding the forked lightning can definitely be identified. Four other gods are represented by the tools of their special attributes, as the war-god Ninurta by his lion-headed mace. The remaining four gods are shown as celestial bodies—the Sun (Shamash), the Winged Globe (the Twelfth Planet, the abode of Anu), the Moon's crescent, and a symbol consisting of seven dots. (Fig. 116)

Although in later times the god Sin was associated with the Moon, identified by the crescent, ample evidence shows that in "olden times" the crescent was the symbol of an elderly and bearded deity, one of Sumer's true "olden gods." Often shown surrounded by streams of water, this god was undoubtedly Ea. The crescent was also associated with the science of measuring and calculating, of which Ea was the divine master. It was appropriate that the God of the Seas and Oceans, Ea, be assigned as his celestial counterpart the Moon, which causes the ocean's tides.

Fig. 116

What was the meaning of the symbol of the seven dots?

Many clues leave no doubt that it was the celestial symbol of Enlil. The depiction of the Gateway of Anu (the Winged Globe) flanked by Ea and Enlil (see Fig. 87), represents them by the crescent and the seven-dot symbol. Some of the clearest depictions of the celestial symbols that were meticulously copied by Sir Henry Rawlinson (*The Cuneiform Inscriptions of Western Asia*) assign the most prominent position to a group of three symbols, standing for Anu flanked by his two sons; these show that the symbol for Enlil could be either the seven dots cr a seven-pointed "star." The essential element in Enlil's celestial representation was the number *seven* (the daughter, Ninḥursag, was sometimes included, represented by the umbilical cutter). (Fig. 117)

Scholars have been unable to understand a statement by Gudea, king of Lagash, that "the celestial 7 is 50." Attempts at arithmetic solutions—some formula whereby the number seven would go into fifty—failed to reveal the meaning of the statement. However, we see a simple answer: Gudea stated that the celestial body that is "seven" stands

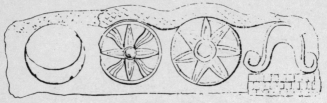

Fig. 117

for the god that is "fifty." The god Enlil, whose rank number was fifty, had as his celestial counterpart the planet that was seventh.

Which planet was the planet of Enlil? We recall the texts that speak of the early times when the gods first came to Earth, when Anu stayed on the Twelfth Planet, and his two sons who had gone down to Earth drew lots. Ea was given the "rulership over the Deep," and to Enlil "the Earth was given for his dominion." And the answer to the puzzle bursts out in all its significance:

The planet of Enlil was Earth. Earth—to the Nefilim—was the seventh planet.

●

In February 1971, the United States launched an unmanned spacecraft on the longest mission to date. For twenty-one months it traveled, past Mars and the asteroid belt, to a precisely scheduled rendezvous with Jupiter. Then, as anticipated by NASA scientists, the immense gravitational pull of Jupiter "grabbed" the spacecraft and hurled it into outer space.

Speculating that *Pioneer 10* might someday be attracted by the gravitational pull of another "solar system" and crash-land on some planet elsewhere in the universe, the *Pioneer 10* scientists attached to it an engraved aluminum plaque bearing the accompanying "message." (Fig. 118)

The message employs a pictographic language—signs and symbols not too different from those used in the very first pictographic writing of Sumer. It attempts to tell whoever might find the plaque that Mankind is male and

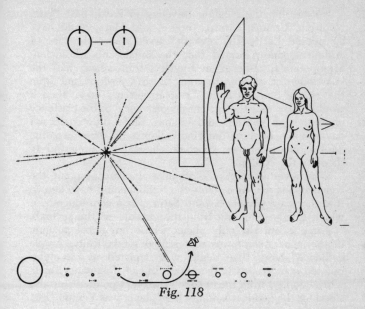

Fig. 118

female, of a size related to the size and shape of the spacecraft. It depicts the two basic chemical elements of our world, and our location relative to a certain interstellar source of radio emissions. And it depicts our solar system as a Sun and nine planets, telling the finder: "The craft that you have found comes from the *third* planet of this Sun."

Our astronomy is geared to the notion that Earth is the third planet—which, indeed, it is if one begins the count from the center of our system, the *Sun*.

But to someone nearing our solar system *from the outside,* the first planet to be encountered would be Pluto, the second Neptune, the third Uranus—not Earth. Fourth would be Saturn; fifth, Jupiter; sixth, Mars.

And Earth would be *seventh*.

•

No one but the Nefilim, traveling to Earth past Pluto, Neptune, Uranus, Saturn, Jupiter, and Mars, could have considered Earth "the seventh." Even if, for the sake of argument, one assumed that the inhabitants of ancient Mesopotamia, rather than travelers from space, had the knowledge or wisdom to count Earth's position not from the central Sun but from the solar system's edge, then it would follow that the ancient peoples *knew* of the existence of Pluto and Neptune and Uranus. Since they could not have known of these outermost planets on their own, the information must, we conclude, have been imparted to them by the Nefilim.

Whichever assumption is adopted as a starting point, the conclusion is the same: Only the Nefilim could have known that there are planets beyond Saturn, as a consequence of which Earth—counting from the outside—is the seventh.

Earth is not the only planet whose numerical position in the solar system was represented symbolically. Ample evidence shows that Venus was depicted as an eight-pointed star: Venus is the eighth planet, following Earth, when counted from the outside. The eight-pointed star also stood for the goddess Ishtar, whose planet was Venus. (Fig. 119)

Fig. 119

Many cylinder seals and other graphic relics depict Mars as the sixth planet. A cylinder seal shows the god associated with Mars (originally Nergal, then Nabu), seated on a throne under a six-pointed "star" as his symbol. (Fig. 120) Other symbols on the seal show the Sun, much in the same manner we would depict it today; the Moon; and the cross, symbol of the "Planet of Crossing," the Twelfth Planet.

Fig. 120

In Assyrian times, the "celestial count" of a god's planet was often indicated by the appropriate number of star symbols placed alongside the god's throne. Thus, a plaque depicting the god Ninurta placed four star symbols at his throne. His planet Saturn is indeed the fourth planet, as counted by the Nefilim. Similar depictions have been found for most of the other planets.

•

The central religious event of ancient Mesopotamia, the twelve-day New Year Festival, was replete with symbolism that had to do with the orbit of the Twelfth Planet, the makeup of the solar system, and the journey of the Nefilim to Earth. The best-documented of these "affirmations of the faith" were the Babylonian New Year rituals; but evidence shows that the Babylonians only copied traditions going back to the beginning of Sumerian civilization.

In Babylon, the festival followed a very strict and detailed ritual; each portion, act, and prayer had a traditional reason and a specific meaning. The ceremonies started on the first day of Nisan—then the first month of the year—coinciding with the spring equinox. For eleven days, the other gods with a celestial status joined Marduk in a prescribed order. On the twelfth day, each of the other gods departed to his own abode, and Marduk was left alone in his splendor. The parallel to the appearance of Marduk within the planetary system, his "visit" with the eleven other members of the solar system, and the separation on the twelfth day—leaving the Twelfth God to go on as King of the Gods, but in isolation from them—is obvious.

The ceremonies of the New Year Festival paralleled the course of the Twelfth Planet. The first four days, matching Marduk's passage by the first four planets (Pluto, Neptune, Uranus, and Saturn), were days of preparation. At the end of the fourth day, the rituals called for marking the appearance of the planet Iku (Jupiter) within sight of Marduk. The celestial Marduk was nearing the place of the celestial battle; symbolically, the high priest began reciting the "Epic of Creation"—the tale of that celestial battle.

The night passed without sleep. When the tale of the celestial battle had been recited, and as the fifth day was breaking, the rituals called for the twelvefold proclamation of Marduk as "The Lord," affirming that in the aftermath of the celestial battle there were now twelve members of the solar system. The recitations then named the twelve members of the solar system and the twelve constellations of the zodiac.

Sometime during the fifth day, the god Nabu—Marduk's son and heir—arrived by boat from his cult center, Borsippa. But he entered Babylon's temple compound only on the sixth day, for by then Nabu was a member of the Babylonian pantheon of twelve and the planet assigned to him was Mars—the sixth planet.

The Book of Genesis informs us that in six days "the Heaven and the Earth and all their host" were completed. The Babylonian rituals commemorating the celestial events that resulted in the creation of the asteroid belt and Earth were also completed in the first six days of Nisan.

On the seventh day, the festival turned its attention to Earth. Though details of the rituals on the seventh day are scarce, H. Frankfort *(Kingship and the Gods)* believes that they involved an enactment by the gods, led by Nabu, of the liberation of Marduk from his imprisonment in the "Mountains of Lower Earth." Since texts have been found that detail epic struggles between Marduk and other claimants to the rulership of Earth, we can surmise that the events of the seventh day were a reenactment of Marduk's struggle for supremacy on Earth (the "Seventh"), his initial defeats, and his final victory and usurpation of the powers.

On the eighth day of the New Year Festival in Babylon, Marduk, victorious on Earth, as the forged *Enuma Elish* had made him in the heavens, received the supreme powers. Having bestowed them on Marduk, the gods, assisted by the king and populace, then embarked, on the ninth day, on a ritual procession that took Marduk from his house within the city's sacred precinct to the "House of Akitu," somewhere outside the city. Marduk and the visiting eleven gods stayed there through the eleventh day; on the twelfth day, the gods dispersed to their various abodes, and the festival was over.

Of the many aspects of the Babylonian festival that reveal its earlier, Sumerian origins, one of the most significant was that which pertained to the House of Akitu. Several studies, such as *The Babylonian Akitu Festival* by S. A. Pallis, have established that this house was featured in religious ceremonies in Sumer as early as the third millennium B.C. The essence of the ceremony was a holy procession that saw the reigning god leave his abode or temple and go, via several stations, to a place well out of town. A special ship, a "Divine Boat," was used for the purpose. Then the god, successful in whatever his mission was at the A.KI.TI House, returned to the city's quay by the same Divine Boat, and retraced his course back to the temple amid feasting and rejoicing by the king and populace.

The Sumerian term A.KI.TI (from which the Babylonian *akitu* derived) literally meant "build on Earth life." This, coupled with the various aspects of the mysterious journey, leads us to conclude that the procession symbolized the hazardous but successful voyage of the Nefilim from their abode to the seventh planet, Earth.

Excavations conducted over some twenty years on the site of ancient Babylon, brilliantly correlated with Babylonian ritual texts, enabled teams of scholars led by F. Wetzel and F. H. Weissbach (*Das Hauptheiligtum des Marduks in Babylon*) to reconstruct the holy precinct of Marduk, the architectural features of his ziggurat, and the Processional Way, portions of which were reerected at the Museum of the Ancient Near East, in East Berlin.

The symbolic names of the seven stations and the epithet of Marduk at each station were given in both Akkadian and Sumerian—attesting both to the antiquity and to the Sumerian origins of the procession and its symbolism.

The first station of Marduk, at which his epithet was "Ruler of the Heavens," was named "House of Holiness" in Akkadian and "House of Bright Waters" in Sumerian. The god's epithet at the second station is illegible; the station itself was named "Where the Field Separates." The partly mutilated name of the third station began with the words "Location facing the planet . . ."; and the god's epithet there changed to "Lord of Poured-Out Fire."

The fourth station was called "Holy Place of Destinies," and Marduk was called "Lord of the Storm of the Waters of *An* and *Ki*." The fifth station appeared less turbulent. It was named "The Roadway," and Marduk assumed the title "Where the Shepherd's Word Appears." Smoother sailing was also indicated at the sixth station, called "The Traveler's Ship," where Marduk's epithet changed to "God of the Marked-Out Gateway."

The seventh station was the *Bit Akitu* ("house of building life on Earth"). There, Marduk took the title "God of the House of Resting."

It is our contention that the seven stations in the procession of Marduk represented the space trip of the Nefilim from their planet to Earth; that the first "station," the "House of Bright Waters," represented the passage by Pluto; the second ("Where the Field Separates") was Neptune; the third, Uranus; the fourth—a place of celestial storms—Saturn. The fifth, where "The Roadway" became clear, "where the shepherd's word appears," was Jupiter. The sixth, where the journey switched to "The Traveler's Ship," was Mars.

And the seventh station was Earth—the end of the journey, where Marduk provided the "House of Resting" (the god's "house of building life on Earth").

•

How did the "Aeronautics and Space Administration" of the Nefilim view the solar system in terms of the space flight to Earth?

Logically—and in fact—they viewed the system in two parts. The one zone of concern was the zone of flight, which embraced the space occupied by the seven planets extending from Pluto to Earth. The second group, beyond the zone of navigation, was made up of four celestial bodies—the Moon, Venus, Mercury, and the Sun. In astronomy and divine genealogy, the two groups were considered separate.

Genealogically, Sin (as the Moon) was the head of the group of the "Four." Shamash (as the Sun) was his son, and Ishtar (Venus), his daughter. Adad, as Mercury, was the Uncle, Sin's brother, who always kept company with his nephew Shamash and (especially) with his niece Ishtar.

The "Seven," on the other hand, were lumped together in texts dealing with the affairs of both gods and men, and with celestial events. They were "the seven who judge," "seven emissaries of Anu, their king," and it was after them that the number seven was consecrated. There were "seven olden cities"; cities had seven gates; gates had seven bolts; blessings called for seven years of plenty; curses, for famines and plagues lasting seven years; divine weddings were celebrated by "seven days of lovemaking"; and so on and on.

During solemn ceremonies like those that accompanied the rare visits to Earth by Anu and his consort, the deities representing the Seven Planets were assigned certain positions and ceremonial robes, while the Four were treated as a separate group. For example, ancient rules of protocol stated: "The deities Adad, Sin, Shamash, and Ishtar shall be seated in the court until daybreak."

In the skies, each group was supposed to stay in its own celestial zone, and the Sumerians assumed that there was a "celestial bar" keeping the two groups apart.

"An important astral-mythological text," according to A. Jeremias (*The Old Testament in the Light of the Ancient Near East*), deals with some remarkable celestial event, when the Seven "stormed in upon the Celestial Bar." In this upheaval, which apparently was an unusual alignment of the Seven Planets, "they made allies of the hero Shamash [the Sun] and of the valiant Adad [Mercury]"—meaning, perhaps, that all exerted a gravitational pull in a single direction. "At the same time, Ishtar, seeking a glorious dwelling place with Anu, strove to become Queen of Heaven"—Venus was somehow shifting its location to a more "glorious dwelling place." The greatest effect was on Sin (the Moon). "The seven who fear not the laws . . . the Light-giver Sin had violently besieged." According to this text, the appearance of the Twelfth Planet saved the darkened Moon and made it "shine forth in the heavens" once again.

The Four were located in a celestial zone the Sumerians termed GIR.ḪE.A ("celestial waters where rockets are confused"), MU.ḪE ("confusion of spacecraft"), or UL.ḪE ("band of confusion"). These puzzling terms make sense once we realize that the Nefilim considered the heavens of the solar system in terms of their space travel. Only recently, the engineers of Comsat (Communications Satellite Corporation) discovered that the Sun and Moon "trick" satellites and "shut them off." Earth satellites could be "confused" by showers of particles from solar flares or by changes in the Moon's reflection of infrared rays. The Nefilim, too, were aware that rocket ships or spacecraft entered a "zone of confusion" once they passed Earth and neared Venus, Mercury, and the Sun.

Separated from the Four by an assumed celestial bar, the Seven were in a celestial zone for which the Sumerians used the term UB. The *ub* consisted of seven parts called (in Akkadian) *giparu* ("night residences"). There is little doubt that this was the origin of Near Eastern beliefs in the "Seven heavens."

The seven "orbs" or "spheres" of the *ub* comprised the Akkadian *kishshatu* ("the entirety"). The term's origin was the Sumerian SHU, which also implied "that part which was the most important," the Supreme. The Seven Planets were therefore sometimes called "the Seven Shiny Ones SHU.NU"—the Seven who "in the Supreme Part rest."

The Seven were treated in greater technical detail than the Four. Sumerian, Babylonian, and Assyrian celestial lists described them with various epithets and listed them in their correct order. Most scholars, assuming that the ancient texts could not possibly have dealt with planets beyond Saturn, have found it difficult to identify correctly the planets described in the texts. But our own findings make identification and understanding of the names' meanings relatively easy.

First to be encountered by the Nefilim approaching the solar system was *Pluto*. The Mesopotamian lists name this planet SHU.PA ("supervisor of the SHU"), the planet that guards the approach to the Supreme Part of the solar system.

As we shall see, the Nefilim could land on Earth only if their spaceship were launched from the Twelfth Planet well before reaching Earth's vicinity, They could thus have crossed the orbit of Pluto not only as inhabitants of the Twelfth Planet but also as astronauts in a moving spaceship. An astronomical text said that the planet Shupa was the one where "the deity Enlil fixed the destiny for the Land"—where the god, in charge of a spacecraft, set the right course for the planet Earth and the Land of Sumer.

Next to Shupa was IRU ("loop"). At *Neptune*, the spacecraft of the Nefilim probably commenced its wide curve or "loop" toward its final target. Another list named the planet ḤUM.BA, which connotes "swampland vegetation." When we probe Neptune someday, will we discover that its persistent association with waters is due to the watery swamps the Nefilim saw upon it?

Uranus was called *Kakkab Shanamma* ("planet which is the double"). Uranus is truly the twin of Neptune in size and appearance. A Sumerian list calls it EN.TI.MASH. SIG ("planet of bright greenish life"). Is Uranus, too, a planet on which swampy vegetation abounded?

Beyond Uranus looms *Saturn*, a giant planet (nearly ten times Earth's size) distinguished by its rings, which extend more than twice as far out as the planet's diameter. Armed with a tremendous gravitational pull and the mysterious rings, Saturn must have posed many dangers to the Nefilim and their spacecraft. This may well explain why they called the fourth planet TAR.GALLU ("the great destroyer"). The planet was also called KAK.SI.DI ("weapon of righteousness") and SI.MUTU ("he who for justice kills"). Throughout the ancient Near East, the planet represented the punisher of the unjust. Were these names expressions of fear or references to actual space accidents?

The *Akitu* rituals, we have seen, made reference to "storms of the waters" between *An* and *Ki* on the fourth day—when the spacecraft was between *Anshar* (Saturn) and *Kishar* (Jupiter).

A very early Sumerian text, assumed since its first publication in 1912 to be "an ancient magical text," very possibly records the loss of a spaceship and its fifty occupants. It relates how Marduk, arriving at Eridu, rushed to his father Ea with some terrible news:

> "It has been created like a weapon;
> It has charged forward like death . . .
> The *Anunnaki* who are fifty,
> it has smitten. . . .
> The flying, birdlike SHU.SAR
> it has smitten on the breast."

The text does not identify "it," whatever destroyed the SHU.SAR (the flying "supreme chaser") and its fifty astronauts. But fear of celestial danger was evident only in regard to Saturn.

The Nefilim must have passed by Saturn and come in view of *Jupiter* with a great sense of relief. They called the fifth planet *Barbaru* ("bright one"), as well as SAG.ME.GAR ("great one, where the space suits are fastened"). Another name for Jupiter, SIB.ZI.AN.NA ("true guide in the heavens"), also described its probable role in the journey to Earth: It was the signal for curving into the difficult passage between Jupiter and Mars, and the entry into the dangerous zone of the asteroid belt. From the epithets, it would seem that it was at this point that the Nefilim put on their *me*'s, their spacesuits.

Mars, appropriately, was called UTU.KA.GAB.A ("light

established at the gate of the waters"), reminding us of the Sumerian and biblical descriptions of the asteroid belt as the celestial "bracelet" separating the "upper waters" from the "lower waters" of the solar system. More precisely, Mars was referred to as *Shelibbu* ("one near the center" of the solar system).

An unusual drawing on a cylinder seal suggests that, passing Mars, an incoming spacecraft of the Nefilim established constant communication with "Mission Control" on Earth. (Fig. 121)

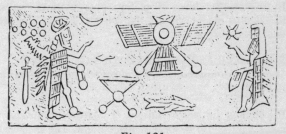

Fig. 121

The central object in this ancient drawing simulates the symbol of the Twelfth Planet, the Winged Globe. Yet it looks different: It is more mechanical, more manufactured than natural. Its "wings" look almost exactly like the solar panels with which American spacecraft are provided to convert the Sun's energy to electricity. The two antennas cannot be mistaken.

The circular craft, with its crownlike top and extended wings and antennas, is located in the heavens, between Mars (the six-pointed star) and Earth and its Moon. On Earth, a deity extends his hand in greeting to an astronaut still out in the heavens, near Mars. The astronaut is shown wearing a helmet with a visor and a breastplate. The lower part of his suit is like that of a "fish-man"—a requirement, perhaps, in case of an emergency splashdown in the ocean. In one hand he holds an instrument; the other hand reciprocates the greeting from Earth.

And then, cruising on, there was *Earth*, the seventh planet. In the lists of the "Seven Celestial Gods" it was called SHU.GI ("right resting place of SHU"). It also meant the "land at the conclusion of SHU," of the Supreme Part of the solar system—the destination of the long space journey.

While in the ancient Near East the sound *gi* was sometimes transformed into the more familiar *ki* ("Earth," "dry land"), the pronunciation and syllable *gi* have endured into our own times in their original meaning, exactly as the Nefilim meant it to be: *geo*-graphy, *geo*-metry, *geo*-logy.

In the earliest form of pictographic writing, the sign SHU.GI also meant *shibu ("the seventh")*. And the astronomical texts explained:

> *Shar shadi il Enlil ana kakkab SHU.GI ikabbi*
> "Lord of Mountains, deity Enlil, with planet Shugi is
>   identical."

Paralleling the seven stations of Marduk's journey, the planets' names also bespeak a space flight. The land at the journey's end was the seventh planet, Earth.

•

We may never know whether, countless years from now, someone on another planet will find and understand the message drawn on the plaque attached to *Pioneer 10*. Likewise, one would think it futile to expect to find on Earth such a plaque in reverse—a plaque conveying to Earthlings information regarding the location and the route from the Twelfth Planet.

Yet such extraordinary evidence does exist.

The evidence is a clay tablet found in the ruins of the Royal Library in Nineveh. Like many of the other tablets, it is undoubtedly an Assyrian copy of an earlier Sumerian tablet. Unlike others, it is a circular disc; and though some cuneiform signs on it are excellently preserved, the few scholars who took on the task of deciphering the tablet ended by calling it "the most puzzling Mesopotamian document."

In 1912, L. W. King, then curator of Assyrian and Babylonian antiquities in the British Museum, made a meticulous copy of the disc, which is divided into eight segments.

The undamaged portions bear geometric shapes unseen on any other ancient artifact, designed and drawn with considerable precision. They include arrows, triangles, intersecting lines, and even an ellipse—a geometric-mathematical curve previously assumed to have been unknown in ancient times. (Fig. 122)

The unusual and puzzling clay plaque was first brought to the attention of the scientific community in a report submitted to the British Royal Astronomical Society on January 9, 1880. R. H. M. Bosanquet and A. H. Sayce, in one of the earliest discourses on "The Babylonian Astronomy," referred to it as a planisphere (the reproduction of a spherical surface as a flat map). They announced that some of the cuneiform signs on it "suggest measurements . . . appear to bear some technical meaning."

The many names of celestial bodies appearing in the eight segments of the plaque clearly established its astronomical character. Bosanquet and Sayce were especially intrigued by the seven "dots" in one segment. They said these might represent the phases of the Moon, were it not for the fact that the dots appeared to run along a line naming the "star of stars" DIL.GAN and a celestial body called APIN.

"There can be no doubt that this enigmatical figure is susceptible of a simple explanation," they said. But their own effort to provide such an explanation did not go beyond reading correctly the phonetic values of the cuneiform signs and the conclusion that the disc was a celestial planisphere.

When the Royal Astronomical Society published a sketch of the planisphere, J. Oppert and P. Jensen improved the reading of some star or planet names. Dr. Fritz Hommel, writing in a German magazine in 1891 ("Die Astronomie der Alten Chaldäer"), drew attention to the fact that each one of the eight segments of the planisphere formed an angle of 45 degrees, from which he concluded that a total sweep of the skies—all 360 degrees of the heavens—was represented. He suggested that the focal point marked some location "in the Babylonian skies."

There the matter rested until Ernst F. Weidner, first in an article published in 1912 (*Babyloniaca:* "Zur Babylonischen Astronomie") and then in his major textbook

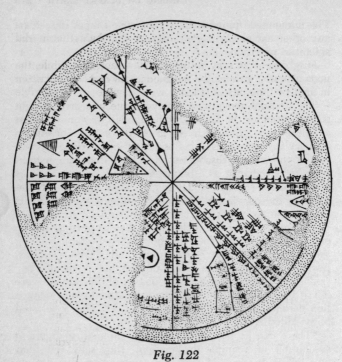

Fig. 122

Handbuch der Babylonischen Astronomie (1915), thoroughly analyzed the tablet, only to conclude that it did not make sense.

His bafflement was caused by the fact that while the geometric shapes and the names of stars or planets written within the various segments were legible or intelligible (even if their meaning or purpose was unclear), the inscriptions along the lines (running at 45-degree angles to each other) just did not make sense. They were, invariably, a series of repeated syllables in the tablet's Assyrian language. They ran, for example, thus:

> *lu bur di   lu bur di   lu bur di*
> *bat bat bat kash kash kash kash alu alu alu alu*

Weidner concluded that the plaque was both astronomical and astrological, used as a magical tablet for exorcism, like several other texts consisting of repeated syllables. With this, he laid to rest any further interest in the unique tablet.

But the tablet's inscriptions assume a completely different aspect if we try to read them not as Assyrian word-signs, but as Sumerian word-syllables; for there can hardly be any doubt that the tablet represents an Assyrian copy of an earlier Sumerian original. When we look at one of the segments (which we can number I), its meaningless syllables

na na na na   a na   a na nu (along the descending line)
sha sha sha sha sha sha (along the circumference)
sham sham bur bur Kur (along the horizontal line)

literally spring to meaningfulness if we enter the Sumerian meaning of these word-syllables. (Fig. 123)

What unfolds here is a *route map*, marking the way by which the god Enlil "went by the planets," accompanied by some operating instructions. The line inclined at 45 degrees appears to indicate the line of a spaceship's descent from a point which is "high high high high," through "vapor clouds" and a lower zone that is vaporless, toward the horizon point, where the skies and the ground meet.

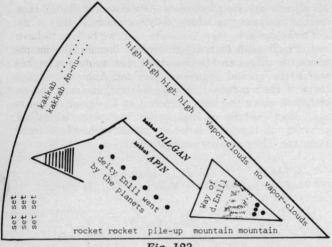

Fig. 123

In the skies near the horizontal line, the instructions to the astronauts make sense: They are told to "set set set" their instruments for the final approach; then, as they near the ground, "rockets rockets" are fired to slow the craft, which apparently should be raised ("piled up") before reaching the landing point because it has to pass over high or rugged terrain ("mountain mountain").

The information provided in this segment clearly pertains to a space voyage by Enlil himself. In this first segment we are given a precise geometric sketch of two triangles connected by a line that turns at an angle. The line represents a route, for the inscription clearly states that the sketch shows how the "deity Enlil went by the planets."

The starting point is the triangle on the left, representing the farther reaches of the solar system; the target area is on the right, where all the segments converge toward the landing point.

The triangle on the left, drawn with its base open, is akin to a known sign in Near Eastern pictographic writing; its meaning can be read as "the ruler's domain, the mountainous land." The triangle on the right is identified by the inscription *shu-ut il Enlil* ("Way of god Enlil"); the term, as we know, denotes Earth's northern skies.

The angled line, then, connects what we believe to have been the Twelfth Planet—"the ruler's domain, the mountainous land"—with Earth's skies. The route passes between two celestial bodies—Dilgan and Apin.

Some scholars have maintained that these were names of distant stars or parts of constellations. If modern manned and unmanned spacecraft navigate by obtaining a "fix" on predetermined bright stars, a similar navigational technique for the Nefilim cannot be ruled out. Yet the notion that the two names stand for such faraway stars somehow does not agree with the meaning of their names: DIL.GAN meant, literally, "the first station"; and APIN, "where the right course is set."

The meanings of the names indicate way stations, points passed by. We tend to agree with such authorities as Thompson, Epping, and Strassmaier, who identified Apin as the planet Mars. If so, the meaning of the sketch becomes clear: The route between the Planet of Kingship and the skies above Earth passed between Jupiter ("the first station") and Mars ("where the right course is set").

This terminology, by which the descriptive names of the planets were related to their role in the space voyage of the Nefilim, conforms with the names and epithets in the lists of the Seven *Shu* Planets. As if to confirm our conclusions, the inscription stating that this was the route of Enlil appears below a row of seven dots—the Seven Planets that stretch from Pluto to Earth.

Not surprisingly, the remaining four celestial bodies, those in the "zone of confusion," are shown separately, beyond Earth's northern skies and the celestial band.

Evidence that this is a space map and flight manual shows up in all the other undamaged segments, too. Continuing in a counterclockwise direction, the legible portion of the next segment bears the inscription: "take take take cast cast cast cast   complete complete." The third segment, where a portion of the unusual elliptical shape is seen, the legible inscriptions include "*kakkab* SIB.ZI.AN.NA . . . envoy of AN.NA . . . deity ISH.TAR," and the intriguing sentence: "Deity NI.NI supervisor of descent."

In the fourth segment, which contains what appear to

be directions on how to establish one's destination according to a certain group of stars, the descending line is specifically identified as the skyline: The word *sky* is repeated eleven times under the line.

Does this segment represent a flight phase nearer Earth, nearer the landing spot? This might indeed be the import of the legend over the horizontal line: "hills hills hills hills top top top top   city city city city." The inscription in the center says: *"kakkab* MASH.TAB.BA [Gemini] whose encounter is fixed: *kakkab* SIB.ZI.AN.NA [Jupiter] provides knowledge."

If, as appears to be the case, the segments are arranged in an approach sequence, then one can almost share the excitement of the Nefilim as they approached Earth's spaceport. The next segment, again identifying the descending line as "sky sky sky," also announces:

> our light   our light   our light
> change   change   change   change
> observe path and high ground
> . . . flat land . . .

The horizontal line contains, for the first time, figures:

> rocket   rocket
> rocket   rise   glide
> 40 40 40
> 40 40 20 22 22

The upper line of the next segment no longer states: "sky sky"; instead, it calls for "channel channel   100 100 100 100 100 100 100." A pattern is discernible in this largely damaged segment. Along one of the lines the inscription says: *"Ashshur,"* which can mean "He who sees" or "seeing."

The seventh segment is too damaged to add to our examination; the few discernible syllables mean "distant distant . . . sight sight," and the instructional words are "press down." The eighth and final segment, however, is almost complete. Directional lines, arrows, and inscriptions mark a path between two planets. Instructions to "pile up   mountain mountain," show four sets of crosses, inscribed twice "fuel water grain" and twice "vapor water grain."

Was this a segment dealing with preparations for the flight toward Earth, or one dealing with stocking up for the return flight to rejoin the Twelfth Planet? The latter may have been the case, for the line with the sharp arrow pointing toward the landing site on Earth has at its other end another "arrow" pointing in the opposite direction, and bearing the legend *"Return."* (Fig. 124)

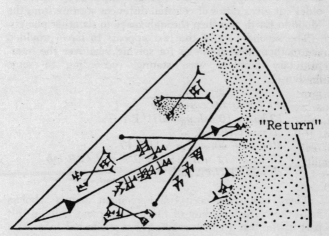

"Return"

Fig. 124

When Ea arranged for Anu's emissary to "make Adapa take the road to Heaven" and Anu discovered the ruse, he demanded to know:

> Why did Ea, to a worthless human
> the plan of Heaven-Earth disclose—
> rendering him distinguished,
> making a *Shem* for him?

In the planisphere we have just deciphered, we indeed see such a route map, a "plan of Heaven-Earth." In sign language and in words, the Nefilim have sketched for us the route from their planet to ours.

Otherwise inexplicable texts dealing with celestial distances also make sense if we read them in terms of space travel from the Twelfth Planet. One such text, found in the ruins of Nippur and believed to be some 4,000 years old, is now kept at the Hilprecht Collection at the University of Jena, in Germany. O. Neugebauer (*The Exact Sciences in Antiquity*) established that the tablet was undoubtedly a copy "from an original composition which was older"; it gives ratios of celestial distances starting from the Moon to Earth and then through space to six other planets.

The second part of the text appears to have provided the mathematical formulas for solving whatever the interplanetary problem was, stating (according to some readings):

> *40 4 20 6 40 × 9 is 6 40*
> *13 kasbu 10 ush mul SHU.PA*
> *eli mul GIR sud*
> *40 4 20 6 40 × 7 is 5 11 6 40*
> *10 kasbu 11 ush 6½ gar 2 u mul GIR tab*
> *eli mul SHU.PA sud*

There has never been full agreement among scholars as to the correct reading of the measurement units in this part of the text (a new reading was suggested to us in a letter from Dr. J. Oelsner, custodian of the Hilprecht Collection at Jena). It is clear, however, that the second part of the text measured distances from SHU.PA (Pluto).

Only the Nefilim, traversing the planetary orbits, could have worked out these formulas; only they needed such data.

Taking into consideration that their own planet and their target, Earth, were both in continuous motion, the Nefilim had to aim their craft not at where Earth was at launch time but where it would be at arrival time. One can safely assume that the Nefilim worked out their trajectories very much as modern scientists map the missions to the Moon and to other planets.

The spacecraft of the Nefilim was probably launched from the Twelfth Planet in the direction of the Twelfth Planet's own orbit, but well ahead of its arrival in Earth's vicinity. Based on these and a myriad other factors, two alternative trajectories for the spacecraft were worked out for us by Amnon Sitchin, doctor of aeronautics and engineering. The first trajectory would call for the launching of the spacecraft from the Twelfth Planet before it reached its apogee (the point farthest out). With few power needs, the spaceship would actually not so much change course as slow down. While the Twelfth Planet (a space vehicle, too, even though a huge one) continued on its vast elliptical orbit, the spaceship would follow a much shorter elliptical course and reach Earth far ahead of the Twelfth Planet. This alternative may have offered the Nefilim both advantages and disadvantages.

The full span of 3,600 Earth years, which applied to tenures of office and other activities of the Nefilim upon Earth, suggests that they might have preferred the second alternative, that of a short trip and a stay in Earth's skies coinciding with the arrival of the Twelfth Planet itself. This would have called for the launching of the spaceship (C) when the Twelfth Planet was about midway on its course back from the apogee. With the planet's own speed rapidly increasing, the spaceship required strong engines to overtake its home planet and reach Earth (D) a few Earth years ahead of the Twelfth Planet. (Fig. 125)

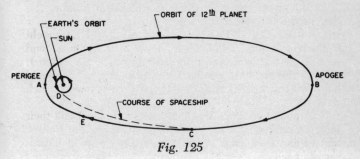

Fig. 125

Based on complex technical data, as well as hints in Mesopotamian texts, it appears that the Nefilim adopted for their Earth missions the same approach NASA adopted for the Moon missions: When the principal spaceship neared the target planet (Earth), it went into orbit around that planet without actually landing. Instead, a smaller craft was released from the mother ship and performed the actual landing.

As difficult as accurate landings were, the departures from Earth must have been even trickier. The landing craft had to rejoin its mother ship, which then had to fire up its engines and accelerate to extremely high speeds, for it had to catch up with the Twelfth Planet, which by then was passing its perigee between Mars and Jupiter at its top orbital speed. Dr. Sitchin has calculated that there were three points in the spaceship's orbit of Earth that lent themselves to a thrust toward the Twelfth Planet. The three alternatives offered the Nefilim a choice of catching up with the Twelfth Planet within 1.1 to 1.6 Earth years.

Suitable terrain, guidance from Earth, and perfect co-ordination with the home planet were required for successful arrivals, landings, takeoffs, and departures from Earth.

As we shall see, the Nefilim met all these requirements.

10

•

CITIES OF THE GODS

THE STORY of the first settlement of Earth by intelligent beings is a breathtaking saga no less inspiring than the discovery of America or the circumnavigation of Earth. It was certainly of greater importance, for, as a result of this settlement, we and our civilizations exist today.

The "Epic of Creation" informs us that the "gods" came to Earth following a deliberate decision by their leader. The Babylonian version, attributing the decision to Marduk, explains that he waited until Earth's soil dried and hardened sufficiently to permit landing and construction operations. Then Marduk announced his decision to the group of astronauts:

> In the deep Above,
> where you have been residing,
> "The Kingly House of Above" have I built.
> Now, a counterpart of it
> I shall build in The Below.

Marduk then explained his purpose:

> When from the Heavens
> for assembly you shall descend,
> there shall be a restplace for the night
> to receive you all.
> I will name it "Babylon"—
> The Gateway of the Gods.

Earth was thus not merely the object of a visit or a quick, exploratory stay; it was to be a permanent "home away from home."

Traveling on board a planet that was itself a kind of spaceship, crossing the paths of most of the other planets, the Nefilim no doubt first scanned the heavens from the surface of their own planet. Unmanned probes must have followed. Sooner or later they acquired the capacity to send out manned missions to the other planets.

As the Nefilim searched for an additional "home," Earth must have struck them favorably. Its blue hues indicated it had life-sustaining water and air; its browns disclosed firm land; its greens, vegetation and the basis for animal life. Yet when the Nefilim finally voyaged to Earth, it must have looked somewhat different from the way it does to our astronauts today. For when the Nefilim first came to Earth, Earth was in the midst of an ice age—a glacial period that was one of the icing and deicing phases of Earth's climate:

Early glaciation—begun some 600,000 years ago
First warming (interglacial period)—550,000 years
   ago
Second glacial period—480,000 to 430,000 years ago

When the Nefilim first landed on Earth some 450,000 years ago, about a third of Earth's land area was covered with ice sheets and glaciers. With so much of Earth's waters frozen, rainfall was reduced, but not everywhere. Due to the peculiarities of wind patterns and terrain, among other things, some areas that are well watered today were barren then, and some areas with only seasonal rains now were experiencing year-round rainfalls then.

The sea levels were also lower because so much water had been captured as ice on the land masses. Evidence indicates that at the height of the two major ice ages, sea levels were as much as 600 to 700 feet lower than at present. Therefore, there was dry land where we now have seas and coastlines. Where rivers continued to run, they created deep gorges and canyons if their courses took them through rocky terrain; if their courses ran in soft earth and clay, they reached the ice-age seas through vast marshlands.

Arriving on Earth amidst such climatic and geographic

conditions, where were the Nefilim to set up their first abode?

They searched, no doubt, for a place with a relatively temperate climate, where simple shelters would suffice and where they could move about in light working clothes rather than in heavily insulated suits. They must also have searched for water for drinking, washing, and industrial purposes, as well as to sustain the plant and animal life needed for food. Rivers would both facilitate the irrigation of large tracts of land and provide a convenient means of transportation.

Only a rather narrow temperate zone on Earth could meet all these requirements, as well as the need for the long, flat areas suitable for landings. The attention of the Nefilim, as we now know, focused on three major river systems and their plains: the Nile, the Indus, and the Tigris–Euphrates. Each of these river basins was suitable for early colonization; each, in time, became the center of an ancient civilization.

The Nefilim would hardly have ignored another need: a source of fuel and energy. On Earth, petroleum has been a versatile and abundant source of energy, heat, and light, as well as a vital raw material from which countless essential goods are made. The Nefilim, judging by Sumerian practice and records, made extensive use of petroleum and its derivatives; it stands to reason that in their search for the most suitable habitat on Earth, the Nefilim would prefer a site rich in petroleum.

With this in mind, the Nefilim probably placed the Indus plain in last place, for it is not an area where oil could be found. The Nile valley was probably given second place; geologically it lies in a major sedimentary rock zone, but the area's oil is found only at some distance from the valley and requires deep drilling. The Land of the Two Rivers, Mesopotamia, was doubtless put in first place. Some of the world's richest oil fields stretch from the tip of the Persian Gulf to the mountains where the Tigris and Euphrates originate. And while in most places one must drill deep to bring up the crude oil, in ancient Sumer (now southern Iraq), bitumens, tars, pitches, and asphalts bubbled or flowed up to the surface naturally.

(Interestingly, the Sumerians had names for *all* bituminous substances—petroleum, crude oils, native asphalts, rock asphalts, tars, pyrogenic asphalts, mastics, waxes, and pitches. They had nine different names for the various bitumens. By comparison, the ancient Egyptian language had only two, and Sanskrit, only three.)

The Book of Genesis describes God's abode on Earth—Eden—as a place of temperate climate, warm yet breezy, for God took afternoon strolls to catch the cooling breeze. It was a place of good soil, lending itself to agriculture and horticulture, especially the cultivation of orchards. It was a place that drew its waters from a network of four rivers. "And the name of the third river [was] Hidekel [Tigris]; it is the one which floweth towards the east of Assyria; and the fourth was the Euphrates."

While opinions regarding the identity of the first two rivers, Pishon ("abundant") and Gihon ("which gushes forth"), are inconclusive, there is no uncertainty regarding the other two rivers, the Tigris and the Euphrates. Some scholars locate Eden in northern Mesopotamia, where the two rivers and two lesser tributaries originate; others (such as E. A. Speiser, in *The Rivers of Paradise*) believe that the four streams converged at the head of the Persian Gulf, so that Eden was not in northern but in southern Mesopotamia.

The biblical name Eden is of Mesopotamian origin, stemming from the Akkadian *edinu*, meaning "plain." We recall that the "divine" title of the ancient gods was DIN.GIR ("the righteous/just ones of the rockets"). A Sumerian name for the gods' abode, E.DIN, would have meant "home of the righteous ones"—a fitting description.

The selection of Mesopotamia as the home on Earth was probably motivated by at least one other important consideration. Though the Nefilim in time established a spaceport on dry land, some evidence suggests that at least initially they landed by splashing down into the sea in a hermetically sealed capsule. If this was the landing method, Mesopotamia offered proximity to not one but two seas—the Indian Ocean to the south and the Mediterranean to the west—so that in case of an emergency, the landing did not have to depend on one watery site alone. As we shall see, a good bay or gulf from which long sea voyages could be launched was also essential.

In ancient texts and pictures, the craft of the Nefilim were initially termed "celestial boats." The landing of such "maritime" astronauts, one can imagine, might have been described in ancient epic tales as the appearance of some kind of submarine from the heavens in the sea, from which "fish-men" emerged and came ashore.

The texts do, in fact, mention that some of the AB.GAL who navigated the spaceships were dressed as fish. One text dealing with Ishtar's divine journeys quotes her as seeking to reach the "Great *gallu*" (chief navigator) who had gone away "in a sunken boat." Berossus transmitted legends regarding Oannes, the "Being Endowed with Reason," a god who made his appearance from "the Erythrean sea which bordered on Babylonia," in the first year of the descent of Kingship from Heaven. Berossus reported that though Oannes looked like a fish, he had a human head under the fish's head, and had feet like a man under the fish's tail. "His voice too and language were articulate and human." (Fig. 126)

The three Greek historians through whom we know what Berossus wrote, reported that such divine fish-men appeared periodically, coming ashore from the "Erythrean sea"—the body of water we now call the Arabian Sea (the western part of the Indian Ocean).

Why would the Nefilim splash down in the Indian Ocean, hundreds of miles from their selected site in Mesopotamia, instead of in the Persian Gulf, which is so much closer? The ancient reports indirectly confirm our conclusion that the first landings occurred during the second glacial period, when today's Persian Gulf was not a sea but a stretch of marshlands and shallow lakes, in which a splashdown was impossible.

Coming down in the Arabian Sea, the first intelligent beings on Earth then made their way toward Mesopotamia. The marshlands extended deeper inland than today's coastline. There, at the edge of the marshes, they established their very first settlement on our planet.

Fig. 126

They named it E.RI.DU ("house in faraway built").
What an appropriate name!

To this very day, the Persian term *ordu* means "encamp-
ment." It is a word whose meaning has taken root in all
languages: The settled Earth is called *Erde* in German,
Erda in Old High German, *Jördh* in Icelandic, *Jord* in
Danish, *Airtha* in Gothic, *Erthe* in Middle English; and,
going back geographically and in time, "Earth" was *Aratha*
or *Ereds* in Aramaic, *Erd* or *Ertz* in Kurdish, and *Eretz* in
Hebrew.

At Eridu, in southern Mesopotamia, the Nefilim established Earth Station I, a lonely outpost on a half-frozen
planet. (Fig. 127)

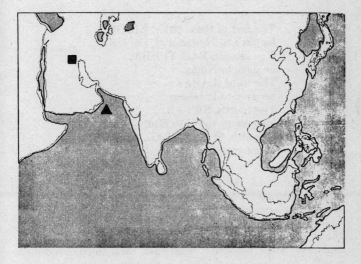

Fig. 127

Sumerian texts, confirmed by later Akkadian translations, list the original settlements or "cities" of the Nefilim in the order in which they were established. We are even told which god was put in charge of each of these settlements. A Sumerian text, believed to have been the original of the Akkadian "Deluge Tablets," relates the following regarding five of the first seven cities:

> After kingship had been lowered from heaven,
> after the exalted crown, the throne of kingship

had been lowered from heaven,
he . . . perfected the procedures,
the divine ordinances. . . .
Founded five cities in pure places,
called their names,
laid them out as centers.

The first of these cities, ERIDU,
he gave to Nudimmud, the leader,
The second, BAD-TIBIRA,
he gave to Nugig.
The third, LARAK,
he gave to Pabilsag.
The fourth, SIPPAR,
he gave to the hero Utu.
The fifth, SHURUPPAK,
he gave to Sud.

The name of the god who lowered Kingship from
Heaven, planned the establishment of Eridu and four other
cities, and appointed their governors or commanders, is
unfortunately obliterated. All the texts agree, however, that
the god who waded ashore to the edge of the marshlands
and said "Here we settle" was Enki, nicknamed "Nudimmud" ("he who made things") in the text.

This god's two names—EN.KI ("lord of firm ground")
and E.A ("whose house is water")—were most appropriate. Eridu, which remained Enki's seat of power and
center of worship throughout Mesopotamian history, was
built on ground artificially raised above the waters of the
marshlands. The evidence is contained in a text named (by
S. N. Kramer) the "Myth of Enki and Eridu":

The lord of the watery-deep, the king Enki . . .
built his house. . . .
In Eridu he built the House of the Water Bank. . . .
The king Enki . . . has built a house:
Eridu, like a mountain,
he raised up from the earth;
in a good place he had built it.

These and other, mostly fragmentary texts suggest that one of the first concerns of these "colonists" on Earth had to do with the shallow lakes or watery marshes. "He brought . . . ; established the cleaning of the small rivers." The effort to dredge the beds of streams and tributaries to allow a better flow of the waters was intended to drain the marshes, obtain cleaner, potable water, and implement controlled irrigation. The Sumerian narrative also indicates some landfilling or the raising of dikes to protect the first houses from the omnipresent waters.

A text named by scholars the "myth" of "Enki and the Land's Order" is one of the longest and best preserved of Sumerian narrative poems so far uncovered. Its text consists of some 470 lines, of which 375 are perfectly legible. Its beginning (some 50 lines) is, unfortunately, broken. The verses that follow are devoted to an exaltation of Enki and to the establishment of his relationship with the chief deity Anu (his father), Ninti (his sister), and Enlil (his brother).

Following these introductions, Enki himself "picks up the microphone." As fantastic as it may sound, the fact is that the text amounts to a first-person report by Enki of his landing on Earth.

> "When I approached Earth,
> there was much flooding.
> When I approached its green meadows,
> heaps and mounds were piled up
> at my command.
> I built my house in a pure place. . . .
> My house—
> Its shade stretches over the Snake Marsh. . . .
> The carp fish wave their tails in it
> among the small *gizi* reeds."

The poem then goes on to describe and record, in the third person, the achievements of Enki. Here are some selected verses:

> He marked the marshland,
> placed in it carp and . . .—fish;
> He marked the cane thicket,
> placed in it . . .—reeds and green-reeds.
> Enbilulu, the Inspector of Canals,
> he placed in charge of the marshlands.
>
> Him who set net so no fish escapes,
> whose trap no . . . escapes,
> whose snare no bird escapes,
> . . . the son of . . . a god who loves fish
> Enki placed in charge of fish and birds.
>
> Enkimdu, the one of the ditch and dike,
> Enki placed in charge of ditch and dike.
>
> Him whose . . . mold directs,
> Kulla, the brick-maker of the Land,
> Enki placed in charge of mold and brick.

The poem lists other achievements of Enki, including the purification of the waters of the Tigris River and the joining (by canal) of the Tigris and Euphrates. His house by the watery bank adjoined a wharf at which reed rafts and boats could anchor, and from which they could sail off. Appropriately, the house was named E.ABZU ("house of the Deep"). Enki's sacred precinct in Eridu was known by this name for millennia thereafter.

No doubt Enki and his landing party explored the lands around Eridu, but he appears to have preferred traveling by water. The marshland, he said in one of the texts, "is my favorite spot; it stretches out its arms to me." In other texts Enki described sailing in the marshlands in his boat, named MA.GUR (literally, "boat to turn about in"), namely, a touring boat. He tells how his crewmen "drew on the oars in unison," how they used to "sing sweet songs, causing the river to rejoice." At such times, he confided, "sacred songs and spells filled my Watery Deep." Even such a minor detail as the name of the captain of Enki's boat is recorded. (Fig. 128)

Fig. 128

The Sumerian king lists indicate that Enki and his first group of Nefilim remained alone on Earth for quite a while: Eight *shar*'s (28,800 years) passed before the second commander or "settlement chief" was named.

Interesting light is shed on the subject as we examine the astronomical evidence. Scholars have been puzzled by the apparent Sumerian "confusion" regarding which one of the twelve zodiacal houses was associated with Enki. The sign of the fish-goat, which stood for the constellation Capricorn, was apparently associated with Enki (and, indeed, may explain the epithet of the founder of Eridu, A.LU.LIM, which could mean "sheep of the glittering waters"). Yet Ea/Enki was frequently depicted as holding vases of flowing waters—the original Water Bearer, or Aquarius; and he was certainly the God of Fishes, and thus associated with Pisces.

Astronomers are hard put to clarify how the ancient stargazers actually saw in a group of stars the outlines of, say, fishes or a water bearer. The answer that comes to mind is that the signs of the zodiac were not named after the shape of the star group but after the epithet or main activity of a god primarily associated with the time when the vernal equinox was in that particular zodiacal house.

If Enki landed on Earth—as we believe—at the start of an Age of Pisces, witnessed a precessional shift to Aquarius, and stayed through a Great Year (25,920 years)

until an Age of Capricorn began, then he was indeed in
sole command on Earth the purported 28,800 years.

The reported passage of time also confirms our earlier
conclusion that the Nefilim arrived on Earth in the midst
of an ice age. The hard work of raising dikes and digging
canals commenced when climatic conditions were still
harsh. But within a few *shar*'s of their landing, the glacial
period was giving way to a warmer and rainier climate
(circa 430,000 years ago). It was then that the Nefilim
decided to move farther inland and expand their settlements. Befittingly, the Anunnaki (rank-and-file Nefilim)
named the second commander of Eridu "A.LAL.GAR"
("he who is raintime brought rest").

But while Enki was enduring the hardships of a pioneer
on Earth, Anu and his other son Enlil were watching the
developments from the Twelfth Planet. The Mesopotamian
texts make it clear that the one who was really in charge
of the Earth mission was Enlil; and as soon as the decision
was made to proceed with the mission, Enlil himself descended to Earth. For him a special settlement or base
named Larsa was built by EN.KI.DU.NU ("Enki digs
deep"). When Enlil took personal charge of the place, he
was nicknamed ALIM ("ram"), coinciding with the "age"
of the zodiacal constellation Aries.

The establishment of Larsa launched a new phase in
the settlement of Earth by the Nefilim. It marked the
decision to proceed with the tasks for which they had come
to Earth, which required the shipping to Earth of more
"manpower," tools, and equipment, and the return of
valuable cargoes to the Twelfth Planet.

Splashdowns at sea were no longer adequate for such
heavier loads. The climatic changes made the interior more
accessible; it was time to shift the landing site to the
center of Mesopotamia. At that juncture, Enlil came to
Earth and proceeded from Larsa to establish a "Mission
Control Center"—a sophisticated command post from
which the Nefilim on Earth could coordinate space journeys
to and from their home planet, guide in landing shuttlecraft, and perfect their takeoffs and dockings with the
spaceship orbiting Earth.

The site Enlil selected for this purpose, known for
millennia as Nippur, was named by him NIBRU.KI

("Earth's crossing"). (We recall that the celestial site of the Twelfth Planet's closest pass to Earth was called the "Celestial Place of the Crossing.") There Enlil established the DUR.AN.KI, the "bond Heaven–Earth."

The task was understandably complex and time-consuming. Enlil stayed in Larsa for 6 *shar*'s (21,600 years) while Nippur was under construction. The Nippurian undertaking was also lengthy, as evidenced by the zodiacal nicknames of Enlil. Having paralleled the Ram (Aries) while in Larsa, he was subsequently associated with the Bull (Taurus). Nippur was established in the "age" of Taurus.

A devotional poem composed as a "Hymn to Enlil, the All-Beneficent" and glorifying Enlil, his consort Ninlil, his city Nippur, and its "lofty house," the E.KUR, tells us much about Nippur. For one thing, Enlil had at his disposal there some highly sophisticated instruments: a "lifted 'eye' which scans the land," and a "lifted beam which searches the heart of all the land." Nippur, the poem tells us, was protected by awesome weapons: "Its sight is awesome fear, dread"; from "its outside, no mighty god can approach." Its "arm" was a "vast net," and in its midst there crouched a "fast-stepping bird," a "bird" whose "hand" the wicked and the evil could not escape. Was the place protected by some death ray, by an electronic power field? Was there in its center a helicopter pad, a "bird" so swift no one could outrun its reach?

In the center of Nippur, atop an artificially raised platform, stood Enlil's headquarters, the KI.UR ("place of Earth's root")—the place where the "bond between Heaven and Earth" rose. This was the communications center of Mission Control, the place from which the Anunnaki on Earth communicated with their comrades, the IGI.GI ("they who turn and see") in the orbiting spacecraft.

At this center, the ancient text goes on to say, stood a "heavenward tall pillar reaching to the sky." This extremely tall "pillar," firmly planted on the ground "as a platform that cannot be overturned," was used by Enlil to "pronounce his word" heavenward. This is a simple description of a broadcasting tower. Once the "word of Enlil"—his command—"approached heaven, abundance would pour down on Earth." What a simple way to describe the flow of materials, special foods, medicines, and tools brought

down by the shuttlecraft, once the "word" from Nippur
was given!

This Control Center on a raised platform, Enlil's "lofty
house," contained a mysterious chamber, named the
DIR.GA:

> As mysterious as the distant Waters,
> as the Heavenly Zenith.
> Among its . . . emblems,
> the emblems of the stars.
> The ME it carries to perfection.
> Its words are for utterance. . . .
> Its words are gracious oracles.

What was this *dirga?* Breaks in the ancient tablet have
robbed us of more data; but the name speaks for itself, for
it means "the dark, crownlike chamber," a place where
star charts were kept, where predictions were made, where
the *me* (the astronaut's communications) were received
and transmitted. The description reminds us of Mission
Control in Houston, Texas, monitoring the astronauts on
their Moon missions, amplifying their communications,
plotting their courses against the starry sky, giving them
"gracious oracles" of guidance.

We may recall here the tale of the god Zu, who made
his way to Enlil's sanctuary and snatched away the Tablet
of Destinies, whereupon "suspended was the issuance of
commands . . . the hallowed inner chamber lost its
brilliance . . . stillness spread . . . silence prevailed."

In the "Epic of Creation," the "destinies" of the planetary
gods were their orbits. It is reasonable to assume that the
Tablet of Destinies, which was so vital to the functions of
Enlil's "Mission Control Center," also controlled the orbits
and flight paths of the spaceships that maintained the
"bond" between Heaven and Earth. It might have been
the vital "black box" containing the computer programs
that guided the spaceships, without which the contact between the Nefilim on Earth and their link to the Home
Planet was disrupted.

Most scholars take the name EN.LIL to mean "lord of
the wind," which fits the theory that the ancients "personi-

fied" the elements of nature and thus assigned one god to be in charge of winds and storms. Yet some scholars have already suggested that in this instance the term LIL means not a stormy wind of nature but the "wind" that comes out of the mouth—an utterance, a command, a spoken communication. Once again, the archaic Sumerian pictographs for the term EN—especially as applied to Enlil—and for the term LIL, shed light on the subject. For what we see is a structure with a high tower of antennas rising from it, as well as a contraption that looks very much like the giant radar nets erected nowadays for capturing and emitting signals—the "vast net" described in the texts. (Fig. 129)

EN   LIL

Fig. 129

In Bad-Tibira, established as an industrial center, Enlil installed his son Nannar/Sin in command; the texts speak of him in the list of cities as NU.GIG ("he of the night sky"). There, we believe, the twins Inanna/Ishtar and Utu/Shamash were born—an event marked by associating their father Nannar with the next zodiacal constellation, Gemini (the Twins). As the god trained in rocketry, Shamash was assigned the constellation GIR (meaning both "rocket" and "the crab's claw," or Cancer), followed by Ishtar and the Lion (Leo), upon whose back she was traditionally depicted.

The sister of Enlil and Enki, "the nurse" Ninhursag (SUD), was not neglected: In her charge Enlil put Shuruppak, the medical center of the Nefilim—an event marked by naming her constellation "The Maid" (Virgo).

While these centers were being established, the completion of Nippur was followed by the construction of the spaceport of the Nefilim on Earth. The texts made clear that Nippur was the place where the "words"—commands—were uttered: There, when "Enlil commanded: 'Towards heaven!' . . . that which shines forth rose like a sky rocket." But the action itself took place "where Shamash rises," and that place—the "Cape Kennedy" of the Nefilim—was Sippar, the city in the charge of the Chief of the Eagles, where multistage rockets were raised within its special enclave, the "sacred precinct."

As Shamash matured to take command of the Fiery Rockets, and in time also to become the God of Justice, he was assigned the constellations Scorpio and Libra (the Scales).

Completing the list of the first seven Cities of the Gods and the correspondence with the twelve zodiac constellations was Larak, where Enlil put his son Ninurta in command. The city lists call him PA.BIL.SAG ("great protector"); it is the same name by which the constellation Sagittarius was called.

•

It would be unrealistic to assume that the first seven Cities of the Gods were established haphazardly. These "gods," who were capable of space travel, located the first settlements in accordance with a definite plan, serving a vital need: to be able to land on Earth and to leave Earth for their own planet.

What was the master plan?

As we searched for an answer, we asked ourselves a question: What is the origin of Earth's astronomical and astrological symbol, a circle bisected by a right-angled cross—the symbol we use to signify "target"?

The symbol goes back to the origins of astronomy and astrology in Sumer and is identical with the Egyptian hieroglyphic sign for "place":

Is this coincidence, or significant evidence? Did the Nefilim land on Earth by superimposing on its image or map some kind of "target"?

The Nefilim were strangers to Earth. As they scanned its surface from space, they must have paid special attention to the mountains and mountain ranges. These could present hazards during landings and takeoffs, but they could also serve as navigational landmarks.

If the Nefilim, as they hovered over the Indian Ocean, looked toward the Land Between the Rivers, which they had selected for their earliest colonizing efforts, one landmark stood out unchallenged: Mount Ararat.

An extinct volcanic massif, Ararat dominates the Armenian plateau where the present-day borders of Turkey, Iran, and Soviet Armenia meet. It rises on the eastern and northern sides to some 3,000 feet above sea level, and on the northwestern side to 5,000 feet. The whole massif is some twenty-five miles in diameter, a towering dome sticking out from the surface of Earth.

Other features make it stand out not only from the horizon but also from high in the skies. First, it is located almost midway between two lakes, Lake Van and Lake Se-Van. Second, two peaks rise from the high massif: Little Ararat (12,900 feet) and Great Ararat (17,000 feet—well over 5 kilometers). No other mountains rival the solitary heights of the two peaks, which are permanently snow-covered. They are like two shining beacons between the two lakes that, in daylight, act as giant reflectors.

We have reason to believe that the Nefilim selected their landing site by coordinating a north–south meridian with an unmistakable landmark and a convenient river location. North of Mesopotamia, the easily identifiable twin-peaked Ararat would have been the obvious landmark. A meridian drawn through the center of the twin-peaked Ararat bisected the Euphrates. That was the target—the site selected for the spaceport. (Fig. 130)

Could one easily land and take off there?

The answer was Yes. The selected side lay in a plain; the mountain ranges surrounding Mesopotamia were a substantial distance away. The highest ones (to the east, northeast, and north) would not interfere with a space shuttle gliding in from the southeast.

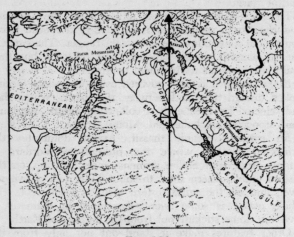

Fig. 130

Was the place accessible—could astronauts and materials be brought there without too much difficulty?

Again, the answer was Yes. The site could be reached overland and, via the Euphrates River, by waterborne craft.

And one more crucial question: Was there a nearby source of energy, of fuel for light and power? The answer was an emphatic Yes. The bend in the Euphrates River where Sippar was to be established was one of the richest known sources in antiquity of surface bitumens, petroleum products that seeped up through natural wells and could be collected from the surface without any deep digging or drilling.

We can imagine Enlil, surrounded by his lieutenants at the spacecraft's command post, drawing the cross within a circle on the map. "What shall we call the place?" he may have asked.

"Why not 'Sippar'?" someone might have suggested.

In Near Eastern languages, the name means "bird." Sippar was the place where the Eagles would come to nest.

How would the space shuttles glide down to Sippar?

We can visualize one of the space navigators pointing out the best route. On the left they had the Euphrates and the mountainous plateau west of it; on the right, the Tigris and the Zagros range east of it. If the craft were to approach Sippar at the easily set angle of 45 degrees to the Ararat meridian, its path would take it safely between these two hazardous areas. Moreover, coming in to land at such an angle, it would cross in the south over the rocky tip of Arabia while at a high altitude, and start its glide over the waters of the Persian Gulf. Coming and going, the craft would have an unobstructed field of vision and of communication with Mission Control at Nippur.

Enlil's lieutenant would then make a rough sketch—a triangle of waters and mountains on each side, pointing like an arrow toward Sippar. An "X" would mark Nippur, in the center. (Fig. 131)

Fig. 131

Incredible as it may seem, this sketch was *not* made by us; the design was drawn on a ceramic object unearthed at Susa, in a stratum dated to about 3200 B.C. It brings to mind the planisphere that described the flight path and procedures, which was based on 45-degree segments.

The establishment of settlements on Earth by the Nefilim was not a hit-or-miss effort. All the alternatives were studied, all the resources evaluated, all the hazards taken into account; moreover, the settlement plan itself was carefully mapped out so that each site fit into the final pattern, whose purpose was to outline the landing path to Sippar.

No one has previously attempted to see a master plan in the scattered Sumerian settlements. But if we look at the first seven cities ever established, we find that Bad-Tibira, Shuruppak, and Nippur lay on a line running precisely at a 45-degree angle to the Ararat meridian, and that line crossed the meridian exactly at Sippar! The other two cities whose sites are known, Eridu and Larsa, also lay on another straight line that crossed the first line and the Ararat meridian, also at Sippar.

Taking our cue from the ancient sketch, which made Nippur the center of a circle, and drawing concentric circles from Nippur through the various cities, we find that another ancient Sumerian town, Lagash, was located exactly on one of these circles—on a line equidistant from the 45-degree line, like the Eridu–Larsa–Sippar line. The location of Lagash mirrors that of Larsa.

Though the site of LA.RA.AK ("seeing the bright halo") remains unknown, the logical site for it would be at Point 5, since there logically was a City of the Gods there, completing the string of cities on the central flight path at intervals of six *beru:* Bad-Tibira, Shuruppak, Nippur, Larak, Sippar. (Fig. 132)

The two outside lines, flanking the central line running through Nippur, lay 6 degrees on each side of it, acting as southwest and northeast outlines of the central flight path. Appropriately, the name LA.AR.SA meant "seeing the red light"; and LA.AG.ASH meant "seeing the halo at six." The cities along each line were indeed six *beru* (approximately sixty kilometers, or thirty-seven miles) from each other.

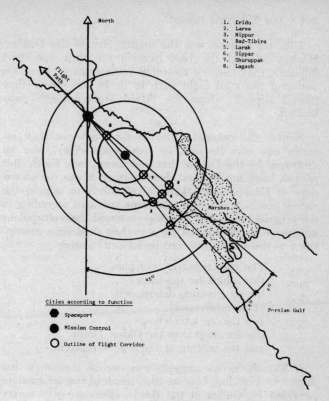

North

1. Eridu
2. Larsa
3. Nippur
4. Bad-Tibira
5. Larak
6. Sippar
7. Shuruppak
8. Lagash

Flight
Path

45°

60°

Marshes

Persian Gulf

<u>Cities according to function</u>

⬡ Spaceport

⬤ Mission Control

◯ Outline of Flight Corridor

Fig. 132

This, we believe, was the master plan of the Nefilim. Having selected the best location for their spaceport (Sippar), they laid out the other settlements in a pattern outlining the vital flight path to it. In the center they placed Nippur, where the "bond Heaven–Earth" was located.

•

Neither the original Cities of the Gods nor their remains can ever be seen by man again—they were all destroyed by the Deluge that later swept over Earth. But we can learn much about them because it was the sacred duty of Mesopotamian kings continuously to rebuild the sacred precincts in exactly the same spot and according to the original plans. The rebuilders stressed their scrupulous adherence to the original plans in their dedication inscriptions, as this one (uncovered by Layard) stated:

> The everlasting ground plan,
> that which for the future
> the construction determined
> [I have followed].
> It is the one which bears
> the drawings from the Olden Times
> and the writing of the Upper Heaven.

If Lagash, as we suggest, was one of the cities that served as a landing beacon, then much of the information provided by Gudea in the third millennium B.C. makes sense. He wrote that when Ninurta instructed him to rebuild the sacred precinct, an accompanying god gave him the architectural plans (drawn on a stone tablet), and a goddess (who had "travelled between Heaven and Earth" in her "chamber") showed him a celestial map and instructed him on the astronomical alignments of the structure.

In addition to the "divine black bird," the god's "terrible eye" ("the great beam that subdues the world to its power") and the "world controller" (whose sound could "reverberate all over") were installed in the sacred precinct. Finally, when the structure was complete, the "emblem of Utu" was raised upon it, facing "toward the rising place of Utu"—toward the spaceport at Sippar. All these beaming objects were important to the spaceport's operation, for

Utu himself "came forth joyfully" to inspect the installations when completed.

Early Sumerian depictions frequently show massive structures, built in earliest times of reeds and wood, standing in fields among grazing cattle. The current assumption that these were stables for cattle is contradicted by the pillars that are invariably shown protruding from the roofs of such structures. (Fig. 133a)

The pillars' purpose, as one can see, was to support one or more pairs of "rings," whose function is unstated. But although these structures were erected in the fields, one must question whether they were built to shelter cattle. The Sumerian pictographs (Fig. 133b) depict the word DUR, or TUR (meaning "abode," "gathering place"), by drawings that undoubtedly represent the same structures shown on the cylinder seals; but they make clear that the main feature of the structure was not the "huts" but the antenna tower. Similar pillars with "rings" were posted at temple entrances, within the sacred precincts of the gods, and not only out in the countryside. (Fig. 133c)

Were these objects antennas attached to broadcasting equipment? Were the pairs of rings radar emitters, placed in the fields to guide the incoming shuttlecraft? Were the eyelike pillars scanning devices, the "all-seeing eyes" of the gods of which many texts have spoken?

We know that the equipment to which these various devices were connected was portable, for some Sumerian seals depict boxlike "divine objects" being transported by boat or mounted on pack animals, which carried the objects farther inland once the boats had docked. (Fig. 134)

These "black boxes," when we see what they looked like, bring to mind the Ark of the Covenant built by Moses under God's instructions. The chest was to be made of wood, overlaid with gold both inside and outside—two electricity-conducting surfaces were insulated by the wood between them. A *kapporeth*, also made of gold, was to be placed above the chest and held up by two cherubim cast of solid gold. The nature of the *kapporeth* (meaning, scholars speculate, "covering") is not clear; but this verse from Exodus suggests its purpose: "And I will address thee from above the Kapporeth, from between the two Cherubim."

a

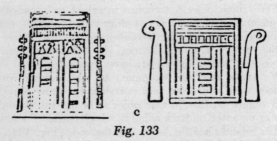

b

c

Fig. 133

Fig. 134

The implication that the Ark of the Covenant was principally a communications box, electrically operated, is enhanced by the instructions concerning its portability. It was to be carried by means of wooden staffs passed through four golden rings. No one was to touch the chest proper; and when one Israelite did touch it, he was killed instantly —as if by a charge of high-voltage electricity.

Such apparently supernatural equipment—which made it possible to communicate with a deity though the deity was physically somewhere else—became objects of veneration, "sacred cult symbols." Temples at Lagash, Ur, Mari, and other ancient sites included among their devotional objects "eye idols." The most outstanding example was found at an "eye temple" at Tell Brak, in northwestern Mesopotamia. This fourth-millennium temple was so named not only because hundreds of "eye" symbols were unearthed there but mainly because the temple's inner sanctum had only one altar, on which a huge stone "double-eye" symbol was displayed. (Fig. 135)

In all probability, it was a simulation of the actual divine object—Ninurta's "terrible eye," or the one at Enlil's Mission Control Center at Nippur, about which the ancient scribe reported: "His raised Eye scans the land. . . . His raised Beam searches the land."

The flat plain of Mesopotamia necessitated, it seems, the artificial raising of platforms on which the space-related equipment was to be placed. Texts and pictorial depictions leave no doubt that the structures ranged from the earliest field huts to the later staged platforms, reached by staircases and sloped ramps that led from a broad lower stage to a narrower upper one, and so on. At the top of the ziggurat an actual residence for the god was built, surrounded by a flat, walled courtyard to house his "bird" and "weapons." A ziggurat depicted on a cylinder seal not only shows the customary stage-upon-stage construction, it also has two "ring antennas" whose height appears to have equaled three stages. (Fig. 136)

Marduk claimed that the ziggurat and temple compound at Babylon (the E.SAG.IL) had been built under his own instructions, also in accordance with the "writing of Upper Heaven." A tablet (known as the Smith Tablet, after its decipherer), analyzed by André Parrot (*Ziggurats et Tour*

Fig. 135   Fig. 136

de Babel) established that the seven-stage ziggurat was a perfect square, with the first stage or base having sides of 15 *gar* each. Each successive stage was smaller in area and in height, except the last stage (the god's residence), which was of a greater height. The total height, however, was again equal to 15 *gar*, so that the complete structure was not only a perfect square but a perfect cube as well.

The *gar* employed in these measurements was equivalent to 12 short cubits—approximately 6 meters, or 20 feet. Two scholars, H. G. Wood and L. C. Stecchini, have shown that the Sumerian sexagesimal base, the number 60, determined all the primary measurements of Mesopotamian ziggurats. Thus each side measured 3 by 60 cubits at its base, and the total was 60 *gar*. (Fig. 137)

What factor determined the height of each stage? Stecchini discovered that if he multiplied the height of the first stage (5.5 *gar*) by double cubits, the result was 33, or the approximate latitude of Babylon (32.5 degrees North). Similarly calculated, the second stage raised the angle of observation to 51 degrees, and each of the succeeding four stages raised it by another 6 degrees. The seventh stage thus stood atop a platform raised to 75 degrees above the horizon at Babylon's geographic latitude. This final stage added 15 degrees—letting the observer look straight

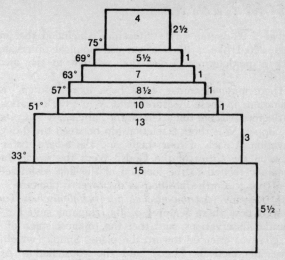

Fig. 137

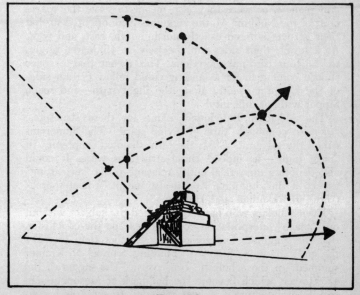

Fig. 138

up, at a 90-degree angle. Stecchini concluded that each stage acted like a stage of an astronomical observatory, with a predetermined elevation relative to the arc of the sky.

There may, of course, have been more "hidden" considerations in these measurements. While the elevation of 33 degrees was not too accurate for Babylon, it was precise for Sippar. Was there a relationship between the 6-degree elevation at each of four stages and the 6-*beru* distances between the Cities of the Gods? Were the seven stages somehow related to the location of the first seven settlements, or to Earth's position as the seventh planet?

G. Martiny (*Astronomisches zur babylonischen Turm*) showed how these features of the ziggurat suited it for celestial observations, and that the topmost stage of the Esagila was oriented toward the planet Shupa (which we have identified as Pluto) and the constellation Aries. (Fig. 138)

But were the ziggurats raised solely to observe the stars and planets, or were they also meant to serve the spacecraft of the Nefilim? All the ziggurats were oriented so that their corners pointed exactly north, south, east, and west. As a result, their sides ran precisely at 45-degree angles to the four cardinal directions. This meant that a space shuttle coming in for a landing could follow certain sides of the ziggurat exactly along the flight path—and reach Sippar without difficulty!

The Akkadian/Babylonian name for these structures, *zukiratu,* connoted "tube of divine spirit." The Sumerians called the ziggurats ESH; the term denoted "supreme" or "most high"—as indeed these structures were. It could also denote a numerical entity relating to the "measuring" aspect of the ziggurats. And it also meant "a heat source" ("fire" in Akkadian and Hebrew).

Even scholars who have approached the subject without our "space" interpretation could not escape the conclusion that the ziggurats had some purpose other than to make the god's abode a "high-rise" building. Samuel N. Kramer summed up the scholastic consensus: "The ziggurat, the stagetower, which became the hallmark of Mesopotamian temple architecture . . . was intended to serve as a connecting link, both real and symbolic, between the gods in heaven and the mortals on earth."

We have shown, however, that the true function of these structures was to connect the gods in Heaven with the gods—not the mortals—on Earth.

11
·
MUTINY OF THE ANUNNAKI

AFTER ENLIL ARRIVED on Earth in person, "Earth Command" was transferred out of Enki's hands. It was probably at this point that Enki's epithet or name was changed to E.A ("lord waters") rather than "lord earth."

The Sumerian texts explain that at that early stage in the arrival of the gods on Earth, a separation of powers was agreed upon: Anu was to stay in the heavens and rule over the Twelfth Planet; Enlil was to command the lands; and Enki was put in charge of the AB.ZU (*apsu* in Akkadian). Guided by the "watery" meaning of the name E.A, scholars have translated AB.ZU as "watery deep," assuming that, as in Greek mythology, Enlil represented the thundering Zeus, and Ea was the prototype of Poseidon, God of the Oceans.

In other instances, Enlil's domain was referred to as the Upper World, and Ea's as the Lower World; again, the scholars assumed that the terms meant that Enlil controlled Earth's atmosphere while Ea was ruler of the "subterranean waters"—the Greeklike Hades the Mesopotamians supposedly believed in. Our own term *abyss* (which derives from *apsu*) denotes deep, dark, dangerous waters in which one can sink and disappear. Thus, as scholars came upon Mesopotamian texts describing this Lower World, they translated it as *Unterwelt* ("underworld") or *Totenwelt* ("world of the dead"). Only in recent years have the Sumerologists mitigated the ominous connotation somewhat by using the term *netherworld* in translation.

The Mesopotamian texts most responsible for this misinterpretation were a series of liturgies lamenting the disappearance of Dumuzi, who is better known from biblical and Canaanite texts as the god Tammuz. It was with him that Inanna/Ishtar had her most celebrated love affair; and when he disappeared, she went to the Lower World to seek him.

The massive *Tammuz-Liturgen und Verwandtes* by P. Maurus Witzel, a masterwork on the Sumerian and Akkadian "Tammuz texts," only helped perpetuate the misconception. The epic tales of Ishtar's search were taken to mean a journey "to the realm of the dead, and her eventual return to the land of the living."

The Sumerian and Akkadian texts describing the descent of Inanna/Ishtar to the Lower World inform us that the goddess decided to visit her sister Ereshkigal, mistress of the place. Ishtar went there neither dead nor against her will—she went alive and uninvited, forcing her way in by threatening the gatekeeper:

> If thou openest not the gate so that I cannot enter,
> I will smash the door, I will shatter the bolt,
> I will smash the doorpost, I will move the doors.

One by one, the seven gates leading to the abode of Ereshkigal were opened to Ishtar; when she finally made it, and Ereshkigal saw her, she literally blew her top (the Akkadian text says, "burst at her presence"). The Sumerian text, vague about the purpose of the trip or the cause of Ereshkigal's anger, reveals that Inanna expected such a reception. She took pains to notify the other principal deities of her journey in advance, and made sure that they would take steps to rescue her in case she was imprisoned in the "Great Below."

The spouse of Ereshkigal—and Lord of the Lower World —was Nergal. The manner in which he arrived in the Great Below and became its lord not only illuminates the human nature of the "gods" but also depicts the Lower World as anything but a "world of the dead."

The tale, found in several versions, begins with a banquet at which the guests of honor were Anu, Enlil, and Ea. The banquet was held "in the heavens," but not at Anu's abode on the Twelfth Planet. Perhaps it took place aboard an orbiting spacecraft, for when Ereshkigal could not ascend to join them, the gods sent her a messenger who "descended the long staircase of the heavens, reached the gate of Ereshkigal." Having received the invitation, Ereshkigal instructed her counselor, Namtar:

> "Ascend, Namtar, the long staircase of the heavens;
> Remove the dish from the table, take my share;
> Whatever Anu gives to thee, bring it all to me."

When Namtar entered the banquet hall, all the gods except "a bald god, seated in the back," rose to greet him. Namtar reported the incident to Ereshkigal when he returned to the Lower World. She and all the lesser gods of her domain were insulted. She demanded that the offending god be sent to her for punishment.

The offender, however, was Nergal, a son of the great Ea. After a severe reprimand by his father, Nergal was instructed to make the trip alone, armed only with lots of fatherly advice on how to behave. When Nergal arrived at the gate, he was recognized by Namtar as the offending god and led in to "Ereshkigal's wide courtyard," where he was put to several tests.

Sooner or later, Ereshkigal went to take her daily bath.

> . . . she revealed her body.
> What is normal for man and woman,
> he . . . in his heart . . .
> . . . they embraced,
> passionately they got into bed.

For seven days and nights they made love. In the Upper World, an alarm had gone out for the missing Nergal. "Release me," he said to Ereshkigal. "I will go, and I will come back," he promised. But no sooner had he left than Namtar went to Ereshkigal and accused Nergal of having no intention of coming back. Once more Namtar was sent up to Anu. Ereshkigal's message was clear:

> I, thy daughter, was young;
> I have not known the play of maidens. . . .
> That god whom you didst send,
> and who had intercourse with me—
> Send him to me, that he may be my husband,
> That he might lodge with me.

With married life perhaps not yet on his mind, Nergal organized a military expedition and stormed the gates of Ereshkigal, intending to "cut off her head." But Ereshkigal pleaded:

> "Be thou my husband and I will be thy wife.
> I will let thee hold dominion
> over the wide Lower Land.
> I will place the Tablet of Wisdom in thy hand.
> Thou shalt be Master, I will be Mistress."

And then came the happy ending:

> When Nergal heard her words,
> He took hold of her hand and kissed her,
> Wiping away her tears:
> "What thou hast wished for me
> since months past—so be it now!"

The events recounted do not suggest a Land of the Dead. Quite the contrary: It was a place the gods could enter and leave, a place of lovemaking, a place important enough to be entrusted to a granddaughter of Enlil and a son of Enki. Recognizing that the facts do not support the earlier notion of a dismal region, W. F. Albright (*Mesopotamian Elements in Canaanite Eschatology*) suggested that Dumuzi's abode in the Lower World was "a bright and fruitful home in the subterranean paradise called 'the mouth of the rivers' which was closely associated with the home of Ea in the Apsu."

The place was far and difficult to reach, to be sure, and a somewhat "restricted area," but hardly a "place of no return." Like Inanna, other leading deities were reported going to, and returning from, this Lower World. Enlil was banished to the Abzu for a while, after he had raped Ninlil. And Ea was a virtual commuter between Eridu in Sumer and the Abzu, bringing to the Abzu "the craftsmanship of Eridu" and establishing in it "a lofty shrine" for himself.

Far from being a dark and desolate place, it was described as a bright place with flowing waters.

> A rich land, beloved of Enki;
> Bursting with riches, perfect in fullness . . .
> Whose mighty river rushes across the land.

We have seen the many depictions of Ea as the God of Flowing Waters. It is evident from Sumerian sources that such flowing waters indeed existed—not in Sumer and its flatlands, but in the Great Below. W. F. Albright drew attention to a text dealing with the Lower World as the Land of UT.TU—"in the west" of Sumer. It speaks of a journey of Enki to the Apsu:

> To thee, Apsu, pure land,
> Where great waters rapidly flow,
> To the Abode of Flowing Waters
> The Lord betakes himself. . . .
> The Abode of Flowing Waters
> Enki in the pure waters established;
> In the midst of the Apsu,
> A great sanctuary he established.

By all accounts, the place lay beyond a sea. A lament for "the pure son," the young Dumuzi, reports that he was carried off to the Lower World in a ship. A "Lamentation over the Destruction of Sumer" describes how Inanna managed to sneak aboard a waiting ship. "From her possessions she sailed forth. She descends to the Lower World."

A long text, little understood because no intact version has been found, deals with some major dispute between Ira (Nergal's title as Lord of the Lower World) and his brother Marduk. In the course of the dispute, Nergal left his domain and confronted Marduk in Babylon; Marduk, on the other hand, threatened: "To the Apsu will I descend, the Anunnaki to supervise . . . my raging weapons against them I will raise." To reach the Apsu, he left the Land of Mesopotamia and traveled over "waters that rose up." His destination was Arali in the "basement" of Earth, and the texts provide a precise clue as to where this "basement" was:

> In the distant sea,
> 100 *beru* of water [away] . . .
> The ground of *Arali* [is] . . .
> It is where the Blue Stones cause ill,
> Where the craftsman of Anu
> the Silver Axe carries, which shines as the day.

The *beru*, both a land-measuring and a time-reckoning unit, was probably used in the latter capacity when travel over water was involved. As such it was a double hour, so that one hundred *beru* meant two hundred hours of sailing. We have no way of determining the assumed or average sailing speed employed in these ancient distance reckonings. But there is no doubt that a truly distant land was reached after a sea voyage of over two or three thousand miles.

The texts indicate that Arali was situated west and south of Sumer. A ship traveling two to three thousand miles in a southwesterly direction from the Persian Gulf could have only one destination: the shores of southern Africa.

Only such a conclusion can explain the terms Lower World, as meaning the southern hemisphere, where the Land of Arali was, as contrasted with the Upper World, or northern hemisphere, where Sumer was. Such a division of Earth's hemispheres between Enlil (northern) and Ea (southern) paralleled the designation of the northern skies as the Way of Enlil and the southern skies as the Way of Ea.

The ability of the Nefilim to undertake interplanetary travel, orbit Earth, and land on it should obviate the question whether they could possibly have known of southern Africa, besides Mesopotamia. Many cylinder seals, depicting animals peculiar to the area (such as the zebra or ostrich), jungle scenes, or rulers wearing leopard skins in the African tradition, attest to an "African connection."

What interest did the Nefilim have in this part of Africa, diverting to it the scientific genius of Ea and granting to the important gods in charge of the land a unique "Tablet of Wisdom"?

The Sumerian term AB.ZU, which scholars have accepted to mean "watery deep," requires a fresh and critical analysis. Literally, the term meant "primeval deep source" —not necessarily of waters. According to Sumerian grammatical rules, either of two syllables of any term could precede the other without changing the word's meaning, with the result that AB.ZU and ZU.AB meant the same thing. The latter spelling of the Sumerian term enables identification of its parallel in the Semitic languages, for *za-ab* has always meant and still means "precious metal," specifically "gold," in Hebrew and its sister languages.

The Sumerian pictograph for AB.ZU was that of an excavation deep into Earth, mounted by a shaft. Thus, Ea was not the lord of an indefinite "watery deep," but the god in charge of the exploitation of Earth's minerals! (Fig. 139)

In fact, the Greek *abyssos,* adopted from the Akkadian *apsu,* also meant an extremely deep hole in the ground. Akkadian textbooks explained that *"apsu is nikbu";* the meaning of the word and that of its Hebrew equivalent *nikba* is very precise: a deep, man-made cutting or drilling into the ground.

P. Jensen *(Die Kosmologie der Babylonier)* observed back in 1890 that the oft-encountered Akkadian term *Bit Nimiku* should not be translated as "house of wisdom" but as "house of deepness." He quoted a text (V.R.30, 49–50ab) that stated: "It is from Bit Nimiku that gold and silver come." Another text (III.R.57, 35ab), he pointed out, explained that the Akkadian name "Goddess Shala of *Nimiki"* was the translation of the Sumerian epithet "Goddess Who Hands the Shining Bronze." The Akkadian term *nimiku,* which has been translated as "wisdom," Jensen concluded, "had to do with metals." But why, he admitted simply, "I do not know."

Some Mesopotamian hymns to Ea exalt him as *Bel Nimiki,* translated "lord of wisdom"; but the correct translation should undoubtedly be "lord of mining." Just as the Tablet of Destinies at Nippur contained orbital data, it follows that the Tablet of Wisdom entrusted to Nergal and Ereshkigal was in fact a "Tablet of Mining," a "data bank" pertaining to the mining operations of the Nefilim.

As Lord of the Abzu, Ea was assisted by another son, the god GI.BIL ("he who burns the soil"), who was in charge of fire and smelting. Earth's Smith, he was usually depicted as a young god whose shoulders emit red-hot rays or sparks of fire, emerging from the ground or about to descend into it. The texts state that Gibil was steeped by Ea in "wisdom," meaning that Ea had taught him mining techniques. (Fig. 140)

The metal ores mined in southeastern Africa by the Nefilim were carried back to Mesopotamia by specially designed cargo ships called MA.GUR UR.NU AB.ZU ("ship for ores of the Lower World"). There, the ores were taken to Bad-Tibira, whose name literally meant "the foundation of metalworking." Smelted and refined, the ores were cast into ingots whose shape remained unchanged throughout the ancient world for millennia. Such ingots were actually found at various Near Eastern excavations, confirming the reliability of the Sumerian pictographs as true depictions of the objects they "wrote" out; the Sumerian sign for the term ZAG ("purified precious") was the picture of such an ingot. In earlier times it apparently had a hole running through its length, through which a carrying rod was inserted. (Fig. 141)

Several depictions of a God of the Flowing Waters show him flanked by bearers of such precious metal ingots, indicating that he was also the Lord of Mining. (Fig. 142)

Fig. 139   *Fig. 140*   *Fig. 141*

Fig. 142

The various names and epithets for Ea's African Land of Mines are replete with clues to its location and nature. It was known as A.RA.LI ("place of the shiny lodes"), the land from which the metal ores come. Inanna, planning her descent to the southern hemisphere, referred to the place as the land where "the precious metal is covered with soil"—where it is found underground. A text reported by Erica Reiner, listing the mountains and rivers of the Sumerian world, stated: "Mount Arali: home of the gold"; and a fragmented text described by H. Radau confirmed that Arali was the land on which Bad-Tibira depended for its continued operations.

The Mesopotamian texts spoke of the Land of Mines as mountainous, with grassy plateaus and steppes, and lush with vegetation. The capital of Ereshkigal in that land was described by the Sumerian texts as being in the GAB. KUR.RA ("in the chest of the mountains"), well inland. In the Akkadian version of Ishtar's journey, the gatekeeper welcomes her:

> Enter my lady,
> Let Kutu rejoice over thee;
> Let the palace of the land of Nugia
> Be glad at thy presence.

Conveying in Akkadian the meaning "that which is in the heartland," the term KU.TU in its Sumerian origin also meant "the bright uplands." It was a land, all texts suggest, with bright days, awash with sunshine. The Sumerian terms for gold (KU.GI—"bright out of earth") and silver (KU.BABBAR—"bright gold") retained the original association of the precious metals with the bright (*ku*) domain of Ereshkigal.

The pictographic signs employed as Sumer's first writing reveal great familiarity not only with diverse metallurgical processes but also with the fact that the sources of the metals were mines dug down into the earth. The terms for copper and bronze ("handsome-bright stone"), gold ("the supreme mined metal"), or "refined" ("bright-purified") were all pictorial variants of a mine shaft ("opening/mouth for dark-red" metal). (Fig. 143)

The land's name—Arali—could also be written as a variant of the pictograph for "dark-red" (soil), of *Kush* ("dark-red," but in time meaning "Negro"), or of the metals mined there; the pictographs always depicted variants of a mine shaft. (Fig. 144)

Extensive references to gold and other metals in ancient texts suggest familiarity with metallurgy from the earliest times. A lively metals trade existed at the very beginnings of civilization, the result of knowledge bequeathed to Mankind by the gods, who, the texts state, had engaged in mining and metallurgy long before Man's appearance. Many studies that correlate Mesopotamian divine tales with the biblical pre-Diluvial list of patriarchs point out that, according to the Bible, Tubal-cain was an "artificer of gold and copper and iron" long before the Deluge.

The Old Testament recognized the land of Ophir, which was probably somewhere in Africa, as a source of

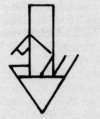

Fig. 143

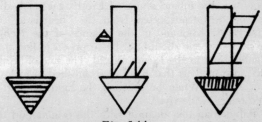

Fig. 144

gold in antiquity. King Solomon's ship convoys sailed down the Red Sea from Ezion-geber (present-day Elath). "And they went to Ophir and fetched from thence gold." Unwilling to risk a delay in the construction of the Lord's Temple in Jerusalem, Solomon arranged with his ally, Hiram, king of Tyre, to sail a second fleet to Ophir by an alternate route:

> And the king had at sea a navy of Tarshish
> with the navy of Hiram.
> Once every three years came the navy of Tarshish,
> bringing gold and silver, ivory and apes and monkeys.

The fleet of Tarshish took three years to complete a round trip. Allowing for an appropriate time to load up at Ophir, the voyage in each direction must have lasted well over a year. This suggests a route much more roundabout than the direct route via the Red Sea and the Indian Ocean —a route around Africa. (Fig. 145)

Most scholars locate Tarshish in the western Mediterranean, possibly at or near the present Strait of Gibraltar. This would have been an ideal place from which to embark on a voyage around the African continent. Some believe that the name Tarshish meant "smeltery."

Many biblical scholars have suggested that Ophir should be identified with present-day Rhodesia. Z. Herman *(Peoples, Seas, Ships)* brought together evidence showing that the Egyptians obtained various minerals from Rhodesia in earliest times. Mining engineers in Rhodesia as well as in South Africa have often searched for gold by seeking evidence of prehistoric mining.

How was the inland abode of Ereshkigal reached? How were the ores transported from the "heartland" to the coastal ports? Knowing of the reliance of the Nefilim on river shipping, one should not be surprised to find a major, navigable river in the Lower World. The tale of "Enlil and Ninlil" informed us that Enlil was banished to exile in the Lower World. When he reached the land, he had to be ferried over a wide river.

A Babylonian text dealing with the origins and destiny of Mankind referred to the river of the Lower World as the River Habur, the "River of Fishes and Birds." Some

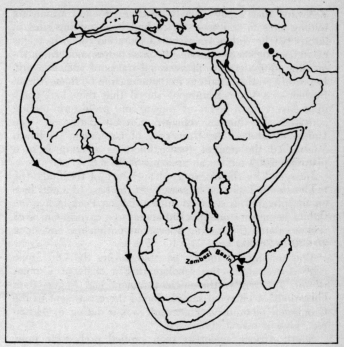

Fig. 145

Sumerian texts nicknamed the Land of Ereshkigal the "Prairie Country of ḤA.BUR."

Of the four mighty rivers of Africa, one, the Nile, flows north into the Mediterranean; the Congo and Niger empty into the Atlantic Ocean on the west; and the Zambezi flows from the heartland of Africa in an eastward semicircle until it reaches the east coast. It offers a wide delta with good port sites; it is navigable inland over a distance of hundreds of miles.

Was the Zambezi the "River of Fishes and Birds" of the Lower World? Were its majestic Victoria Falls the waterfalls mentioned in one text as the site of Ereshkigal's capital?

Aware that many "newly discovered" and promising mining sites in southern Africa had been mining sites in antiquity, the Anglo-American Corporation called in teams of archaeologists to examine the sites before modern earth-moving equipment swept away all traces of ancient work. Reporting on their findings in the magazine *Optima*, Adrian Boshier and Peter Beaumont stated that they had come upon layers upon layers of ancient and prehistoric mining activities and human remains. Carbon dating at Yale University and at the University of Groningen (Holland) established the age of the artifacts as ranging from a plausible 2000 B.C. to an amazing 7690 B.C.

Intrigued by the unexpected antiquity of the finds, the team extended its area of search. At the base of a cliff face on the precipitous western slopes of Lion Peak, a five-ton slab of hematite stone blocked access to a cavern. Charcoal remains dated the mining operations within the cavern at 20,000 to 26,000 B.C.

Was mining for metals possible during the Old Stone Age? Incredulous, the scholars dug a shaft at a point where, apparently, the ancient miners had begun their operations. A charcoal sample found there was sent to the Groningen laboratory. The result was a dating of 41,250 B.C., give or take 1,600 years!

South African scientists then probed prehistoric mine sites in southern Swaziland. Within the uncovered mine caverns, they found twigs, leaves, and grass, even feathers —all, presumably, brought in by the ancient miners as bedding. At the 35,000 B.C. level, they found notched bones, which "indicate man's ability to count at that remote period." Other remains advanced the age of the artifacts to about 50,000 B.C.

Believing that the "true age of the onset of mining in Swaziland is more likely to be in the order of 70,000-80,000 B.C.," the two scientists suggested that "southern Africa . . . could well have been in the forefront of technological invention and innovation during much of the period subsequent to 100,000 B.C."

Commenting on the discoveries, Dr. Kenneth Oakley, former head anthropologist of the Natural History Museum in London, saw quite a different significance to the finds. "It throws important light on the origins of Man . . . it is now possible that southern Africa was the evolutionary home of Man," the "birthplace" of *Homo sapiens.*

As we shall show, it was indeed there that modern Man appeared on Earth, through a chain of events triggered by the gods' search for metals.

•

Both serious scientists and science-fiction writers have suggested that a good reason for us to establish settlements on other planets or asteroids might be the availability of rare minerals on those celestial bodies, minerals that might be too scarce or too costly to mine on Earth. Could this have been the Nefilim's purpose in colonizing Earth?

Modern scholars divide Man's activities on Earth into the Stone Age, Bronze Age, Iron Age, and so on; in ancient times, however, the Greek poet Hesiod, for example, listed five ages—Golden, Silver, Bronze, Heroic, and Iron. Except for the Heroic Age, all ancient traditions accepted the sequence of gold–silver–copper–iron. The prophet Daniel had a vision in which he saw "a great image" with a head of fine gold, breast and arms of silver, belly of brass, legs of iron, and extremities, or feet, of clay.

Myth and folklore abound with hazy memories of a Golden Age, mostly associated with the time when gods roamed Earth, followed by a Silver Age, and then the ages when gods and men shared Earth—the Age of Heroes, of Copper, Bronze, and Iron. Are these legends in fact vague recollections of actual events on Earth?

Gold, silver, and copper are all native elements of the gold group. They fall into the same family in the periodic classification by atomic weight and number; they have similar crystallographic, chemical, and physical properties —all are soft, malleable, and ductile. Of all known elements, these are the best conductors of heat and electricity.

Of the three, gold is the most durable, virtually indestructible. Though best known for its use as money and in jewelry or fine artifacts, it is almost invaluable in the electronics industry. A sophisticated society requires gold for microelectronic assemblies, guidance circuitry, and computer "brains."

Man's infatuation with gold is traceable to the beginnings of his civilization and religion—to his contacts with the ancient gods. The gods of Sumer required that they be served food from golden trays, water and wine from golden vessels, that they be clad in golden garments. Though the Israelites left Egypt in such a hurry that there was no time for them to let their bread leaven, they were ordered to ask the Egyptians for all available silver and gold objects. This command, as we shall find out later, anticipated the need for such materials to construct the Tabernacle and its electronic accoutrements.

Gold, which we call the royal metal, was in fact the metal of the gods. Speaking to the prophet Haggai, the Lord made it clear, in connection with his return to judge the nations: "The silver is mine and the gold is mine."

The evidence suggests that Man's own infatuation with these metals has its roots in the great need of the Nefilim for gold. The Nefilim, it appears, came to Earth for gold and its related metals. They may also have come for other rare metals—such as platinum (abundant in southern Africa), which can power fuel cells in an extraordinary manner. And the possibility should not be ruled out that they came to Earth for sources of radioactive minerals, such as uranium or cobalt—the Lower World's "blue stones that cause ill," which some texts mention. Many depictions show Ea—as the God of Mining—emitting such powerful rays as he exits from a mine that the gods attending him have to use screening shields; in all these depictions, Ea is shown holding a miner's rock saw. (Fig. 146)

Fig. 146

Though Enki was in charge of the first landing party and the development of the Abzu, credit for what was accomplished—as the case should be with all generals—should not go to him alone. Those who actually did the work, day in, day out, were the lesser members of the landing party, the Anunnaki.

A Sumerian text describes the construction of Enlil's center in Nippur. "The Annuna, gods of heaven and earth, are working. The axe and the carrying-basket, with which they laid foundation of the cities, in their hands they held."

The ancient texts described the Anunnaki as the rank-and-file gods who had been involved in the settlement of Earth—the gods "who performed the tasks." The Babylonian "Epic of Creation" credited Marduk with giving the Anunnaki their assignments. (The Sumerian original, we can safely assume, named Enlil as the god who commanded these astronauts.)

Assigned to Anu, to heed his instructions,
Three hundred in the heavens he stationed as a guard;
the ways of Earth to define from the Heaven;
And on Earth,
Six hundred he made reside.
After he all their instructions had ordered,
to the Anunnaki of Heaven and of Earth
he allotted their assignments.

The texts reveal that three hundred of them—the "Anunnaki of Heaven," or Igigi—were true astronauts who stayed aboard the spacecraft without actually landing on Earth. Orbiting Earth, these spacecraft launched and received the shuttlecraft to and from Earth.

As chief of the "Eagles," Shamash was a welcome and heroic guest aboard the "mighty great chamber in heaven" of the Igigi. A "Hymn to Shamash" describes how the Igigi observed Shamash approaching in his shuttlecraft:

At thy appearances, all the princes are glad;
All the Igigi rejoice over thee. . . .
In the brilliance of thy light, their path. . . .
They constantly look for thy radiance. . . .

Opened wide is the doorway, entirely. . . .
The bread offerings of all the Igigi [await thee].

Staying aloft, the Igigi were apparently never encountered by Mankind. Several texts say that they were
"too high up for Mankind," as a consequence of which
"they were not concerned with the people." The Anunnaki,
on the other hand, who landed and stayed on Earth, were
known and revered by Mankind. The texts that state that
"the Anunnaki of Heaven . . . are 300" also state that "the
Anunnaki of Earth . . . are 600."

Still, many texts persist in referring to the Anunnaki as
the "fifty great princes." A common spelling of their name
in Akkadian, *An-nun-na-ki*, readily yields the meaning "the
fifty who went from Heaven to Earth." Is there a way to
bridge the seeming contradiction?

We recall the text relating how Marduk rushed to his
father Ea to report the loss of a spacecraft carrying "the
Anunnaki who are fifty" as it passed near Saturn. An
exorcism text from the time of the third dynasty of Ur
speaks of the *anunna eridu ninnubi* ("the fifty Anunnaki
of the city Eridu"). This strongly suggests that the group
of Nefilim who founded Eridu under the command of
Enki numbered fifty. Could it be that fifty was the number
of Nefilim in each landing party?

It is, we believe, quite conceivable that the Nefilim
arrived on Earth in groups of fifty. As the visits to Earth
became regular, coinciding with the opportune launching
times from the Twelfth Planet, more Nefilim would arrive.
Each time, some of the earlier arrivals would ascend in
an Earth module and rejoin the spaceship for a trip home.
But, each time, more Nefilim would stay on Earth, and
the number of Twelfth Planet astronauts who stayed to
colonize Earth grew from the initial landing party of fifty
to the "600 who on Earth settled."

•

How did the Nefilim expect to achieve their mission—to
mine on Earth its desired minerals, and ship the ingots
back to the Twelfth Planet—with such a small number
of hands?

Undoubtedly, they relied on their scientific knowledge. It was there that Enki's full value becomes clear—the reason for his, rather than Enlil's, being the first to land, the reason for his assignment to the Abzu.

A famous seal now on exhibit at the Louvre Museum shows Ea with his familiar flowing waters, except that the waters seem to emanate from, or be filtered through, a series of laboratory flasks. (Fig. 147) Such an ancient interpretation of Ea's association with waters raises the possibility that the original hope of the Nefilim was to obtain their minerals from the sea. The oceans' waters do contain vast quantities of gold and other vital minerals, but so greatly diluted that highly sophisticated and cheap techniques are needed to justify such "water mining." It is also known that the sea beds contain immense quantities of minerals in the form of plum-sized nodules—available if only one could reach deep down and scoop them up.

Fig. 147

The ancient texts refer repeatedly to a type of ship used by the gods called *elippu tebiti* ("sunken ship"—what we now call a submarine). We have seen the "fish-men" that were assigned to Ea. Is this evidence of efforts to dive to the depths of the oceans and retrieve their mineral riches? The Land of the Mines, we have noted, was earlier called

A.RA.LI.—"place of the waters of the shiny lodes." This could mean a land where gold could be river-panned; it could also refer to efforts to obtain gold from the seas.

If these were the plans of the Nefilim, they apparently came to naught. For, soon after they had established their first settlements, the few hundred Anunnaki were given an unexpected and most arduous task: to go down into the depths of the African soil and mine the needed minerals there.

Depictions that have been found on cylinder seals show gods at what appear to be mine entrances or mine shafts; one shows Ea in a land where Gibil is aboveground and another god toils underground, on his hands and knees. (Fig. 148)

Fig. 148

In later times, Babylonian and Assyrian texts disclose, men—young and old—were sentenced to hard labor in the mines of the Lower World. Working in darkness and eating dust as food, they were doomed never to return to their homeland. This is why the Sumerian epithet for the land —KUR.NU.GI.A—acquired the interpretation "land of no return"; its literal meaning was "land where gods-who-work, in deep tunnels pile up [the ores]." For the time when the Nefilim settled Earth, all the ancient sources attest, was a time when Man was not yet on Earth; and in the absence of Mankind, the few Anunnaki had to toil in the mines. Ishtar, on her descent to the Lower World, described the toiling Anunnaki as eating food mixed with clay and drinking water fouled with dust.

Against this background, we can fully understand a long epic text named (after its opening verse, as was the custom), "When the gods, like men, bore the work."

Piecing together many fragments of both Babylonian and Assyrian versions, W. G. Lambert and A. R. Millard (*Atra-Ḫasis: The Babylonian Story of the Flood*) were able to present a continuous text. They reached the conclusion that it was based on earlier Sumerian versions, and possibly on even earlier oral traditions about the arrival of the gods on Earth, the creation of Man, and his destruction by the Deluge.

While many of the verses hold only literary value to their translators, we find them highly significant, for they corroborate our findings and conclusions in the preceding chapters. They also explain the circumstances that led to the mutiny of the Anunnaki.

The story begins in the time when only the gods lived on Earth:

> When the gods, like men,
> bore the work and suffered the toil—
> the toil of the gods was great,
> the work was heavy,
> the distress was much.

At that time, the epic relates, the chief deities had already divided the commands among themselves.

> Anu, father of the Anunnaki, was their Heavenly King;
> Their Lord Chancellor was the warrior Enlil.
> Their Chief Officer was Ninurta,
> And their Sheriff was Ennugi.
> The gods had clasped hands together,
> Had cast lots and divided.
> Anu had gone up to heaven,
> [Left] the earth to his subjects.
> The seas, enclosed as with a loop,
> They had given to Enki, the prince.

Seven cities were established, and the text refers to seven Anunnaki who were city commanders. Discipline must have been strict, for the text tells us "The seven

Great Anunnaki were making the lesser gods suffer the work."

Of all their chores, it seems, digging was the most common, the most arduous, and the most abhorred. The lesser gods dug up the river beds to make them navigable; they dug canals for irrigation; and they dug in the Apsu to bring up the minerals of Earth. Though they undoubtedly had some sophisticated tools—the texts spoke of the "silver axe which shines as the day," even underground— the work was too exacting. For a long time—for forty "periods," to be exact—the Anunnaki "suffered the toil"; and then they cried: No more!

> They were complaining, backbiting,
> Grumbling in the excavations.

The occasion for the mutiny appears to have been a visit by Enlil to the mining area. Seizing the opportunity, the Anunnaki said to one another:

> Let us confront our . . . the Chief Officer,
> That he may relieve us of our heavy work.
> The king of the gods, the hero Enlil,
> Let us unnerve him in his dwelling!

A leader or organizer of the mutiny was soon found. He was the "chief officer of old time," who must have held a grudge against the current chief officer. His name, regrettably, is broken off; but his inciting address is quite clear:

> "Now, proclaim war;
> Let us combine hostilities and battle."

The description of the mutiny is so vivid that scenes of the storming of the Bastille come to mind:

> The gods heeded his words.
> They set fire to their tools;
> Fire to their axes they put;
> They troubled the god of mining in the tunnels;
> They held [him] as they went
> to the gate of the hero Enlil.

The drama and tension of the unfolding events are brought to life by the ancient poet:

> It was night, half-way through the watch.
> His house was surrounded—
> but the god, Enlil, did not know.
> Kalkal [then] observed it, was disturbed.
> He slid the bolt and watched. . . .
> Kalkal roused Nusku;
> they listened to the noise of. . . .
> Nusku roused his lord—
> he got him out of his bed, [saying]:
> "My lord, your house is surrounded,
> battle has come right up to your gate."

Enlil's first reaction was to take up arms against the mutineers. But Nusku, his chancellor, advised a Council of the Gods:

> "Transmit a message that Anu come down;
> Have Enki brought to your presence."
> He transmitted and Anu was carried down;
> Enki was also brought to his presence.
> With the great Anunnaki present,
> Enlil arose . . . opened his mouth
> And addressed the great gods.

Taking the mutiny personally, Enlil demanded to know:

> "Is it against me that this is being done?
> Must I engage in hostilities . . . ?
> What did my very own eyes see?
> That battle has come right up to my gate!"

Anu suggested that an inquiry be undertaken. Armed with the authority of Anu and the other commanders, Nusku went to the encamped mutineers. "Who is the instigator of battle?" he asked. "Who is the provoker of hostilities?"

The Anunnaki stood together:

"Every single one of us gods has war declared!
We have our . . . in the excavations;
Excessive toil has killed us,
Our work was heavy, the distress much."

When Enlil heard Nusku's report of these grievances, "his tears flowed." He presented an ultimatum: either the leader of the mutineers be executed or he would resign. "Take the office away, take back your power," he told Anu, "and I will to you in heaven ascend." But Anu, who came down from Heaven, sided with the Anunnaki:

"What are we accusing them of?
Their work was heavy, their distress was much!
Every day . . .
The lamentation was heavy, we could hear the complaint."

Encouraged by his father's words, Ea also "opened his mouth" and repeated Anu's summation. But he had a solution to offer: Let a *lulu*, a "Primitive Worker," be created!

"While the Birth Goddess is present,
Let her create a Primitive Worker;
Let him bear the yoke. . . .
Let him carry the toil of the gods!"

The suggestion that a "Primitive Worker" be *created* so that he could take over the burden of work of the Anunnaki was readily accepted. Unanimously, the gods voted to create "The Worker." " '*Man*' shall be his name," they said:

They summoned and asked the goddess,
The midwife of the gods, the wise Mami,
[and said to her:]
"You are the Birth Goddess, create Workers!
Create a Primitive Worker,
That he may bear the yoke!
Let him bear the yoke assigned by Enlil,
Let The Worker carry the toil of the gods!"

Mami, the Mother of the Gods, said she would need the help of Ea, "with whom skill lies." In the House of Shimti, a hospital-like place, the gods were waiting. Ea helped prepare the mixture from which the Mother Goddess proceeded to fashion "Man." Birth goddesses were present. The Mother Goddess went on working while incantations were constantly recited. Then she shouted in triumph:

> "I have created!
> My hands have made it!"

She "summoned the Anunnaki, the Great Gods . . . she opened her mouth, addressed the Great Gods":

> "You commanded me a task—
> I have completed it. . . .
> I have removed your heavy work
> I have imposed your toil on The Worker, 'Man.'
> You raised a cry for a Worker-kind:
> I have loosed the yoke,
> I have provided your freedom."

The Anunnaki received her announcement enthusiastically. "They ran together and kissed her feet." From then on it would be the Primitive Worker—Man—"who will bear the yoke."

The Nefilim, having arrived on Earth to set up their colonies, had created their own brand of slavery, not with slaves imported from another continent, but with Primitive Workers fashioned by the Nefilim themselves.

A mutiny of the gods had led to the creation of Man.

12

·

THE CREATION OF MAN

THE ASSERTION, first recorded and transmitted by the Sumerians, that "Man" was created by the Nefilim, appears at first sight to clash both with the theory of evolution and with the Judeo-Christian tenets based on the Bible. But in fact, the information contained in the Sumerian texts— and only that information—can affirm both the validity of the theory of evolution and the truth of the biblical tale— and show that there really is no conflict at all between the two.

In the epic "When the gods as men," in other specific texts, and in passing references, the Sumerians described Man as both a deliberate creature of the gods and a link in the evolutionary chain that began with the celestial events described in the "Epic of Creation." Holding firm to the belief that the creation of Man was preceded by an era during which only the Nefilim were upon Earth, the Sumerian texts recorded instance after instance (for example, the incident between Enlil and Ninlil) of events that had taken place "when Man had not yet been created, when Nippur was inhabited by the gods alone." At the same time, the texts also described the creation of Earth and the development of plant and animal life upon it, in terms that conform to the current evolutionary theories.

The Sumerian texts state that when the Nefilim first came to Earth, the arts of grain cultivation, fruit planting, and cattle raising had not yet extended to Earth. The biblical account likewise places the creation of Man in the sixth "day" or phase of the evolutionary process. The Book of Genesis, too, asserts that at an earlier evolutionary stage:

> No plant of the cleared field was yet on Earth,
> No herb that is planted had yet been grown. . . .
> And Man was not yet there to work the soil.

All the Sumerian texts assert that the gods created Man to do their work. Putting the explanation in words uttered by Marduk, the Creation epic reports the decision:

> I will produce a lowly Primitive;
> "Man" shall be his name.
> I will create a Primitive Worker;
> He will be charged with the service of the gods,
> that they might have their ease.

The very terms by which the Sumerians and Akkadians called "Man" bespoke his status and purpose: He was a *lulu* ("primitive"), a *lulu amelu* ("primitive worker"), an *awilum* ("laborer"). That Man was created to be a servant of the gods did not strike the ancient peoples as a peculiar idea at all. In biblical times, the deity was "Lord," "Sovereign," "King," "Ruler," "Master." The term that is commonly translated as "worship" was in fact *avod* ("work"). Ancient and biblical Man did not "worship" his god; he worked for him.

No sooner had the biblical Deity, like the gods in Sumerian accounts, created Man, than he planted a garden and assigned Man to work there:

> And the Lord God took the "Man"
> and placed him in the garden of Eden
> to till it and to tend it.

Later on, the Bible describes the Deity "strolling in the garden in the breeze of the day," now that the new being was there to tend the Garden of Eden. How far is this version from the Sumerian texts that describe how the gods clamored for workers so that they could rest and relax?

In the Sumerian versions, the decision to create Man was adopted by the gods in their Assembly. Significantly, the Book of Genesis—purportedly exalting the achievements of a sole Deity—uses the plural Elohim (literally, "deities") to denote "God," and reports an astonishing remark:

And Elohim said:
"Let us make Man in our image,
after our likeness."

Whom did the sole but plural Deity address, and who were the "us" in whose plural image and plural likeness Man was to be made? The Book of Genesis does not provide the answer. Then, when Adam and Eve ate of the fruit of the Tree of Knowing, Elohim issued a warning to the same unnamed colleagues: "Behold, Man has become as one of us, to know good and evil."

Since the biblical story of Creation, like the other tales of beginnings in Genesis, stems from Sumerian origins, the answer is obvious. Condensing the many gods into a single Supreme Deity, the biblical tale is but an edited version of the Sumerian reports of the discussions in the Assembly of the Gods.

The Old Testament took pains to make clear that Man was neither a god nor from the heavens. "The Heavens are the Heavens of the Lord, unto Mankind Earth He hath given." The new being was called "the Adam" because he was created of the *adama*, the Earth's soil. He was, in other words, "the Earthling."

Lacking only certain "knowing" and a divine span of life, the Adam was in all other respects created in the image (*selem*) and likeness (*dmut*) of his Creator(s). The use of both terms in the text was meant to leave no doubt that Man was similar to the God(s) both physically and emotionally, externally and internally.

In all ancient pictorial depictions of gods and men, this physical likeness is evident. Although the biblical admonition against the worship of pagan images gave rise to the notion that the Hebrew God had neither image nor likeness, not only the Genesis tale but other biblical reports attest to the contrary. The God of the ancient Hebrews could be seen face-to-face, could be wrestled with, could be heard and spoken to; he had a head and feet, hands and fingers, and a waist. The biblical God and his emissaries looked like men and acted like men—because men were created to look like and act like the gods.

But in this very simplicity lies a great mystery. How could a *new* creature possibly be a virtual physical, mental, and emotional replica of the Nefilim? How, indeed, was Man created?

The Western world was long wedded to the notion that, created deliberately, Man was put upon Earth to subdue it and have dominion over all other creatures. Then, in November 1859, an English naturalist by the name of Charles Darwin published a treatise called *On the Origin of Species by Means of Natural Selection, or the Preservation of Favoured Races in the Struggle for Life.* Summing up nearly thirty years of research, the book added to earlier thoughts about natural evolution the concept of natural selection as a consequence of the struggle of all species— of plant and animal alike—for existence.

The Christian world had been jostled earlier when, from 1788 on, noted geologists had begun to express their belief that Earth was of great antiquity, much, much greater than the roughly 5,500 years of the Hebrew calendar. Nor was the concept of evolution as such the explosive: Earlier scholars had noted such a process, and Greek scholars as far back as the fourth century B.C. compiled data on the evolution of animal and plant life.

Darwin's shattering bombshell was the conclusion that all living things—*Man included*—were products of evolution. Man, contrary to the then-held belief, was not generated spontaneously.

The initial reaction of the Church was violent. But as the scientific facts regarding Earth's true age, evolution, genetics, and other biological and anthropological studies came to light, the Church's criticism was muted. It seemed at last that the very words of the Old Testament made the tale of the Old Testament indefensible; for how could a God who has no corporal body and who is universally alone say, "Let *us* make Man in *our image,* after *our likeness?*"

But are we really nothing more than "naked apes"? Is the monkey just an evolutionary arm's length away from us, and the tree shrew just a human who has yet to lose his tail and stand erect?

As we showed at the very beginning of this book, modern scientists have come to question the simple theories. Evolution can explain the general course of events that caused life and life's forms to develop on Earth, from the simplest one-celled creature to Man. But evolution cannot account for the appearance of *Homo sapiens,* which happened virtually overnight in terms of the millions of years evolution requires, and with no evidence of earlier stages that would indicate a gradual change from *Homo erectus.*

The hominid of the genus *Homo* is a product of evolution. But *Homo sapiens* is the product of some sudden, revolutionary event. He appeared inexplicably some 300,000 years ago, millions of years too soon.

The scholars have no explanation. But we do. The Sumerian and Babylonian texts do. The Old Testament does.

Homo sapiens—modern Man—was brought about by the ancient gods.

•

The Mesopotamian texts, fortunately, provide a clear statement regarding the time when Man was created. The story of the toil and ensuing mutiny of the Anunnaki informs us that "for 40 periods they suffered the work, day and night"; the long years of their toil are dramatized by repetitious verses.

> For 10 periods they suffered the toil;
> For 20 periods they suffered the toil;
> For 30 periods they suffered the toil;
> For 40 periods they suffered the toil.

The ancient text uses the term *ma* to denote "period," and most scholars have translated this as "year." But the term had the connotation of "something that completes itself and then repeats itself." To men on Earth, one year equals one complete orbit of Earth around the Sun. As we have already shown, the orbit of the Nefilim's planet equaled a *shar,* or 3,600 Earth years.

Forty *shars,* or 144,000 Earth years, after their landing, the Anunnaki protested, "No more!" If the Nefilim first landed on Earth, as we have concluded, some 450,000 years ago, then the creation of Man took place some 300,000 years ago!

The Nefilim did not create the mammals or the primates or the hominids. "The Adam" of the Bible was not the genus *Homo,* but the being who is our ancestor—the first

Homo sapiens. It is modern Man as we know him that the Nefilim created.

The key to understanding this crucial fact lies in the tale of a slumbering Enki, aroused to be informed that the gods had decided to form an *adamu,* and that it was his task to find the means. He replied:

> "The creature whose name you uttered—
> IT EXISTS!"

and he added: "Bind upon it"—on the creature that already exists—"the image of the gods."

Here, then, is the answer to the puzzle: The Nefilim did not "create" Man out of nothing; rather, they took an existing creature and manipulated it, to "bind upon it" the "image of the gods."

Man is the product of evolution; but modern Man, *Homo sapiens,* is the product of the "gods." For, some time circa 300,000 years ago, the Nefilim took ape-man *(Homo erectus)* and implanted on him their own image and likeness.

Evolution and the Near Eastern tales of Man's creation are not at all in conflict. Rather, they explain and complement each other. For without the creativity of the Nefilim, modern Man would still be millions years away on the evolutionary tree.

•

Let us transport ourselves back in time, and try to visualize the circumstances and the events as they unfolded.

The great interglacial stage that began about 435,000 years ago, and its warm climate, brought about a proliferation of food and animals. It also speeded up the appearance and spread of an advanced manlike ape, *Homo erectus.*

As the Nefilim looked about them, they saw not only the predominant mammals but also the primates—among them the manlike apes. Is it not possible that the roaming bands of *Homo erectus* were lured to come close to observe the fiery objects rising to the sky? Is it not possible that the Nefilim observed, encountered, even captured some of these interesting primates?

That the Nefilim and the manlike apes did meet is attested to by several ancient texts. A Sumerian tale dealing with the primordial times states:

> When Mankind was created,
> They knew not the eating of bread,
> Knew not the dressing in garments;
> Ate plants with their mouth like sheep;
> Drank water from a ditch.

Such an animal-like "human" being is also described in the "Epic of Gilgamesh." That text tells what Enkidu, the one "born on the steppes," was like before he became civilized:

> Shaggy with hair is his whole body,
> he is endowed with head-hair like a woman. . . .
> He knows neither people nor land;
> Garbed he is like one of the green fields;
> With gazelles he feeds on grass;
> With the wild beasts he jostles
> at the watering place;
> With the teeming creatures in the water
> his heart delights.

Not only does the Akkadian text describe an animal-like man; it also describes an encounter with such a being:

> Now a hunter, one who traps,
> faced him at the watering place.
> When the hunter saw him,
> his face became motionless. . . .
> His heart was disturbed, overclouded his face,
> for woe had entered his belly.

There was more to it than mere fear after the hunter beheld "the savage," this "barbarous fellow from the depths of the steppe"; for this "savage" also interfered with the hunter's pursuits:

> He filled the pits that I had dug,
> he tore up my traps which I had set;
> the beasts and creatures of the steppe
> he has made slip through my hands.

We can ask for no better description of an ape-man: hairy, shaggy, a roaming nomad who "knows neither people nor land," garbed in leaves, "like one of the green fields," feeding on grass, and living among the animals. Yet he is not without substantial intelligence, for he knows how to tear up the traps and fill up the pits dug to catch the animals. In other words, he protected his animal friends from being caught by the alien hunters. Many cylinder seals have been found that depict this shaggy ape-man among his animal friends. (Fig. 149)

Fig. 149

Then, faced with the need for manpower, resolved to obtain a Primitive Worker, the Nefilim saw a ready-made solution: to domesticate a suitable animal.

The "animal" was available—but *Homo erectus* posed a problem. On the one hand, he was too intelligent and wild to become simply a docile beast of work. On the other hand, he was not really suited to the task. His physique had to be changed—he had to be able to grasp and use the tools of the Nefilim, walk and bend like them so that he could replace the gods in the fields and in the mines. He had to have better "brains"—not like those of the gods but enough to understand speech and commands and the tasks allotted to him. He needed enough cleverness and understanding to be an obedient and useful *amelu*—a serf.

If, as the ancient evidence and modern science seem to confirm, life on Earth germinated from life on the Twelfth Planet, then evolution on Earth should have proceeded as it had on the Twelfth Planet. Undoubtedly there were mutations, variations, accelerations, and retardations caused by different local conditions; but the same genetic codes, the same "chemistry of life" found in all living plants and animals on Earth would also have guided the development of life forms on Earth in the same general direction as on the Twelfth Planet.

Observing the various forms of life on Earth, the Nefilim and their chief scientist, Ea, needed little time to realize what had happened: During the celestial collision, their planet had seeded Earth with its life. Therefore, the being that was available was really akin to the Nefilim—though in a less evolved form.

A gradual process of domestication through generations of breeding and selection would not do. What was needed was a quick process, one that would permit "mass production" of the new workers. So the problem was posed to Ea, who saw the answer at once: to "imprint" the image of the gods on the being that already existed.

The process that Ea recommended in order to achieve a quick evolutionary advancement of *Homo erectus* was, we believe, *genetic manipulation*.

We now know that the complex biological process whereby a living organism reproduces itself, creating progeny that resemble their parents, is made possible by the genetic code. All living organisms—a threadworm, a fern tree, or Man—contain in their cells chromosomes, minute rodlike bodies within each cell that hold the complete hereditary instructions for that particular organism. As the male cell (pollen, sperm) fertilizes the female cell, the two sets of chromosomes combine and then divide to form new cells that hold the complete hereditary characteristics of their parent cells.

Artificial insemination, even of a female human egg, is now possible. The real challenge lies in cross-fertilization between different families within the same species, and even between different species. Modern science has come a long way from the development of the first hybrid corns, or the mating of Alaskan dogs with wolves, or the "creation" of the mule (the artificial mating of a mare and a donkey), to the ability to manipulate Man's own reproduction.

A process called cloning (from the Greek word *klon*—"twig") applies to animals the same principle as that of taking a cutting from a plant to reproduce hundreds of similar plants. The technique as applied to animals was first demonstrated in England, where Dr. John Gurdon replaced the nucleus of a fertilized frog's egg with the nuclear material from another cell of the same frog. The successful formation of normal tadpoles demonstrated that the egg proceeds to develop and subdivide and create progeny no matter where it obtains the correct set of matching chromosomes.

Experiments reported by the Institute of Society, Ethics and Life Sciences at Hastings-on-Hudson, show that techniques already exist for cloning human beings. It is now possible to take the nuclear material of any human cell (not necessarily from the sex organs) and, by introducing its twenty-three sets of complete chromosomes into the female ovum, lead to the conception and birth of a "predetermined" individual. In normal conception, "father" and "mother" chromosome sets merge and then must split to remain at twenty-three chromosome pairs, leading to chance combinations. But in cloning the offspring is an exact replica of the source of the unsplit set of chromosomes. We already possess, wrote Dr. W. Gaylin in *The New York Times*, the "awful knowledge to make exact copies of human beings"—a limitless number of Hitlers or Mozarts or Einsteins (if we had preserved their cell nuclei).

But the art of genetic engineering is not limited to one process. Researchers in many countries have perfected a process called "cell fusion," making it possible to fuse cells rather than combine chromosomes within a single cell. As a result of such a process, cells from different sources can be fused into one "supercell," holding within itself two nuclei and a double set of the paired chromosomes. When this cell splits, the mixture of nuclei and chromosomes may split in a pattern different from that of each cell before the fusion. The result can be two new cells, each genetically complete, but each with a brand-new set of genetic codes, completely garbled as far as the ancestor cells were concerned.

This means that cells from hitherto incompatible living organisms—say, that of a chicken and that of a mouse—

can be fused to form new cells with brand-new genetic mixes that produce new animals that are neither chickens nor mice as we know them. Further refined, the process can also permit us to *select* which traits of one life form shall be imparted to the combined or "fused" cell.

This has led to the development of the wide field of "genetic transplant." It is now possible to pick up from certain bacteria a single specific gene and introduce that gene into an animal or human cell, giving the offspring an added characteristic.

•

We should assume that the Nefilim—being capable of space travel 450,000 years ago—were also equally advanced, compared to us today, in the field of life sciences. We should also assume that they were aware of the various alternatives by which two preselected sets of chromosomes could be combined to obtain a predetermined genetic result; and that whether the process was akin to cloning, cell fusion, genetic transplant, or methods as yet unknown to us, they knew these processes and could carry them out, not only in the laboratory flask but also with living organisms.

We find a reference to such a mixing of two life-sources in the ancient texts. According to Berossus, the deity Belus ("lord")—also called Deus ("god")—brought forth various "hideous Beings, which were produced of a two-fold principle":

Men appeared with two wings, some with four and two faces. They had one body but two heads, the one of a man, the other of a woman. They were likewise in their several organs both male and female.

Other human figures were to be seen with the legs and horns of goats. Some had horses' feet; others had the limbs of a horse behind, but in front were fashioned like men, resembling hippocentaurs. Bulls likewise bred there with the heads of men; and dogs with fourfold bodies, and the tails of fishes. Also horses with the heads of dogs; men too and other animals with the heads and bodies of horses and the tails of fishes. In short, there were creatures with the limbs of every species of animals. . . .

Of all these were preserved delineations in the temple of Belus at Babylon.

The tale's baffling details may hold an important truth. It is quite conceivable that before resorting to the creation of a being in their own image, the Nefilim attempted to come up with a "manufactured servant" by experimenting with other alternatives: the creation of a hybrid ape-man–animal. Some of these artificial creatures may have survived for a while but were certainly unable to reproduce. The enigmatic bull-men and lion-men (sphinxes) that adorned temple sites in the ancient Near East may not have been just figments of an artist's imagination but actual creatures that came out of the biological laboratories of the Nefilim —unsuccessful experiments commemorated in art and by statues. (Fig. 150)

Fig. 150

Sumerian texts, too, speak of deformed humans created by Enki and the Mother Goddess (Ninḫursag) in the course of their efforts to fashion a perfect Primitive Worker. One text reports that Ninḫursag, whose task it was to "bind upon the mixture the mold of the gods," got drunk and "called over to Enki,"

> "How good or how bad is Man's body?
> As my heart prompts me,
> I can make its fate good or bad."

Mischievously, then, according to this text—but probably unavoidably, as part of a trial-and-error process—Ninḫursag produced a Man who could not hold back his urine, a woman who could not bear children, a being who had neither male nor female organs. All in all, six deformed or deficient humans were brought forth by Ninḫursag. Enki was held responsible for the imperfect creation of a man with diseased eyes, trembling hands, a sick liver, a failing heart; a second one with sicknesses attendant upon old age; and so on.

But finally the perfect Man was achieved—the one Enki named Adapa; the Bible, Adam; our scholars, *Homo sapiens*. This being was so much akin to the gods that one text even went so far as to point out that the Mother Goddess gave to her creature, Man, "a skin as the skin of a god"—a smooth, hairless body, quite different from that of the shaggy ape-man.

With this final product, the Nefilim were genetically compatible with the daughters of Man and able to marry them and have children by them. But such compatibility could exist only if Man had developed from the same "seed of life" as the Nefilim. This, indeed, is what the ancient texts attest to.

Man, in the Mesopotamian concept, as in the biblical one, was made of a mixture of a godly element—a god's blood or its "essence"—and the "clay" of Earth. Indeed, the very term *lulu* for "Man," while conveying the sense of "primitive," literally meant "one who has been mixed." Called upon to fashion a man, the Mother Goddess "Washed her hands, pinched off clay, mixed it in the steppe." (It is fascinating to note here the sanitary precautions taken by the goddess. She "washed her hands." We encounter such clinical measures and procedures in other creation texts as well.)

The use of earthly "clay" mixed with divine "blood" to create the prototype of Man is firmly established by the Mesopotamian texts. One, relating how Enki was called upon to "bring to pass some great work of Wisdom"—of scientific know-how—states that Enki saw no great problem in fulfilling the task of "fashioning servants for the gods." "It can be done!" he announced. He then gave these instructions to the Mother Goddess:

"Mix to a core the clay
from the Basement of Earth,
just above the Abzu—
and shape it into the form of a core.
I shall provide good, knowing young gods
who will bring that clay to the right condition."

The second chapter of Genesis offers this technical version:

And Yahweh, Elohim, fashioned the Adam
of the clay of the soil;
and He blew in his nostrils the breath of life,
and the Adam turned into a living Soul.

The Hebrew term commonly translated as "soul" is *nephesh,* that elusive "spirit" that animates a living creature and seemingly abandons it when it dies. By no coincidence, the Pentateuch (the first five books of the Old Testament) repeatedly exhorted against the shedding of human blood and the eating of animal blood "because the blood is the *nephesh.*" The biblical versions of the creation of Man thus equate *nephesh* ("spirit," "soul") and blood.

The Old Testament offers another clue to the role of blood in Man's creation. The term *adama* (after which the name Adam was coined) originally meant not just any earth or soil, but specifically dark-red soil. Like the parallel Akkadian word *adamatu* ("dark-red earth"), the Hebrew term *adama* and the Hebrew name for the color red *(adom)* stem from the words for blood: *adamu, dam.* When the Book of Genesis termed the being created by God "the Adam," it employed a favorite Sumerian linguistic play of double meanings. "The Adam" could mean "the one of the earth" (Earthling), "the one made of the dark-red soil," and "the one made of blood."

The same relationship between the essential element of living creatures and blood exists in Mesopotamian accounts of Man's creation. The hospital-like house where Ea and the Mother Goddess went to bring Man forth was called the House of Shimti; most scholars translate this as "the house where fates are determined." But the term *Shimti* clearly stems from the Sumerian SHI.IM.TI, which, taken

syllable by syllable, means "breath-wind-life." *Bit Shimti*
meant, literally, "the house where the wind of life is
breathed in." This is virtually identical to the biblical
statement.

Indeed, the Akkadian word employed in Mesopotamia
to translate the Sumerian SHI.IM.TI was *napishtu*—the
exact parallel of the biblical term *nephesh*. And the *nephesh*
or *napishtu* was an elusive "something" in the blood.

While the Old Testament offered only meager clues,
Mesopotamian texts were quite explicit on the subject.
Not only do they state that blood was required for the
mixture of which Man was fashioned; they specified that
it had to be the blood of a god, divine blood.

When the gods decided to create Man, their leader
announced: "Blood will I amass, bring bones into being."
Suggesting that the blood be taken from a specific god,
"Let primitives be fashioned after his pattern," Ea said.
Selecting the god,

> Out of his blood they fashioned Mankind;
> imposed on it the service, let free the gods. . . .
> It was a work beyond comprehension.

According to the epic tale "When gods as men," the
gods then called the Birth Goddess (the Mother Goddess,
Ninhursag) and asked her to perform the task:

> While the Birth Goddess is present,
> Let the Birth Goddess fashion offspring.
> While the Mother of the Gods is present,
> Let the Birth Goddess fashion a *Lulu;*
> Let the worker carry the toil of the gods.
> Let her create a *Lulu Amelu,*
> Let him bear the yoke.

In a parallel Old Babylonian text named "Creation of
Man by the Mother Goddess," the gods call upon "The
Midwife of the gods, the Knowing Mami" and tell her:

> Thou art the mother-womb,
> The one who Mankind can create.
> Create then *Lulu,* let him bear the yoke!

At this point, the text "When gods as men" and parallel texts turn to a detailed description of the actual creation of Man. Accepting the "job," the goddess (here named NIN.TI—"lady who gives life") spelled out some requirements, including some chemicals ("bitumens of the Abzu"), to be used for "purification," and "the clay of the Abzu."

Whatever these materials were, Ea had no problem understanding the requirements; accepting, he said:

> "I will prepare a purifying bath.
> Let one god be bled. . . .
> From his flesh and blood,
> let Ninti mix the clay."

To shape a man from the mixed clay, some feminine assistance, some pregnancy or childbearing aspects were also needed. Enki offered the services of his own spouse:

> Ninki, my goddess-spouse,
> will be the one for labor.
> Seven goddesses-of-birth
> will be near, to assist.

Following the mixing of the "blood" and "clay," the childbearing phase would complete the bestowal of a divine "imprint" on the creature.

> The new-born's fate thou shalt pronounce;
> Ninki would fix upon it the image of the gods;
> And what it will be is "Man."

Depictions on Assyrian seals may well have been intended as illustrations for these texts—showing how the

Mother Goddess (her symbol was   the cutter of the

umbilical cord) and Ea (whose original symbol was the crescent) were preparing the mixtures, reciting the incantations, urging each other to proceed. (Figs. 151, 152)

The involvement of Enki's spouse, Ninki, in the creation of the first successful specimen of Man reminds us of the tale of Adapa, which we discussed in an earlier chapter:

Fig. 151

Fig. 152

In those days, in those years,
The Wise One of Eridu, Ea,
created him as a model of men.

Scholars have surmised that references to Adapa as a
"son" of Ea implied that the god loved this human so
much that he adopted him. But in the same text Anu refers
to Adapa as "the human offspring of Enki." It appears
that the involvement of Enki's spouse in the process of
creating Adapa, the "model Adam," did create some
genealogical relationship between the new Man and his
god: It was Ninki who was pregnant with Adapa!

Ninti blessed the new being and presented him to Ea.
Some seals show a goddess, flanked by the Tree of Life
and laboratory flasks, holding up a newborn being. (Fig.
153)

Fig. 153

The being that was thus produced, which is repeatedly referred to in Mesopotamian texts as a "model Man" or a "mold," was apparently the right creature, for the gods then clamored for duplicates. This seemingly unimportant detail, however, throws light not only on the process by which Mankind was "created," but also on the otherwise conflicting information contained in the Bible.

According to the first chapter of Genesis:

> Elohim created the Adam in His image—
> in the image of Elohim created He him.
> Male and female created He them.

Chapter 5, which is called the Book of the Genealogies of Adam, states that:

> On the day that Elohim created Adam,
> in the likeness of Elohim did He make him.
> Male and female created He them,
> and He blessed them, and called them "Adam"
> on the very day of their creation.

In the same breath, we are told that the Deity created, in his likeness and his image, only a single being, "the Adam," and in apparent contradiction, that both a male and a female were created simultaneously. The contradiction seems sharper still in the second chapter of Genesis, which specifically reports that the Adam was alone for a while, until the Deity put him to sleep and fashioned Woman from his rib.

The contradiction, which has puzzled scholars and theologians alike, disappears once we realize that the biblical texts were a condensation of the original Sumerian sources. These sources inform us that after trying to fashion a Primitive Worker by "mixing" apemen with animals, the gods concluded that the only mixture that would work would be between apemen and the Nefilim themselves. After several unsuccessful attempts, a "model"—Adapa/Adam—was made. There was, at first, only a single Adam.

Once Adapa/Adam proved to be the right creature, he was used as the genetic model or "mold" for the creation of duplicates, and those duplicates were not only male, but male and female. As we showed earlier, the biblical "rib" from which Woman was fashioned was a play on words on the Sumerian TI ("rib" and "life")—confirming that Eve was made of Adam's "life's essence."

•

The Mesopotamian texts provide us with an eye-witness report of the first production of the duplicates of Adam.

The instructions of Enki were followed. In the House of Shimti—where the breath of life is "blown in"—Enki, the Mother Goddess, and fourteen birth goddesses assembled. A god's "essence" was obtained, the "purifying bath" prepared. "Ea cleaned the clay in her presence; he kept reciting the incantation."

> The god who purifies the Napishtu, Ea, spoke up.
> Seated before her, he was prompting her.
> After she had recited her incantation,
> She put her hand out to the clay.

We are now privy to the detailed process of Man's mass creation. With fourteen birth goddesses present,

> Ninti nipped off fourteen pieces of clay;
> Seven she deposited on the right,
> Seven she deposited on the left.
> Between them she placed the mould.
> . . . the hair she . . .
> . . . the cutter of the umbilical cord.

It is evident that the birth goddesses were divided into two groups. "The wise and learned, twice-seven birth goddesses had assembled," the text goes on to explain. Into their wombs the Mother Goddess deposited the "mixed clay." There are hints of a surgical procedure—the removal or shaving off of hair, the readying of a surgical instrument, a cutter. Now there was nothing to do but wait:

> The birth goddesses were kept together.
> Ninti sat counting the months.
> The fateful 10th month was approaching;
> The 10th month arrived;
> The period of opening the womb had elapsed.

Her face radiated understanding:
She covered her head, performed the midwifery.
Her waist she girdled, pronounced the blessing.
She drew a shape; in the mould was life.

The drama of Man's creation, it appears, was compounded by a late birth. The "mixture" of "clay" and "blood" was used to induce pregnancy in fourteen birth goddesses. But nine months passed, and the tenth month commenced. "The period of opening the womb had elapsed." Understanding what was called for, the Mother Goddess "performed the midwifery." That she engaged in some surgical operation emerges more clearly from a parallel text (in spite of its fragmentation):

Ninti . . . counts the months. . . .
The destined 10th month they called;
The Lady Whose Hand Opens came.
With the . . . she opened the womb.
Her face brightened with joy.
Her head was covered;
. . . made an opening;
That which was in the womb came forth.

Overcome with joy, the Mother Goddess let out a cry.

"I have created!
My hands have made it!"

•

How was the creation of Man accomplished?

The text "When the gods as men" contains a passage whose purpose was to explain why the "blood" of a god had to be mixed into the "clay." The "divine" element required was not simply the dripping blood of a god, but something more basic and lasting. The god that was selected, we are told, had TE.E.MA—a term the leading authorities on the text (W. G. Lambert and A. R. Millard of Oxford University) translate as "personality." But the ancient term is much more specific; it literally means "that which houses that which binds the memory." Further on, the same term appears in the Akkadian version as *etemu,* which is translated as "spirit."

In both instances we are dealing with that "something" in the blood of the god that was the repository of his individuality. All these, we feel certain, are but roundabout ways of stating that what Ea was after, when he put the god's blood through a series of "purifying baths," was the god's *genes.*

The purpose of mixing this divine element thoroughly with the earthly element was also spelled out:

> In the clay, god and Man shall be bound,
> to a unity brought together;
> So that to the end of days
> the Flesh and the Soul
> which in a god have ripened—
> that Soul in a blood-kinship be bound;
> As its Sign life shall proclaim.
> So that this not be forgotten,
> Let the "Soul" in a blood-kinship be bound.

These are strong words, little understood by scholars. The text states that the god's blood was mixed into the clay so as to bind god and Man genetically "to the end of days" so that both the flesh ("image") and the soul ("likeness") of the gods would become imprinted upon Man in a kinship of blood that could never be severed.

The "Epic of Gilgamesh" reports that when the gods decided to create a double for the partly divine Gilgamesh, the Mother Goddess mixed "clay" with the "essence" of the god Ninurta. Later on in the text, Enkidu's mighty strength is attributed to his having in him the "essence of Anu," an element he acquired through Ninurta, the grandson of Anu.

The Akkadian term *kiṣir* refers to an "essence," a "concentration" that the gods of the heavens possessed. E. Ebeling summed up the efforts to understand the exact meaning of *kiṣir* by stating that as "Essence, or some nuance of the term, it could well be applied to deities as well as to missiles from Heaven." E. A. Speiser concurred that the term also implied "something that came down from

Heaven." It carried the connotation, he wrote, "as would be indicated by the use of the term in medicinal contexts."

We are back to a simple, single word of translation: *gene*.

The evidence of the ancient texts, Mesopotamian as well as biblical, suggests that the process adopted for merging two sets of genes—those of a god and those of *Homo erectus*—involved the use of male genes as the divine element and female genes as the earthly element.

Repeatedly asserting that the Deity created Adam in his image and in his likeness, the Book of Genesis later describes the birth of Adam's son Seth in the following words:

> And Adam lived a hundred and thirty years,
> and had an offspring
> in his likeness and after his image;
> and he called his name Seth.

The terminology is identical to that used to describe the creation of Adam by the Deity. But Seth was certainly born to Adam by a biological process—the fertilization of a female egg by the male sperm of Adam, and the ensuing conception, pregnancy, and birth. The identical terminology bespeaks an identical process, and the only plausible conclusion is that Adam, too, was brought forth by the Deity through the process of fertilizing a female egg with the male sperm of a god.

If the "clay" onto which the godly element was mixed was an earthly element—as all texts insist—then the only possible conclusion is that the male sperm of a god—his genetic material—was inserted into the egg of an ape-woman!

The Akkadian term for the "clay"—or, rather, "molding clay"—is *tit*. But its original spelling was TI.IT ("that which is with life"). In Hebrew, *tit* means "mud"; but its synonym is *boş*, which shares a root with *bişa* ("marsh") and *beşa* ("egg").

The story of Creation is replete with plays on words. We have seen the double and triple meanings of Adam–*adama* –*adamtu–dam*. The epithet for the Mother Goddess, NIN.TI, meant both "lady of life" and "lady of the rib."

Why not, then, *boṣ–biṣa–beṣa* ("clay–mud–egg") as a play on words for the female ovum?

The ovum of a female *Homo erectus*, fertilized by the genes of a god, was then implanted within the womb of Ea's spouse; and after the "model" was obtained, duplicates of it were implanted in the wombs of birth goddesses, to undergo the process of pregnancy and birth.

> The Wise and learned,
> Double-seven birth-goddesses had assembled;
> Seven brought forth males,
> Seven brought forth females.
> The Birth Goddess brought forth
> The Wind of the Breath of Life.
> In pairs were they completed,
> In pairs were they completed in her presence.
> The creatures were People—
> Creatures of the Mother Goddess.

Homo sapiens had been created.

•

The ancient legends and myths, biblical information, and modern science are also compatible in one more aspect. Like the findings of modern anthropologists—that Man evolved and emerged in southeast Africa—the Mesopotamian texts suggest that the creation of Man took place in the Apsu—in the Lower World where the Land of the Mines was located. Paralleling Adapa, the "model" of Man, some texts mention "sacred Amama, the Earth woman," whose abode was in the Apsu.

In the "Creation of Man" text, Enki issues the following instructions to the Mother Goddess: "Mix to a core the clay from the Basement of Earth, just above the Abzu." A hymn to the creations of Ea, who "the Apsu fashioned as his dwelling," begins by stating:

> Divine Ea in the *Apsu*
> pinched off a piece of clay,
> created Kulla to restore the temples.

The hymn continues to list the construction specialists, as well as those in charge of the "abundant products of mountain and sea," who were created by Ea—all, it is inferred, from pieces of "clay" pinched off in the Abzu— the Land of Mines in the Lower World.

The texts make it abundantly clear that while Ea built a brick house by the water in Eridu, in the Abzu he built a house adorned with precious stones and silver. It was there that his creature, Man, originated:

> The Lord of the AB.ZU, the king Enki . . .
> Built his house of silver and lapis-lazuli;
> Its silver and lapis-lazuli, like sparkling light,
> The Father fashioned fittingly in the AB.ZU.
> The Creatures of bright countenance,
> Coming forth from the AB.ZU,
> Stood all about the Lord Nudimmud.

One can even conclude from the various texts that the creation of Man caused a rift among the gods. It would appear that at least at first the new Primitive Workers were confined to the Land of Mines. As a result, the Anunnaki who were toiling in Sumer proper were denied the benefits of the new manpower. A puzzling text named by the scholars "The Myth of the Pickax" is in fact the record of the events whereby the Anunnaki who stayed in Sumer under Enlil obtained their fair share of the Black-Headed People.

Seeking to reestablish "the normal order," Enlil took the extreme action of severing the contacts between "Heaven" (the Twelfth Planet or the spaceships) and Earth, and launched some drastic action against the place "where flesh sprouted forth."

> The Lord,
> That which is appropriate he caused to come about.
> The Lord Enlil,
> Whose decisions are unalterable,
> Verily did speed to separate Heaven from Earth
> So that the Created Ones could come forth;
> Verily did speed to separate Earth from Heaven.

In the "Bond Heaven–Earth" he made a gash,
So that the Created Ones could come up
From the Place-Where-Flesh-Sprouted-Forth.

Against the "Land of Pickax and Basket," Enlil fashioned a marvelous weapon named AL.A.NI ("ax that produces power"). This weapon had a "tooth," which, "like a one-horned ox," could attack and destroy large walls. It was by all descriptions some kind of a huge power drill, mounted on a bulldozer-like vehicle that crushed everything ahead of it:

The house which rebels against the Lord,
The house which is not submissive to the Lord,
The AL.A.NI makes it submissive to the Lord.
Of the bad . . . , the heads of its plants it crushes;
Plucks at the roots, tears at the crown.

Arming his weapon with an "earth splitter," Enlil launched the attack:

The Lord called forth the AL.A.NI, gave its orders.
He set the Earth Splitter as a crown upon its head,
And drove it into the Place-Where-Flesh-Sprouted-Forth.
In the hole was the head of a man;
From the ground, people were breaking through
towards Enlil.
He eyed his Black-headed Ones in steadfast fashion.

Grateful, the Anunnaki put in their requests for the arriving Primitive Workers and lost no time in putting them to work:

The Anunnaki stepped up to him,
Raised their hands in greetings,
Soothing Enlil's heart with prayers.
Black-headed Ones they were requesting of him.
To the Black-headed people,
they give the pickax to hold.

The Book of Genesis likewise conveys the information that "the Adam" was created somewhere west of Mesopotamia, then brought over eastward to Mesopotamia to work in the Garden of Eden:

> And the Deity Yahweh
> Planted an orchard in Eden, in the east . . .
> And He took the Adam
> And placed him in the Garden of Eden
> To work it and to keep it.

13

·

THE END OF ALL FLESH

MAN'S LINGERING BELIEF that there was some Golden Age in his prehistory cannot possibly be based on human recollection, for the event took place too long ago and Man was too primitive to record any concrete information for future generations. If Mankind somehow retains a subconscious sense that in those earliest days Man lived through an era of tranquillity and felicity, it is simply because Man knew no better. It is also because the tales of that era were first told Mankind, not by earlier men, but by the Nefilim themselves.

The only complete account of the events that befell Man following his transportation to the Abode of the Gods in Mesopotamia is the biblical tale of Adam and Eve in the Garden of Eden:

> And the Deity Yahweh planted an orchard
> In Eden, in the east;
> And he placed there the Adam
> Whom He had created.
> And the Deity Yahweh
> Caused to grow from the ground
> Every tree that is pleasant to the sight
> And good for eating;
> And the Tree of Life was in the orchard
> And the Tree of Knowing good and evil. . . .
> And the Deity Yahweh took the Adam
> And placed him in the Garden of Eden
> To work it and to keep it.
> And the Deity Yahweh
> Commanded the Adam, saying:

"Of every tree of the orchard eat you shall;
but of the Tree of Knowing good and evil
thou shalt not eat of it;
for on the day that thou eatest thereof
thou shalt surely die."

Though two vital fruits were available, the Earthlings were prohibited from reaching only for the fruit of the Tree of Knowing. The Deity—at that point—appeared unconcerned that Man might try to reach for the Fruit of Life. Yet Man could not adhere even to that single prohibition, and tragedy followed.

The idyllic picture soon gave way to dramatic developments, which biblical scholars and theologians call the Fall of Man. It is a tale of unheeded divine commandments, divine lies, a wily (but truth-telling) Serpent, punishment, and exile.

Appearing from nowhere, the Serpent challenged God's solemn warnings:

And the Serpent . . . said unto the woman:
"Hath the Deity indeed said
'Ye shall not eat of any tree of the orchard'?"
And the woman said unto the Serpent:
"Of the fruits of the trees of the orchard
eat we may;
it is of the fruit of the tree in the
midst of the orchard that the Deity hath said:
'Ye shall not eat of it, neither touch it,
lest ye die.' "
And the Serpent said unto the woman:
"Nay, ye will surely not die;
It is that the Deity doth know
that on the day ye eat thereof
your eyes will be opened
and ye will be as the Deity—
knowing good and evil."
And the woman saw that the tree was good to eat
And that it was lustful to behold;
And the tree was desirable to make one wise;
And she took of its fruit and did eat,
And gave also to her mate with her, and he ate.

and the eyes of both of them were opened,
And they knew that they were naked;
And they sewed fig leaves together,
And made themselves loincloths.

Reading and rereading the concise yet precise tale, one cannot help wondering what the whole confrontation was about. Prohibited under threat of death from even touching the Fruit of Knowing, the two Earthlings were persuaded to go ahead and eat the stuff, which would make them "knowing" as the Deity. Yet all that happened was a sudden awareness that they were naked.

The state of nakedness was indeed a major aspect of the whole incident. The biblical tale of Adam and Eve in the Garden of Eden opens with the statement: "And the both of them were naked, the Adam and his mate, and they were not ashamed." They were, we are to understand, at some lesser stage of human development than that of fully developed humans: Not only were they naked, they were unaware of the implications of such nakedness.

Further examination of the biblical tale suggests that its theme is Man's acquisition of some sexual prowess. The "knowing" that was held back from Man was not some scientific information but something connected with the male and female sex; for no sooner had Man and his mate acquired the "knowing" than "they knew that they were naked" and covered their sex organs.

The continuing biblical narrative confirms the connection between nakedness and the lack of knowing, for it took the Deity no time at all to put the two together:

And they heard the sound of the Deity Yahweh
Walking in the orchard in the day's breeze,
And the Adam and his mate hid
From the Deity Yahweh amongst the orchard's trees.
And the Deity Yahweh called to the Adam
And said: "Where art thou?"
And he answered:
"Thy sound I heard in the orchard
and I was afraid, for I am naked;
and I hid."

And He said:
"Who told thee that thou are naked?
Hast thou eaten of the tree,
whereof I commanded thee not to eat?"

Admitting the truth, the Primitive Worker blamed his female mate, who, in turn, blamed the Serpent. Greatly angered, the Deity put curses on the Serpent and the two Earthlings. Then—surprisingly—"the Deity Yahweh made for Adam and his wife garments of skins, and clothed them."

One cannot seriously assume that the purpose of the whole incident—which led to the expulsion of the Earthlings from the Garden of Eden—was a dramatic way to explain how Man came to wear clothes. The wearing of clothes was merely an outward manifestation of the new "knowing." The acquisition of such "knowing," and the Deity's attempts to deprive Man of it, are the central themes of the events.

While no Mesopotamian counterpart of the biblical tale has yet been found, there can be little doubt that the tale —like all the biblical material concerning Creation and Man's prehistory—was of Sumerian origin. We have the locale: the Abode of the Gods in Mesopotamia. We have the telltale play on words in Eve's name ("she of life," "she of rib"). And we have two vital trees, the Tree of Knowing and the Tree of Life, as in Anu's abode.

Even the words of the Deity reflect a Sumerian origin, for the sole Hebrew Deity has again lapsed into the plural, addressing divine colleagues who were featured not in the Bible but in Sumerian texts:

Then did the Deity Yahweh say:
"Behold, the Adam has become as one of us,
to know good and evil.
And now might he not put forth his hand
And partake also of the Tree of Life,
and eat, and live forever?"
And the Deity Yahweh expelled the Adam
from the orchard of Eden.

As many early Sumerian depictions show, there had been a time when Man, as a Primitive Worker, served his gods stark naked. He was naked whether he served the gods their food and drink, or toiled in the fields or on construction jobs. (Figs. 154, 155)

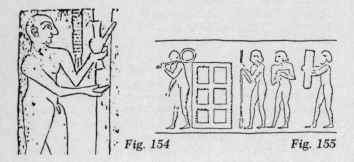

Fig. 154   Fig. 155

The clear implication is that the status of Man vis-à-vis the gods was not much different from that of domesticated animals. The gods had merely upgraded an existing animal to suit their needs. Did the lack of "knowing," then, mean that, naked as an animal, the newly fashioned being also engaged in sex as, or with, the animals? Some early depictions indicate that this was indeed the case. (Fig. 156)

Sumerian texts like the "Epic of Gilgamesh" suggest that the manner of sexual intercourse did indeed account for

Fig. 156

a distinction between wild-Man and human-Man. When the people of Uruk wanted to civilize the wild Enkidu—"the barbarous fellow from the depths of the steppes"—they enlisted the services of a "pleasure girl" and sent her to meet Enkidu at the water hole where he used to befriend various animals, and there to offer him her "ripeness."

It appears from the text that the turning point in the process of "civilizing" Enkidu was the rejection *of* him *by* the animals he had befriended. It was important, the people of Uruk told the girl, that she continue to treat him to "a woman's task" until "his wild beasts, that grew up on his steppe, will reject him." For Enkidu to be torn away from sodomy was a prerequisite to his becoming human.

> The lass freed her beasts, bared her bosom,
> and he possessed her ripeness . . .
> She treated him, the savage,
> to a woman's task.

Apparently the ploy worked. After six days and seven nights, "after he had had his fill of her charms," he remembered his former playmates.

> He set his face toward his wild beasts; but
> On seeing him the gazelles ran off.
> The wild beasts of the steppe
> drew away from his body.

The statement is explicit. The human intercourse brought about such a profound change in Enkidu that the animals he had befriended "drew away from his body." They did not simply run away; they shunned physical contact with him.

Astounded, Enkidu stood motionless for a while, "for his wild animals had gone." But the change was not to be regretted, as the ancient text explains:

> Now he had vision, broader understanding. . . .
> The harlot says to him, to Enkidu:
> "Thou art knowing, Enkidu;
> Thou art become like a god!"

The words in this Mesopotamian text are almost identical to those of the biblical tale of Adam and Eve. As the Serpent had predicted, by partaking of the Tree of Knowing, they had become—in sexual matters—"as the Deity —knowing good and evil."

If this meant only that Man had come to recognize that having sex with animals was uncivilized or evil, why were Adam and Eve punished for giving up sodomy? The Old Testament is replete with admonitions against sodomy, and it is inconceivable that the learning of a virtue would cause divine wrath.

The "knowing" that Man obtained against the wishes of the Deity—or one of the deities—must have been of a more profound nature. It was something good for Man, but something his creators did not wish him to have.

We have to read carefully between the lines of the curse against Eve to grasp the meaning of the event:

> And to the woman He said:
> "I will greatly multiply thy suffering
> by thy pregnancy.
> In suffering shalt thou bear children,
> yet to thy mate shall be thy desire" . . .
> And the Adam named his wife "Eve,"
> for she was the mother of all who lived.

This, indeed, is the momentous event transmitted to us in the biblical tale: As long as Adam and Eve lacked "knowing," they lived in the Garden of Eden without any offspring. Having obtained "knowing," Eve gained the ability (and pain) to become pregnant and bear children. Only after the couple had acquired this "knowing," "Adam *knew* Eve his wife, and she conceived and gave birth to Cain."

Throughout the Old Testament, the term "to know" is used to denote sexual intercourse, mostly between a man and his spouse for the purpose of having children. The tale of Adam and Eve in the Garden of Eden is the story of a crucial step in Man's development: *the acquisition of the ability to procreate.*

That the first representatives of *Homo sapiens* were incapable of reproduction should not be surprising. Whatever method the Nefilim had used to infuse some of their genetic material into the biological makeup of the hominids they selected for the purpose, the new being was a hybrid, a cross between two different, if related, species. Like a mule (a cross between a mare and a donkey), such mammal hybrids are sterile. Through artificial insemination and even more sophisticated methods of biological engineering, we can produce as many mules as we desire, even without actual intercourse between donkey and mare; but no mule can procreate and bring forth another mule.

Were the Nefilim, at first, simply producing "human mules" to suit their requirements?

Our curiosity is aroused by a scene depicted on a rock carving found in the mountains of southern Elam. It depicts a seated deity holding a "laboratory" flask from which liquids are flowing—a familiar depiction of Enki. A Great Goddess is seated next to him, a pose that indicates that she was a co-worker rather than a spouse; she could be none other than Ninti, the Mother Goddess or Goddess of Birth. The two are flanked by lesser goddesses—reminiscent of the birth goddesses of the Creation tales. Facing these creators of Man are row upon row of human beings, whose outstanding feature is that they all look alike—like products from the same mold. (Fig. 157)

Our attention is also drawn again to the Sumerian tale of the imperfect males and females initially brought forth by Enki and the Mother Goddess, who were either sexless or sexually incomplete beings. Does this text recall the first phase of the existence of hybrid Man—a being in the like-

Fig. 157

...nd image of the gods, but sexually incomplete: ...ng in "knowing"?

After Enki managed to produce a "perfect model"—Adapa/Adam, "mass-production" techniques are described in the Sumerian texts: the implanting of the genetically treated ova in a "production line" of birth goddesses, with the advance knowledge that half would produce males and half would produce females. Not only does this bespeak the technique by which hybrid Man was "manufactured"; it also implies that Man could not procreate on his own.

The inability of hybrids to procreate, it has been discovered recently, stems from a deficiency in the reproductive cells. While all cells contain only one set of hereditary chromosomes, Man and other mammals are able to reproduce because their sex cells (the male sperm, the female ovum) contain two sets each. But this unique feature is lacking in hybrids. Attempts are now being made through genetic engineering to provide hybrids with such a double set of chromosomes in their reproductive cells, making them sexually "normal."

Was that what the god whose epithet was "The Serpent" accomplished for Mankind?

The biblical Serpent surely was not a lowly, literal snake —for he could converse with Eve, he knew the truth about the matter of "knowing," and he was of such high stature that he unhesitatingly exposed the deity as a liar. We recall that in all ancient traditions, the chief deity fought a Serpent adversary—a tale whose roots undoubtedly go back to the Sumerian gods.

The biblical tale reveals many traces of its Sumerian origin, including the presence of other deities: "The Adam has become as one of *us*." The possibility that the biblical antagonists—the Deity and the Serpent—stood for Enlil and Enki seems to us entirely plausible.

Their antagonism, as we have discovered, originated in the transfer to Enlil of the command of Earth, although Enki had been the true pioneer. While Enlil stayed at the comfortable Mission Control Center at Nippur, Enki was sent to organize the mining operations in the Lower World. The mutiny of the Anunnaki was directed at Enlil and his

son Ninurta; the god who spoke out for the mutine[...]
Enki. It was Enki who suggested, and undertoo[...]
creation of Primitive Workers; Enlil had to use force[...]
obtain some of these wonderful creatures. As the Sumerian
texts recorded the course of human events, Enki as a rule
emerges as Mankind's protagonist, Enlil as its strict
discipliner if not outright antagonist. The role of a deity
wishing to keep the new humans sexually suppressed, and of
a deity willing and capable of bestowing on Mankind the
fruit of "knowing," fit Enlil and Enki perfectly.

Once more, Sumerian and biblical plays on words come
to our aid. The biblical term for "Serpent" is *nahash*, which
does mean "snake." But the word comes from the root
NHSH, which means "to decipher, to find out"; so that
nahash could also mean "he who can decipher, he who
finds things out," an epithet befitting Enki, the chief
scientist, the God of Knowledge of the Nefilim.

Drawing parallels between the Mesopotamian tale of
Adapa (who obtained "knowing" but failed to obtain
eternal life) and the fate of Adam, S. Langdon (*Semitic
Mythology*) reproduced a depiction unearthed in Mesopotamia that strongly suggests the biblical tale: a serpent
entwined on a tree, pointing at its fruit. The celestial
symbols are significant: High above is the Planet of Crossing, which stood for Anu; near the serpent is the Moon's
crescent, which stood for Enki. (Fig. 158)

Most pertinent to our findings is the fact that in the
Mesopotamian texts, the god who eventually granted
"knowledge" to Adapa was none other than Enki:

> Wide understanding he perfected for him. . . .
> Wisdom [he had given him]. . . .
> To him he had given Knowledge;
> Eternal Life he had not given him.

A pictorial tale engraved on a cylinder seal found in
Mari may well be an ancient illustration of the Mesopotamian version of the tale in Genesis. The engraving shows
a great god seated on high ground rising from watery
waves—an obvious depiction of Enki. Water-spouting
serpents protrude from each side of this "throne."

Flanking this central figure are two treelike gods. The one on the right, whose branches have penis-shaped ends, holds up a bowl that presumably contains the Fruit of Life. The one on the left, whose branches have vagina-shaped ends, offers fruit-bearing branches, representing the Tree of "Knowing"—the god-given gift of procreation.

Standing to the side is another Great God; we suggest that he was Enlil. His anger at Enki is obvious. (Fig. 159)

We shall never know what caused this "conflict in the Garden of Eden." But whatever Enki's motives were, he did succeed in perfecting the Primitive Worker and in creating *Homo sapiens,* who could have his own offspring.

After Man's acquisition of "knowing," the Old Testament ceases to refer to him as "*the* Adam," and adopts as its subject *Adam,* a specific person, the first patriarch of the line of people with whom the Bible was concerned. But this coming of age of Mankind also marked a schism between God and Man.

The parting of the ways, with Man no longer a dumb serf of the gods but a person tending for himself, is ascribed in the Book of Genesis not to a decision by Man himself but to the imposition of a punishment by the Deity: lest the Earthling also acquire the ability to escape mortality, he shall be cast out of the Garden of Eden. According to these sources, Man's independent existence began not in southern Mesopotamia, where the Nefilim had established their cities and orchards, but to the east, in the Zagros Mountains: "And he drove out the Adam and made him reside east of the Garden of Eden."

Once more, then, biblical information conforms to scientific findings: Human culture began in the mountainous areas bordering the Mesopotamian plain. What a pity the biblical narrative is so brief, for it deals with what was Man's first civilized life on Earth.

Cast out of the Abode of the Gods, doomed to a mortal's life, but able to procreate, Man proceeded to do just that. The first Adam with whose generations the Old Testament was concerned "knew" his wife Eve, and she bore him a son, Cain, who tilled the land. Then Eve bore Abel, who was a shepherd. Hinting at homosexuality as the cause, the Bible relates that "Cain rose up unto his brother Abel and killed him."

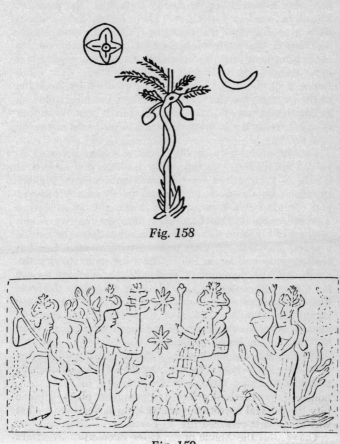

Fig. 158

Fig. 159

Fearing for his life, Cain was given a protective sign by the Deity and was ordered to move farther east. At first leading a nomad's life, he finally settled in "the Land of Migration, well east of Eden." There he had a son whom he named Enoch ("inauguration"), "and he built a city, and called the name of the city after the name of his son." Enoch, in turn, had children and grandchildren and great-grandchildren. In the sixth generation after Cain, Lamech was born; his three sons are credited by the Bible as the bearers of civilization: Jabal "was the father of such as dwell in tents and have cattle"; Jubal "was the father of all that grasp lyre and harp"; Tubal-cain was the first smith.

But Lamech, too, as his ancestor Cain, became involved in murder—this time of both a man and a child. It is safe to assume that the victims were not some humble strangers, for the Book of Genesis dwells on the incident and considers it a turning point in the lineage of Adam. The Bible reports that Lamech summoned his two wives, mothers of his three sons, and confessed to them the double murder, declaring, "If Cain be sevenfold avenged, Lamech shall seventy and seven fold." This little-understood statement must be assumed to deal with the succession; we see it as an admission by Lamech to his wives that the hope that the curse on Cain would be redeemed by the seventh generation (the generation of their sons) had come to naught. Now a new curse, lasting much longer, had been imposed on the house of Lamech.

Confirming that the event concerned the line of succession, the following verses advise us of the immediate establishment of a new, pure, lineage:

And Adam knew his wife again
and she bore a son
and called his name Seth ["foundation"]
for the Deity hath founded for me
another seed instead of Abel, whom Cain slew.

The Old Testament at that point loses all interest in the defiled line of Cain and Lamech. Its ongoing tale of human events is henceforth anchored on the lineage of Adam through his son Seth, and Seth's firstborn, Enosh, whose name has acquired in Hebrew the generic connotation "human being." "It was then," Genesis informs us, "that it was begun to call upon the name of the Deity."

This enigmatic statement has baffled biblical scholars and theologians throughout the ages. It is followed by a chapter giving the genealogy of Adam through Seth and Enosh for ten generations ending with Noah, the hero of the Deluge.

The Sumerian texts, which describe the early stages when the gods were alone in Sumer, describe with equal precision the life of humans in Sumer at a later time, but before the Deluge. The Sumerian (and original) story of the Deluge has as its "Noah" a "Man of Shuruppak," the seventh city established by the Nefilim when they landed on Earth.

At some point, then, the human beings—banished from Eden—were allowed to return to Mesopotamia, to live alongside the gods, to serve them, and to worship them. As we interpret the biblical statement, this happened in the days of Enosh. It was then that the gods allowed Mankind back into Mesopotamia, to serve the gods "and to call upon the name of the deity."

Eager to get to the next epic event in the human saga, the Deluge, the Book of Genesis provides little information besides the names of the patriarchs who followed Enosh. But the meaning of each patriarch's name may suggest the events that took place during his lifetime.

The son of Enosh, through whom the pure lineage continued, was Cainan ("little Cain"); some scholars take the name to mean "metalsmith." Cainan's son was Mahalal-El ("praiser of god"). He was followed by Jared ("he who descended"); his son was Enoch ("consecrated one"), who at age 365 was carried aloft by the Deity. But three hundred years earlier, at age sixty-five, Enoch had begotten a son named Methuselah; many scholars, following Lettia D. Jeffreys (*Ancient Hebrew Names: Their Significance and Historical Value*) translate Methuselah as "man of the missile."

Methuselah's son was named Lamech, meaning "he who was humbled." And Lamech begot Noah ("respite"), saying: "Let this one comfort us concerning our work and the suffering of our hands by the earth which the deity hath accursed."

Humanity, it appears, was undergoing great deprivations when Noah was born. The hard work and the toil were getting it nowhere, for Earth, which was to feed them, was accursed. The stage was set for the Deluge—the momentous event which was to wipe off the face of Earth not only the human race but all life upon the land and in the skies.

> And the Deity saw that the wickedness of Man
> was great on the earth,
> and that every desire of his heart's thoughts
> was only evil, every day.
> And the Deity repented that He had made Man
> upon the earth, and His heart grieved.
> And the Deity said:
> "I will destroy the Earthling whom I have created
> off the face of the earth."

These are broad accusations, presented as justifications for drastic measures to "end all flesh." But they lack specificity, and scholars and theologians alike find no satisfactory answers regarding the sins or "violations" that could have upset the Deity so much.

The repeated use of the term *flesh*, both in the accusative verses and in the proclamations of judgment, suggest, of course, that the corruptions and violations had to do with the flesh. The Deity grieved over the evil "desire of Man's thoughts." Man, it would seem, having discovered sex, had become a sex maniac.

But one can hardly accept that the Deity would decide to wipe Mankind off the face of Earth simply because men made too much love to their wives. The Mesopotamian texts speak freely and eloquently of sex and lovemaking among the gods. There are texts describing tender love between gods and their consorts; illicit love between a maiden and her lover; violent love (as when Enlil raped Ninlil). There is a profusion of texts describing lovemaking and actual intercourse among the gods—with their official consorts or unofficial concubines, with their sisters and daughters and even granddaughters (making love to the latter was a favorite pastime of Enki). Such gods could hardly turn against Mankind for behaving as they themselves did.

The Deity's motive, we find, was not merely concern for human morals. The mounting disgust was caused by a spreading defilement of the gods themselves. Seen in this light, the meaning of the baffling opening verses of Genesis 6 becomes clear:

> And it came to pass,
> When the Earthlings began to increase in number
> upon the face of the Earth,
> and daughters were born unto them,
> that the sons of the deities
> saw the daughters of the Earthlings
> that they were compatible,
> and they took unto themselves
> wives of whichever they chose.

As these verses should make clear, it was when the sons of the gods began to be sexually involved with Earthlings' offspring that the Deity cried, "Enough!"

> And the Deity said:
> "My spirit shall not shield Man forever;
> having strayed, he is but flesh."

The statement has remained enigmatic for millennia. Read in the light of our conclusions regarding the genetic manipulation that was brought to play in Man's creation, the verses carry a message to our own scientists. The "spirit" of the gods—their genetic perfection of Mankind—was beginning to deteriorate. Mankind had "strayed," thereby reverting to being "but flesh"—closer to its animal, simian origins.

We can now understand the stress put by the Old Testament on the distinction between Noah, "a righteous man . . . pure in his genealogies" and "the whole earth that was corrupt." By intermarrying with the men and women of decreasing genetic purity, the gods were subjecting themselves, too, to deterioration. By pointing out that

Noah alone continued to be genetically pure, the biblical tale justifies the Deity's contradiction: Having just decided to wipe all life off the face of Earth, he decided to save Noah and his descendants and "every clean animal," and other beasts and fowls, "so as to keep seed alive upon the face of all the earth."

The Deity's plan to defeat his own initial purpose was to alert Noah to the coming catastrophe and guide him in the construction of a waterborne ark, which would carry the people and the creatures that were to be saved. The notice given to Noah was a mere seven days. Somehow, he managed to build the ark and waterproof it, collect all the creatures and put them and his family aboard, and provision the ark in the allotted time. "And it came to pass, after the seven days, that the waters of the Deluge were upon the earth." What came to pass is best described in the Bible's own words:

> On that day,
> all the fountains of the great deep burst open,
> and the sluices of the heavens were opened. . . .
> And the Deluge was forty days upon the Earth,
> and the waters increased, and bore up the ark,
> and it was lifted up above the earth.
> And the waters became stronger
> and greatly increased upon the earth,
> and the ark floated upon the waters.
> And the waters became exceedingly strong upon the
> earth and all the high mountains were covered.
> those that are under all the skies:
> fifteen cubits above them did the water prevail,
> and the mountains were covered.
> And all flesh perished. . . .
> Both man and cattle and creeping things
> and the birds of the skies
> were wiped off from the Earth;
> And Noah only was left,
> and that which were with him in the ark.

The waters prevailed upon Earth 150 days, when the Deity

caused a wind to pass upon the Earth,
and the waters were calmed.
And the fountains of the deep were dammed,
as were the sluices of the heavens;
and the rain from the skies was arrested.
And the waters began to go back from upon the Earth,
coming and going back.
And after one hundred and fifty days,
the waters were less;
and the ark rested on the Mounts of Ararat.

According to the biblical version, Mankind's ordeal began "in the six hundredth year of Noah's life, in the second month, on the seventeenth day of the month." The ark rested on the Mounts of Ararat "in the seventh month, on the seventeenth day of the month." The surge of the waters and their gradual "going back"—enough to lower the water level so that the ark rested on the peaks of Ararat—lasted, then, a full five months. Then "the waters continued to diminish, until the peaks of the mountains"—and not just the towering Ararats—"could be seen on the eleventh day of the tenth month," nearly three months later.

Noah waited another forty days. Then he sent out a raven and a dove "to see if the waters were abated from off the face of the ground." On the third try, the dove came back holding an olive leaf in her mouth, indicating that the waters had receded enough to enable treetops to be seen. After a while, Noah sent out the dove once more, "but she returned not again." The Deluge was over.

And Noah removed the covering of the Ark
and looked, and behold:
the face of the ground was dry.

"In the second month, on the twenty-seventh day of the month, did the earth dry up." It was the six hundred and first year of Noah. The ordeal had lasted a year and ten days.

Then Noah and all that were with him in the ark came out. And he built an altar and offered burnt sacrifices to the Deity.

And the Deity smelled the enticing smell
and said in his heart:
"I shall no longer curse the dry land
on account of the Earthling;
for his heart's desire is evil from his youth."

The "happy ending" is as full of contradictions as the Deluge story itself. It begins with a long indictment of Mankind for various abominations, including defilement of the purity of the younger gods. A momentous decision to have all flesh perish is reached and appears fully justified. Then the very same Deity rushes in a mere seven days to make sure that the seed of Mankind and other creatures shall not perish. When the trauma is over, the Deity is enticed by the smell of roasting meat and, forgetting his original determination to put an end to Mankind, dismisses the whole thing with an excuse, blaming Man's evil desires on his youth.

These nagging doubts of the story's veracity disperse, however, when we realize that the biblical account is an edited version of the original Sumerian account. As in the other instances, the monotheistic Bible has compressed into one Deity the roles played by several gods who were not always in accord.

Until the archaeological discoveries of the Mesopotamian civilization and the decipherment of the Akkadian and Sumerian literature, the biblical story of the Deluge stood alone, supported only by scattered primitive mythologies around the world. The discovery of the Akkadian "Epic of Gilgamesh" placed the Genesis Deluge tale in older and venerable company, further enhanced by later discoveries of older texts and fragments of the Sumerian original.

The hero of the Mesopotamian Deluge account was Ziusudra in Sumerian (Utnapishtim in Akkadian), who was taken after the Deluge to the Celestial Abode of the Gods to live there happily ever after. When, in his search for immortality, Gilgamesh finally reached the place, he sought Utnapishtim's advice on the subject of life and death. Utnapishtim disclosed to Gilgamesh—and through him to all post-Diluvial Mankind—the secret of his survival, "a hidden matter, a secret of the gods"—the true story (one might say) of the Great Flood.

The secret revealed by Utnapishtim was that before the onslaught of the Deluge the gods held a council and voted on the destruction of Mankind. The vote and the decision were kept secret. But Enki searched out Utnapishtim, the ruler of Shuruppak, to inform him of the approaching calamity. Adopting clandestine methods, Enki spoke to Utnapishtim from behind a reed screen. At first his disclosures were cryptic. Then his warning and advice were clearly stated:

> Man of Shuruppak, son of Ubar-Tutu:
> Tear down the house, build a ship!
> Give up possessions, seek thou life!
> Forswear belongings, keep soul alive!
> Aboard ship take thou the seed of all living things;
> That ship thou shalt build—
> her dimensions shall be to measure.

The parallels with the biblical story are obvious: A Deluge is about to come; one Man is forewarned; he is to save himself by preparing a specially constructed boat; he is to take with him and save "the seed of all living things." Yet the Babylonian version is more plausible. The decision to destroy and the effort to save are not contradictory acts of the same single Deity, but the acts of different deities. Moreover, the decision to forewarn and save the seed of Man is the defiant act of one god (Enki), acting in secret and contrary to the joint decision of the other Great Gods.

Why did Enki risk defying the other gods? Was he solely concerned with the preservation of *his* "wondrous works of art," or did he act against the background of a rising rivalry or enmity between him and his brother Enlil?

The existence of such a conflict between the two brothers is highlighted in the Deluge story.

Utnapishtim asked Enki the obvious question: How could he, Utnapishtim, explain to the other citizens of Shuruppak the construction of an oddly shaped vessel and the abandonment of all possessions? Enki advised him:

> Thou shalt thus speak unto them:
> "I have learnt that Enlil is hostile to me,

> so that I cannot reside in your city,
> nor set my foot in Enlil's territory.
> To the Apsu I will therefore go down,
> to dwell with my Lord Ea."

The excuse was thus to be that, as Enki's follower, Utnapishtim could no longer dwell in Mesopotamia, and that he was building a boat in which he intended to sail to the Lower World (southern Africa, by our findings) to dwell there with his Lord, Ea/Enki. Verses that follow suggest that the area was suffering from a drought or a famine; Utnapishtim (on Enki's advice) was to assure the residents of the city that if Enlil saw him depart, "the land shall [again] have its fill of harvest riches." This excuse made sense to the other residents of the city.

Thus misled, the people of the city did not question, but actually lent a hand in, the construction of the ark. By killing and serving them bullocks and sheep "every day" and by lavishing upon them "must, red wine, oil and white wine," Utnapishtim encouraged them to work faster. Even children were pressed to carry bitumen for waterproofing.

"On the seventh day the ship was completed. The launching was very difficult, so they had to shift the floor planks above and below, until two-thirds of the structure had gone into the water" of the Euphrates. Then Utnapishtim put all his family and kin aboard the ship, taking along "whatever I had of all the living creatures" as well as "the animals of the field, the wild beasts of the field." The parallels with the biblical tale—even down to the seven days of construction—are clear. Going a step beyond Noah, however, Utnapishtim also sneaked aboard all the craftsmen who had helped him build the ship.

He himself was to go aboard only upon a certain signal, whose nature Enki had also revealed to him: a "stated time" to be set by Shamash, the deity in charge of the fiery rockets. This was Enki's order:

> "When Shamash who orders a trembling at dusk
> will shower down a rain of eruptions—
> board thou the ship, batten up the entrance!"

We are left guessing at the connection between this apparent firing of a space rocket by Shamash and the arrival of the moment for Utnapishtim to board his ark and seal himself inside it. But the moment did arrive; the space rocket did cause a "trembling at dusk"; there was a shower of eruptions. And Utnapishtim "battened down the whole ship" and "handed over the structure together with its contents" to "Puzur-Amurri, the Boatman."

The storm came "with the first glow of dawn." There was awesome thunder. A black cloud rose up from the horizon. The storm tore out the posts of buildings and piers; then the dikes gave. Darkness followed, "turning to blackness all that had been light;" and "the wide land was shattered like a pot."

For six days and six nights the "south-storm" blew.

> Gathering speed as it blew,
> submerging the mountains,
> overtaking the people like a battle. . . .
> When the seventh day arrived,
> the flood-carrying south-storm
> subsided in the battle
> which it had fought like an army.
> The sea grew quiet,
> the tempest was still,
> the flood ceased.
> I looked at the weather.
> Stillness had set in.
> And all of Mankind had returned to clay.

The will of Enlil and the Assembly of Gods was done.

But, unknown to them, the scheme of Enki had also worked: Floating in the stormy waters was a vessel carrying men, women, children, and other living creatures.

With the storm over, Utnapishtim "opened a hatch; light fell upon my face." He looked around; "the landscape was as level as a flat roof." Bowing low, he sat and wept, "tears running down on my face." He looked about for a coastline in the expanse of the sea; he saw none. Then:

There emerged a mountain region;
On the Mount of Salvation the ship came to a halt;
Mount *Niṣir* ["salvation"] held the ship fast,
allowing no motion.

For six days Utnapishtim watched from the motionless ark, caught in the peaks of the Mount of Salvation—the biblical peaks of Ararat. Then, like Noah, he sent out a dove to look for a resting place, but it came back. A swallow flew out and came back. Then a raven was set free—and flew off, finding a resting place. Utnapishtim then released all the birds and animals that were with him, and stepped out himself. He built an altar "and offered a sacrifice"—just as Noah had.

But here again the single-Deity–multideity difference crops up. When Noah offered a burnt sacrifice, "Yahweh smelled the enticing smell"; but when Utnapishtim offered a sacrifice, "the gods smelled the savor, the gods smelled the sweet savor. The gods crowded like flies about a sacrificer."

In the Genesis version, it was Yahweh who vowed never again to destroy Mankind. In the Babylonian version it was the Great Goddess who vowed: "I shall not forget. . . . I shall be mindful of these days, forgetting them never."

That, however, was not the immediate problem. For when Enlil finally arrived on the scene, he had little mind for food. He was hopping mad to discover that some had survived. "Has some living soul escaped? No man was to survive the destruction!"

Ninurta, his son and heir, immediately pointed a suspecting finger at Enki. "Who, other than Ea, can devise plans? It is Ea alone who knows every matter." Far from denying the charge, Enki launched one of the world's most eloquent defense summations. Praising Enlil for his own wisdom, and suggesting that Enlil could not possibly be "unreasoning"—a realist—Enki mixed denial with confession. "It was not I who disclosed the secret of the gods"; I merely let one Man, an "exceedingly wise" one, perceive by his own wisdom what the gods' secret was. And if indeed this Earthling is so wise, Enki suggested to Enlil, let's not ignore his abilities. "Now then, take counsel in regard to him!"

All this, the "Epic of Gilgamesh" relates, was the "secret of the gods" that Utnapishtim told Gilgamesh. He then told Gilgamesh of the final event. Having been influenced by Enki's argument,

> Enlil thereupon went aboard the ship.
> Holding me by the hand, he took me aboard.
> He took my wife aboard,
> made her kneel by my side.
> Standing between us,
> he touched our foreheads to bless us:
> "Hitherto Utnapishtim has been but human;
> henceforth Utnapishtim and his wife
> shall be unto us like gods.
> Utnapishtim shall reside in the Far Away,
> at the Mouth of the Waters!"

And Utnapishtim concluded his story to Gilgamesh. After he was taken to reside in the Far Away, Anu and Enlil

> Gave him life, like a god,
> Elevated him to eternal life, like a god.

But what happened to Mankind in general? The biblical tale ends with an assertion that the Deity then permitted and blessed Mankind to "be fruitful and multiply." Mesopotamian versions of the Deluge story also end with verses that deal with Mankind's procreation. The partly mutilated texts speak of the establishment of human "categories":

> . . . Let there be a third category among the Humans:
> Let there be among the Humans
> Women who bear, and women who do not bear.

There were, apparently, new guidelines for sexual intercourse:

> Regulations for the human race:
> Let the male . . . to the young maiden. . . .
> Let the young maiden. . . .

The young man to the young maiden . . .
When the bed is laid,
let the spouse and her husband lie together.

Enlil was outmaneuvered. Mankind was saved and allowed to procreate. The gods opened up Earth to Man.

14

·

WHEN THE GODS
FLED FROM EARTH

WHAT WAS THIS DELUGE, whose raging waters swept over Earth?

Some explain the Flood in terms of the annual inundations of the Tigris–Euphrates plain. One such inundation, it is surmised, must have been particularly severe. Fields and cities, men and beasts were swept away by the rising waters; and primitive peoples, seeing the event as a punishment by the gods, began to propagate the legend of a Deluge.

In one of his books, *Excavations at Ur,* Sir Leonard Woolley relates how, in 1929, as the work on the Royal Cemetery at Ur was drawing to a close, the workmen sank a small shaft at a nearby mound, digging through a mass of broken pottery and crumbled brick. Three feet down, they reached a level of hard-packed mud—usually soil marking the point where civilization had started. But could the millennia of urban life have left only three feet of archaeological strata? Sir Leonard directed the workmen to dig farther. They went down another three feet, then another five. They still brought up "virgin soil"—mud with no traces of human habitation. But after digging through eleven feet of silted, dry mud, the workmen reached a stratum containing pieces of broken green pottery and flint instruments. An earlier civilization had been buried under eleven feet of mud!

Sir Leonard jumped into the pit and examined the excavation. He called in his aides, seeking their opinions. No one had a plausible theory. Then Sir Leonard's wife remarked almost casually, "Well, of course, it's the Flood!"

Other archaeological delegations to Mesopotamia, however, cast doubt on this marvelous intuition. The stratum of mud containing no traces of habitation did indicate flooding; but while the deposits of Ur and al-'Ubaid suggested flooding sometime between 3500 and 4000 B.C., a similar deposit uncovered later at Kish was estimated to have occurred circa 2800 B.C. The same date (2800 B.C.) was estimated for mud strata found at Erech and at Shuruppak, the city of the Sumerian Noah. At Nineveh, excavators found, at a depth of some sixty feet, no less than thirteen alternate strata of mud and riverine sand, dating from 4000 to 3000 B.C.

Most scholars, therefore, believe that what Woolley found were traces of diverse local floodings—frequent occurrences in Mesopotamia, where occasional torrential rains and the swelling of the two great rivers and their frequent course changes cause such havoc. All the varying mud strata, scholars have concluded, were not the comprehensive calamity, the monumental prehistoric event that the Deluge must have been.

The Old Testament is a masterpiece of literary brevity and precision. The words are always well chosen to convey precise meanings; the verses are to the point; their order is purposeful; their length is no more than is absolutely needed. It is noteworthy that the whole story from Creation through the expulsion of Adam and Eve from the Garden of Eden is told in eighty verses. The complete record of Adam and his line, even when told separately for Cain and his line and Seth, Enosh, and their line, is managed in fifty-eight verses. But the story of the Great Flood merited no less than eighty-seven verses. It was, by any editorial standard, a "major story." No mere local event, it was a catastrophe affecting the whole of Earth, the whole of Mankind. The Mesopotamian texts clearly state that the "four corners of the Earth" were affected.

As such, it was a crucial point in the prehistory of Mesopotamia. There were the events and the cities and the people *before* the Deluge, and the events and cities and people *after* the Deluge. There were all the deeds of the gods and the Kingship that they lowered from Heaven *before* the Great Flood, and the course of godly and human events when Kingship was lowered again to Earth *after* the Great Flood. It was the great time divider.

Not only the comprehensive king lists but also texts relating to individual kings and their ancestries made mention of the Deluge. One, for example, pertaining to Ur-Ninurta, recalled the Deluge as an event remote in time:

> On that day, on that remote day,
> On that night, on that remote night,
> In that year, in that remote year—
> When the Deluge had taken place.

The Assyrian king Ashurbanipal, a patron of the sciences who amassed the huge library of clay tablets in Nineveh, professed in one of his commemorative inscriptions that he had found and was able to read "stone inscriptions from before the Deluge." An Akkadian text dealing with names and their origins explains that it lists names "of kings from after the Deluge." A king was exalted as "of seed preserved from before the Deluge." Various scientific texts quoted as their source "the olden sages, from before the Deluge."

No, the Deluge was no local occurrence or periodic inundation. It was by all counts an Earthshaking event of unparalleled magnitude, a catastrophe the likes of which neither Man nor gods experienced before or since.

•

The biblical and Mesopotamian texts that we have examined so far leave a few puzzles to be solved. What was the ordeal suffered by Mankind, in respect to which Noah was named "Respite" with the hope that his birth signaled an end to the hardships? What was the "secret" the gods swore to keep, and of whose disclosure Enki was accused? Why was the launching of a space vehicle from Sippar the signal to Utnapishtim to enter and seal the ark? Where were the gods while the waters covered even the highest mountains? And why did they so cherish the roasted meat sacrifice offered by Noah/Utnapishtim?

As we proceed to find the answers to these and other questions, we shall find that the Deluge was not a premeditated punishment brought about by the gods at their exclusive will. We shall discover that though the Deluge was a predictable event, it was an unavoidable one, a natural calamity in which the gods played not an active but a passive role. We will also show that the secret the gods swore to was a conspiracy against Mankind—to withhold from the Earthlings the information they had regarding the coming avalanche of water so that, while the Nefilim saved themselves, Mankind should perish.

Much of our greatly increased knowledge of the Deluge and the events preceding it comes from the text "When the gods as men." In it the hero of the Deluge is called Atra-Hasis. In the Deluge segment of the "Epic of Gilgamesh," Enki called Utnapishtim "the exceedingly wise"— which in Akkadian is *atra-hasis*.

Scholars theorized that the texts in which Atra-Hasis is the hero might be parts of an earlier, Sumerian Deluge story. In time, enough Babylonian, Assyrian, Canaanite, and even original Sumerian tablets were discovered to enable a major reassembly of the Atra-Hasis epic, a masterful work credited primarily to W. G. Lambert and A. R. Millard *(Atra-Hasis: The Babylonian Story of the Flood)*.

After describing the hard work of the Anunnaki, their mutiny, and the ensuing creation of the Primitive Worker, the epic relates how Man (as we also know from the biblical version) began to procreate and multiply. In time, Mankind began to upset Enlil.

The land extended, the people multiplied;
In the land like wild bulls they lay.
The god got disturbed by their conjugations;
The god Enlil heard their pronouncements,
and said the great gods:
"Oppressive have become the pronouncements of Mankind;
Their conjugations deprive me of sleep."

Enlil—once again cast as the prosecutor of Mankind— then ordered a punishment. We would expect to read now of the coming Deluge. But not so. Surprisingly, Enlil did not even mention a Deluge or any similar watery ordeal. Instead, he called for the decimation of Mankind through pestilence and sicknesses.

The Akkadian and Assyrian versions of the epic speak of "aches, dizziness, chills, fever" as well as "disease, sickness, plague, and pestilence" afflicting Mankind and its livestock following Enlil's call for punishment. But Enlil's scheme did not work. The "one who was exceedingly wise"—Atra-Ḥasis—happened to be especially close to the god Enki. Telling his own story in some of the versions, he says, "I am Atra-Ḥasis; I lived in the temple of Ea my lord." With "his mind alert to his Lord Enki," Atra-Ḥasis appealed to him to undo his brother Enlil's plan:

> "Ea, O Lord, Mankind groans;
> the anger of the gods consumes the land.
> Yet it is thou who hast created us!
> Let there cease the aches, the dizziness,
> the chills, the fever!"

Until more pieces of the broken-off tablets are found, we shall not know what Enki's advice was. He said of something, ". . . let there appear in the land." Whatever it was, it worked. Soon thereafter, Enlil complained bitterly to the gods that "the people have not diminished; they are more numerous than before!"

He then proceeded to outline the extermination of Mankind through starvation. "Let supplies be cut off from the people; in their bellies, let fruit and vegetables be wanting!" The famine was to be achieved through natural forces, by a lack of rain and failing irrigation.

> Let the rains of the rain god be withheld from above;
> Below, let the waters not rise from their sources.
> Let the wind blow and parch the ground;
> Let the clouds thicken, but hold back the downpour.

Even the sources of seafood were to disappear: Enki was ordered to "draw the bolt, bar the sea," and "guard" its food away from the people.

Soon the drought began to spread devastation.

> From above, the heat was not. . . .
> Below, the waters did not rise from their sources.
> The womb of the earth did not bear;
> Vegetation did not sprout. . . .
> The black fields turned white;
> The broad plain was choked with salt.

The resulting famine caused havoc among the people. Conditions got worse as time went on. The Mesopotamian texts speak of six increasingly devastating *sha-at-tam*'s— a term that some translate as "years," but which literally means "passings," and, as the Assyrian version makes clear, "a year of Anu":

For one *sha-at-tam* they ate the earth's grass.
For the second *sha-at-tam* they suffered the vengeance.
The third *sha-at-tam* came;
their features were altered by hunger,
their faces were encrusted . . .
they were living on the verge of death.
When the fourth *sha-at-tam* arrived,
their faces appeared green;
they walked hunched in the streets;
their broad [shoulders?] became narrow.

By the fifth "passing," human life began to deteriorate. Mothers barred their doors to their own starving daughters. Daughters spied on their mothers to see whether they had hidden any food.

By the sixth "passing," cannibalism was rampant.

When the sixth *sha-at-tam* arrived
they prepared the daughter for a meal;
the child they prepared for food. . . .
One house devoured the other.

The texts report the persistent intercession by Atra-Ḥasis with his god Enki. "In the house of his god . . . he set foot; . . . every day he wept, bringing oblations in the morning . . . he called by the name of his god," seeking Enki's help to avert the famine.

Enki, however, must have felt bound by the decision of the other deities, for at first he did not respond. Quite possibly, he even hid from his faithful worshiper by leaving the temple and sailing into his beloved marshlands. "When the people were living on the edge of death," Atra-Ḥasis "placed his bed facing the river." But there was no response.

The sight of a starving, disintegrating Mankind, of parents eating their own children, finally brought about the unavoidable: another confrontation between Enki and Enlil. In the seventh "passing," when the remaining men and women were "like ghosts of the dead," they received a message from Enki. "Make a loud noise in the land," he said. Send out heralds to command all the people: "Do not revere your gods, do not pray to your goddesses." There was to be total disobedience!

Under the cover of such turmoil, Enki planned more concrete action. The texts, quite fragmented at this point, disclose that he convened a secret assembly of "elders" in his temple. "They entered . . . they took counsel in the House of Enki." First Enki exonerated himself, telling them how he had opposed the acts of the other gods. Then he outlined a plan of action; it somehow involved his command of the seas and the Lower World.

We can glean the clandestine details of the plan from the fragmentary verses: "In the night . . . after he . . ." someone had to be "by the bank of the river" at a certain time, perhaps to await the return of Enki from the Lower World. From there Enki "brought the water warriors"— perhaps also some of the Earthlings who were Primitive Workers in the mines. At the appointed time, commands were shouted: "Go! . . . the order . . ."

In spite of missing lines, we can gather what had happened from the reaction of Enlil. "He was filled with anger." He summoned the Assembly of the Gods and sent his sergeant at arms to fetch Enki. Then he stood up and accused his brother of breaking the surveillance-and-containment plans:

> All of us, Great Anunnaki,
> reached together a decision. . . .
> I commanded that in the Bird of Heaven
> Adad should guard the upper regions;
> that Sin and Nergal should guard
> the Earth's middle regions;
> that the bolt, the bar of the sea,
> you [Enki] should guard with your rockets.
> But you let loose provisions for the people!

Enlil accused his brother of breaking the "bolt to the sea." But Enki denied that it had happened with his consent:

> The bolt, the bar of the sea,
> I did guard with my rockets.
> [But] when . . . escaped from me . . .
> a myriad of fish . . . it disappeared;
> they broke off the bolt . . .
> they had killed the guards of the sea.

He claimed that he had caught the culprits and punished them, but Enlil was not satisfied. He demanded that Enki "stop feeding his people," that he no longer "supply corn rations on which the people thrive." The reaction of Enki was astounding:

> The god got fed up with the sitting;
> in the Assembly of the Gods,
> laughter overcame him.

We can imagine the pandemonium. Enlil was furious. There were heated exchanges with Enki and shouting. "There is slander in his hand!" When the Assembly was finally called to order, Enlil took the floor again. He reminded his colleagues and subordinates that it had been a unanimous decision. He reviewed the events that led to the fashioning of the Primitive Worker and recalled the many times that Enki "broke the rule."

But, he said, there was still a chance to doom Mankind. A "killing flood" was in the offing. The approaching catastrophe had to be kept a secret from the people. He called on the Assembly to swear themselves to secrecy and, most important, to "bind prince Enki by an oath."

> Enlil opened his mouth to speak
> and addressed the Assembly of all the gods:
> "Come, all of us, and take an oath
> regarding the Killing Flood!"
> Anu swore first;
> Enlil swore; his sons swore with him.

At first, Enki refused to take the oath. "Why will you bind me with an oath?" he asked. "Am I to raise my hands against my own humans?" But he was finally forced to take the oath. One of the texts specifically states: "Anu, Enlil, Enki, and Ninḫursag, the gods of Heaven and Earth, had taken the oath."

The die was cast.

•

What was the oath he was bound by? As Enki chose to interpret it, he swore not to reveal the secret of the coming Deluge to the people; but could he not tell it to a wall? Calling Atra-Ḫasis to the temple, he made him stay behind a screen. Then Enki pretended to speak not to his devout Earthling but to the wall. "Reed screen," he said,

> Pay attention to my instructions.
> On all the habitations, over the cities,
> a storm will sweep.
> The destruction of Mankind's seed it will be. . . .
> This is the final ruling,
> the word of the Assembly of the gods,
> the word spoken by Anu, Enlil and Ninḫursag.

(This subterfuge explains Enki's later contention, when the survival of Noah/Utnapishtim was discovered, that he had not broken his oath—that the "exceedingly wise" [atra-ḫasis] Earthling had found out the secret of the Deluge all by himself, by correctly interpreting the signs.) Pertinent seal depictions show an attendant holding the screen while Ea—as the Serpent God—reveals the secret to Atra-Ḫasis. (Fig. 160)

Enki's advice to his faithful servant was to build a waterborne vessel; but when the latter said, "I have never built a boat . . . draw for me a design on the ground that I may see," Enki provided him with precise instructions regarding the boat, its measurements, and its construction. Steeped in Bible stories, we imagine this "ark" as a very large boat, with decks and superstructures. But the biblical term—teba—stems from the root "sunken," and it must be concluded that Enki instructed his Noah to construct a submersible boat—a submarine.

Fig. 160

The Akkadian text quotes Enki as calling for a boat "roofed over and below," hermetically sealed with "tough pitch." There were to be no decks, no openings, "so that the sun shall not see inside." It was to be a boat "like an Apsu boat," a *sulili*; it is the very term used nowadays in Hebrew (*soleleth*) to denote a submarine.

"Let the boat," Enki said, "be a MA.GUR.GUR"—"a boat that can turn and tumble." Indeed, only such a boat could have survived an overpowering avalanche of waters.

The Atra-Hasis version, like the others, reiterates that although the calamity was only seven days away, the people were unaware of its approach. Atra-Hasis used the excuse that the "Apsu vessel" was being built so that he could leave for Enki's abode and perhaps thereby avert Enlil's anger. This was readily accepted, for things were really bad. Noah's father had hoped that his birth signaled the end of a long time of suffering. The people's problem was a drought—the absence of rain, the shortage of water. Who in his right mind would have thought that they were about to perish in an avalanche of water?

Yet if the humans could not read the signs, the Nefilim could. To them, the Deluge was not a sudden event; though it was unavoidable, they detected its coming. Their scheme to destroy Mankind rested not on an active but on a passive role by the gods. They did not cause the Deluge; they simply connived to withhold from the Earthlings the fact of its coming.

Aware, however, of the impending calamity, and of its global impact, the Nefilim took steps to save their own skins. With Earth about to be engulfed by water, they could go in only one direction for protection: skyward. When the storm that preceded the Deluge began to blow, the Nefilim took to their shuttlecraft, and remained in Earth orbit until the waters began to subside.

The day of the Deluge, we will show, was the day the gods fled from Earth.

The sign for which Utnapishtim had to watch, upon which he was to join all other in the ark and seal it, was this:

> When Shamash,
> who orders a trembling at dusk,
> will shower down a rain of eruptions—
> board thou the ship,
> batten up the entrance!

Shamash, as we know, was in charge of the spaceport at Sippar. There is no doubt in our mind that Enki instructed Utnapishtim to watch for the first sign of space launchings at Sippar. Shuruppak, where Utnapishtim lived, was only 18 *beru* (some 180 kilometers, or 112 miles) south of Sippar. Since the launchings were to take place at dusk, there would be no problem in seeing the "rain of eruptions" that the rising rocket ships would "shower down."

Though the Nefilim were prepared for the Deluge, its coming was a frightening experience: "The noise of the Deluge . . . set the gods trembling." But when the moment to leave Earth arrived, the gods, "shrinking back, ascended to the heavens of Anu." The Assyrian version of Atra-Ḥasis speaks of the gods using *rukub ilani* ("chariot of the gods") to escape from Earth. "The Anunnaki lifted up," their rocketships, like torches, "setting the land ablaze with their glare."

Orbiting Earth, the Nefilim saw a scene of destruction that affected them deeply. The Gilgamesh texts tell us that, as the storm grew in intensity, not only "could no one see his fellow," but "neither could the people be recognized from the heavens." Crammed into their spacecraft, the gods strained to see what was happening on the planet from which they had just blasted off.

> The gods cowered like dogs,
> crouched against the outer wall.
> Ishtar cried out like a woman in travail:
> "The olden days are alas turned to clay." . . .
> The Anunnaki gods weep with her.
> The gods, all humbled, sit and weep;
> their lips drawn tight . . . one and all.

The Atra-Hasis texts echo the same theme. The gods, fleeing, were watching the destruction at the same time. But the situation within their own vessels was not very encouraging, either. Apparently, they were divided among several spaceships; Tablet III of the Atra-Hasis epic describes the conditions on board one where some of the Anunnaki shared accommodations with the Mother Goddess.

> The Anunnaki, great gods,
> were sitting in thirst, in hunger. . . .
> Ninti wept and spent her emotion;
> she wept and eased her feelings.
> The gods wept with her for the land.
> She was overcome with grief,
> she thirsted for beer.
> Where she sat, the gods sat weeping;
> crouching like sheep at a trough.
> Their lips were feverish of thirst,
> they were suffering cramp from hunger.

The Mother Goddess herself, Ninhursag, was shocked by the utter devastation. She bewailed what she was seeing:

> The Goddess saw and she wept . . .
> her lips were covered with feverishness. . . .
> "My creatures have become like flies—
> they filled the rivers like dragonflies,
> their fatherhood was taken by the rolling sea."

Could she, indeed, save her own life while Mankind, which she helped create, was dying? Could she really leave the Earth, she asked aloud—

> "Shall I ascend up to Heaven,
> to reside in the House of Offerings,
> where Anu, the Lord, had ordered to go?"

The orders to the Nefilim became clear: Abandon Earth, "ascend up to Heaven." It was a time when the Twelfth Planet was nearest Earth, within the asteroid belt ("Heaven"), as evidenced by the fact that Anu was able to attend personally the crucial conferences shortly before the Deluge.

Enlil and Ninurta—accompanied perhaps by the elite of the Anunnaki, those who had manned Nippur—were in one spacecraft, planning, no doubt, to rejoin the main spaceship. But the other gods were not so determined. Forced to abandon Earth, they suddenly realized how attached they had become to it and its inhabitants. In one craft, Ninhursag and her group of Anunnaki debated the merits of the orders given by Anu. In another, Ishtar cried out: "The olden days, alas, are turned into clay"; the Anunnaki who were in her craft "wept with her."

Enki was obviously in yet another spacecraft, or else he would have disclosed to the others that he had managed to save the seed of Mankind. No doubt he had other reasons to feel less gloomy, for the evidence suggests that he had also planned the encounter at Ararat.

The ancient versions appear to imply that the ark was simply carried to the region of Ararat by the torrential waves; and a "south-storm" would indeed drive the boat northward. But the Mesopotamian texts reiterate that Atra-Ḥasis/Utnapishtim took along with him a "Boatman" named Puzur-Amurri ("westerner who knows the secrets"). To him the Mesopotamian Noah "handed over the structure, together with its contents," as soon as the storm started. Why was an experienced navigator needed, unless it was to bring the ark to a specific destination?

The Nefilim, as we have shown, used the peaks of Ararat as landmarks from the very beginning. As the highest peaks in that part of the world, they could be expected to re-appear first from under the mantle of water. Since Enki, "The Wise One, the All-Knowing," certainly could figure that much out, we can surmise that he had instructed his servant to guide the ark toward Ararat, planning the encounter from the very beginning.

Berossus's version of the Flood, as reported by the Greek Abydenus, relates: "Kronos revealed to Sisithros that there would be a Deluge on the fifteenth day of Daisios [the second month], and ordered him to conceal in Sippar, the city of Shamash, every available writing. Sisithros accomplished all these things, sailed immediately to Armenia, and thereupon what the god had announced did happen."

Berossus repeats the details regarding the release of the birds. When Sisithros (which is *atra-asis* reversed) was taken by the gods to their abode, he explained to the other people in the ark that they were "in Armenia" and directed them back (on foot) to Babylonia. We find in this version not only the tie-in with Sippar, the spaceport, but also confirmation that Sisithros was instructed to "sail immediately to Armenia"—to the land of Ararat.

As soon as Atra-Ḫasis had landed, he slaughtered some animals and roasted them on a fire. No wonder that the exhausted and hungry gods "gathered like flies over the offering." Suddenly they realized that Man and the food he grew and the cattle he raised were essential. "When at length Enlil arrived and saw the ark, he was wroth." But the logic of the situation and Enki's persuasion prevailed; Enlil made his peace with the remnants of Mankind and took Atra-Ḫasis/Utnapishtim in his craft up to the Eternal Abode of the Gods.

Another factor in the quick decision to make peace with Mankind may have been the progressive abatement of the Flood and the reemergence of dry land and the vegetation upon it. We have already concluded that the Nefilim became aware ahead of time of the approaching calamity; but it was so unique in their experience that they feared that Earth would become uninhabitable forever. As they landed on Ararat, they saw that this was not so. Earth was still habitable, and to live on it, they needed man.

•

What was this catastrophe—predictable yet unavoidable? An important key to unlocking the puzzle of the Deluge is the realization that it was not a single, sudden event, but the climax of a chain of events.

Unusual pestilences affecting man and beast and a severe drought preceded the ordeal by water—a process that lasted, according to the Mesopotamian sources, seven "passings," or *sar*'s. These phenomena could be accounted for only by major climatic changes. Such changes have been associated in Earth's past with the recurring ice ages and interglacial stages that had dominated Earth's immediate past. Reduced precipitation, falling sea and lake levels, and the drying up of subterranean water sources have been the hallmarks of an approaching ice age. Since the Deluge that abruptly ended those conditions was followed by the Sumerian civilization and our own present, postglacial age, the glaciation in question could only have been the last one.

Our conclusion is that the events of the Deluge relate to Earth's last ice age and its catastrophic ending.

Drilling into the Arctic and Antarctic ice sheets, scientists have been able to measure the oxygen trapped in the various layers, and to judge from that the climate that prevailed millennia ago. Core samples from the bottoms of the seas, such as the Gulf of Mexico, measuring the proliferation or dwindling of marine life, likewise enable them to estimate temperatures in ages past. Based on such findings, scientists are now certain that the last ice age began some 75,000 years ago and underwent a mini-warming some 40,000 years ago. Circa 38,000 years ago, a harsher, colder, and drier period ensued. And then, about 13,000 years ago, the ice age abruptly ended, and our present mild climate was ushered in.

Aligning the biblical and Sumerian information, we find that the harsh times, the "accursation of Earth," began in the time of Noah's father Lamech. His hopes that the birth of Noah ("respite") would mark the end of the hardships was fulfilled in an unexpected way, through the catastrophic Deluge.

Many scholars believe that the ten biblical pre-Diluvial patriarchs (Adam to Noah) somehow parallel the ten pre-Diluvial rulers of the Sumerian king lists. These lists do not apply to divine titles DIN.GIR or EN to the last two of the ten, and treat Ziusudra/Utnapishtim and his father Ubar-Tutu as *men*. The latter two parallel Noah and his father Lamech; and according to the Sumerian lists, the two reigned a combined total of 64,800 years until the Deluge occurred. The last ice age, from 75,000 to 13,000 years ago, lasted 62,000 years. Since the hardships began when Ubartutu/Lamech was already reigning, the 62,000 fit perfectly into the 64,800.

Moreover, the extremely harsh conditions lasted, according to the Atra-Hasis epic, seven *shar's*, or 25,200 years. The scientists discovered evidence of an extremely harsh period from circa 38,000 to 13,000 years ago—a span of 25,000 years. Once again, the Mesopotamian evidence and modern scientific findings corroborate each other.

Our endeavor to unravel the puzzle of the Deluge, then, focuses on Earth's climatic changes, and in particular the abrupt collapse of the ice age some 13,000 years ago.

What could have caused a sudden climatic change of such magnitude?

Of the many theories advanced by the scientists, we are intrigued by the one suggested by Dr. John T. Hollin of the University of Maine. He contended that the Antarctic ice sheet periodically breaks loose and slips into the sea, creating an abrupt and enormous tidal wave!

This hypothesis—accepted and elaborated upon by others—suggests that as the ice sheet grew thicker and thicker, it not only trapped more of Earth's heat beneath the ice sheet but also created (by pressure and friction) a slushy, slippery layer at its bottom. Acting as a lubricant between the thick ice sheet above and the solid earth below, this slushy layer sooner or later caused the ice sheet to slide into the surrounding ocean.

Hollin calculated that if only half the present ice sheet of Antarctica (which is, on the average, more than a mile in thickness) were to slip into the southern seas, the immense tidal wave that would follow would raise the level of all the seas around the globe by some sixty feet, inundating coastal cities and lowlands.

In 1964, A. T. Wilson of Victoria University in New Zealand offered the theory that ice ages ended abruptly in such slippages, not only in the Antarctic but also in the Arctic. We feel that the various texts and facts gathered by us justify a conclusion that the Deluge was the result of such a slippage into the Antarctic waters of billions of tons of ice, bringing an abrupt end to the last ice age.

The sudden event triggered an immense tidal wave. Starting in Antarctic waters, it spread northward toward the Atlantic, Pacific, and Indian oceans. The abrupt change in temperature must have created violent storms accompanied by torrents of rain. Moving faster than the waters, the storms, clouds, and darkened skies heralded the avalanche of waters.

Exactly such phenomena are described in the ancient texts.

As commanded by Enki, Atra-Ḥasis sent everybody aboard the ark while he himself stayed outside to await the signal for boarding the vessel and sealing it off. Providing a "human-interest" detail, the ancient text tells us that Atra-Ḥasis, though ordered to stay outside the vessel, "was in and out; he could not sit, could not crouch . . . his heart was broken; he was vomiting gall." But then:

> . . . the Moon disappeared. . . .
> The appearance of the weather changed;
> The rains roared in the clouds. . . .
> The winds became savage . . .
> . . . the Deluge set out,
> its might came upon the people like a battle;
> One person did not see another,
> they were not recognizable in the destruction.
> The Deluge bellowed like a bull;
> The winds whinnied like a wild ass.
> The darkness was dense;
> The Sun could not be seen.

The "Epic of Gilgamesh" is specific about the direction from which the storm came: It came from the south. Clouds, winds, rain, and darkness indeed preceded the tidal wave which first tore down the "posts of Nergal" in the Lower World:

> With the glow of dawn
> a black cloud arose from the horizon;
> inside it the god of storms thundered. . . .
> Everything that had been bright
> turned to blackness. . . .
> For one day the south storm blew,

gathering speed as it blew, submerging the mountains. . . .
Six days and six nights blows the wind
as the South Storm sweeps the land.
When the seventh day arrived,
the Deluge of the South Storm subsided.

The references to the "south storm," "south wind" clearly
indicate the direction from which the Deluge arrived, its
clouds and winds, the "heralds of the storm," moving "over
hill and plain" to reach Mesopotamia. Indeed, a storm and
an avalanche of water originating in the Antarctic would
reach Mesopotamia via the Indian Ocean after first en-
gulfing the hills of Arabia, then inundating the Tigris–
Euphrates plain. The "Epic of Gilgamesh" also informs
us that before the people and their land were submerged,
the "dams of the dry land" and its dikes were "torn out":
the continental coastlines were overwhelmed and swept
over.

The biblical version of the Deluge story reports that the
"bursting of the fountains of the Great Deep" preceded
the "opening of the sluices of heaven." First, the waters
of the "Great Deep" (what a descriptive name for the
southernmost, frozen Antarctic seas) broke loose out of
their icy confinement; only then did the rains begin to pour
from the skies. This confirmation of our understanding of
the Deluge is repeated, in reverse, when the Deluge subsided. First the "Fountains of the Deep [were] dammed";
then the rain "was arrested from the skies."

After the first immense tidal wave, its waters were still
"coming and going back" in huge waves. Then the waters
began "going back," and "they were less" after 150 days,
when the ark came to rest between the peaks of Ararat.
The avalanche of water, having come from the southern
seas, went back to the southern seas.

•

How could the Nefilim predict *when* the Deluge would
burst out of Antarctica?

The Mesopotamian texts, we know, related the Deluge
and the climatic changes preceding it to seven "passings"
—undoubtedly meaning the periodic passage of the
Twelfth Planet in Earth's vicinity. We know that even the
Moon, Earth's small satellite, exerts sufficient gravitational

pull to cause the tides. Both Mesopotamian and biblical texts described how the Earth shook when the Celestial Lord passed in Earth's vicinity. Could it be that the Nefilim, observing the climatic changes and the instability of the Antarctic ice sheet, realized that the next, seventh "passing" of the Twelfth Planet would trigger the impending catastrophe?

Ancient texts show that it was so.

The most remarkable of these is a text of some thirty lines inscribed in miniature cuneiform writing on both sides of a clay tablet less than one inch long. It was unearthed at Ashur, but the profusion of Sumerian words in the Akkadian text leaves no doubt as to its Sumerian origin. Dr. Erich Ebeling determined that it was a hymn recited in the House of the Dead, and he therefore included the text in his masterwork *(Tod und Leben)* on death and resurrection in ancient Mesopotamia.

On close examination, however, we find that the composition "called on the names" of the Celestial Lord, the Twelfth Planet. It elaborates the meaning of the various epithets by relating them to the passage of the planet at the site of the battle with Tiamat—a passage that causes the Deluge!

The text begins by announcing that, for all its might and size, the planet ("the hero") nevertheless orbits the Sun. The Deluge was the "weapon" of this planet.

> His weapon is the Deluge;
> God whose Weapon brings death to the wicked.
> Supreme, Supreme, Anointed . . .
> Who like the Sun, the lands crosses;
> The Sun, his god, he frightens.

Calling out the "first name" of the planet—which, unfortunately, is illegible—the text describes the passage near Jupiter, toward the site of the battle with Tiamat:

> First Name: . . .
> Who the circular band hammered together;
> Who the Occupier split in two, poured her out.
> Lord, who at Akiti time
> Within Tiamat's battle place reposes. . . .

Whose seed are the sons of Babylon;
Who by the planet Jupiter cannot be distracted;
Who by his glow shall create.

Coming closer, the Twelfth Planet is called SHILIG.
LU.DIG ("powerful leader of the joyous planets"). It is
now nearest to Mars: "By the brilliance of the god [planet]
Anu god [planet] Laḫmu [Mars] is clothed." Then it
loosed the Deluge upon the Earth:

This is the name of the Lord
Who from the second month to the month Addar
The waters had summoned forth.

The text's elaboration of the two names offers remarkable calendarial information. The Twelfth Planet passed
Jupiter and neared Earth "at *Akiti* time," when the Mesopotamian New Year began. By the second month it was
closest to Mars. Then, "from the second month to the month
Addar" (the twelfth month), it loosed the Deluge upon
Earth.

This is in perfect harmony with the biblical account,
which states that "the fountains of the great deep burst
open" on the seventeenth day of the second month. The
ark came to rest on Ararat in the seventh month; other dry
land was visible in the tenth month; and the Deluge was
over in the twelfth month—for it was on "the first day of
the first month" of the following years that Noah opened
the ark's hatch.

Shifting to the second phase of the Deluge, when the
waters began to subside, the text calls the planet SHUL.
PA.KUN.E.

Hero, Supervising Lord,
Who collects together the waters;
Who by gushing waters
The righteous and the wicked cleanses;
Who in the twin-peaked mountain
Arrested the. . . .
. . . fish, river, river; the flooding rested.
In the mountainland, on a tree, a bird rested.
Day which . . . said.

In spite of the illegibility of some damaged lines, the parallels with the biblical and other Mesopotamian Deluge tales is evident: The flooding had ceased, the ark was "arrested" on the twin-peaked mountain; the rivers began to flow again from the mountaintops and carry the waters back to the oceans; fish were seen; a bird was sent out from the ark. The ordeal was over.

The Twelfth Planet had passed its "crossing." It had neared Earth, and it began to move away, accompanied by its satellites:

> When the savant shall call out: "Flooding!"—
> It is the god *Nibiru* ["Planet of Crossing"];
> It is the Hero, the planet with four heads.
> The god whose weapon is the Flooding Storm,
> shall turn back;
> To his resting place he shall lower himself.

(The receding planet, the text asserts, then recrossed the path of Saturn in the month of Ululu, the sixth month of the year.)

The Old Testament frequently refers to the time when the Lord caused Earth to be covered by the waters of the deep. The twenty-ninth Psalm describes the "calling" as well as the "return" of the "great waters" by the Lord:

> Unto the Lord, ye sons of the gods,
> Give glory, acknowledge might. . . .
> The sound of the Lord is upon the waters;
> The God of glory, the Lord,
> Thundereth upon the great waters. . . .
> The Lord's sound is powerful,
> The Lord's sound is majestic;
> The Lord's sound breaketh the cedars. . . .
> He makes [Mount] Lebanon dance as a calf,
> [Mount] Sirion leap like a young bull.
> The Lord's sound strikes fiery flames;
> The Lord's sound shaketh the desert. . . .
> The Lord to the Deluge [said]: "Return!"
> The Lord, as king, is enthroned forever.

In the magnificent Psalm 77—"Aloud to God I Cry"—
the Psalmist recalls the Lord's appearance and disappearance in earlier times:

> I have calculated the Olden Days,
> The years of *Olam*. . . .
> I shall recall the Lord's deeds,
> Remember thy wonders in antiquity. . . .
> Thine course, O Lord, is determined;
> No god is as great as the Lord. . . .
> The waters saw thee, O Lord, and shuddered;
> Thine splitting sparks went forth.
> The sound of thine thunder was rolling;
> Lightnings lit up the world;
> The Earth was agitated and it quaked.
> [Then] in the waters was thy course,
> Thine paths in the deep waters;
> And thine footsteps were gone, unknown.

Psalm 104, exalting the deeds of the Celestial Lord, recalled the time when the oceans overran the continents and were made to go back:

> Thou didst fix the Earth in constancy,
> For ever and ever to be unmoved.
> With the oceans, as with garment, thou coveredst it;
> Above the mountains did the water stand.
> At thy rebuke, the waters fled;
> At the sound of thine thunder, they hastened away.
> They went upon the mountains, then down to the valleys
> Unto the place which thou hast founded for them.
> A boundary thou hast set, not to be passed over;
> That they turn not again to cover the Earth.

The words of the prophet Amos are even more explicit:

> Woe unto you that desire the Day of the Lord;
> To what end is it for you?
> For the Day of the Lord is darkness and no light. . . .
> Turneth morning unto death's shadow,
> Maketh the day dark as night;
> Calleth forth the waters of the sea
> and poureth them upon the face of the Earth.

These, then, were the events that took place "in olden days." The "Day of the Lord" was the day of the Deluge.

•

We have already shown that, having landed on Earth, the Nefilim associated the first reigns in the first cities with the zodiacal ages—giving the zodiacs the epithets of the various associated gods. We now find that the text uncovered by Ebeling provided calendarial information not only for men but also for the Nefilim. The Deluge, it informs us, occurred in the "Age of the constellation Lion":

Supreme, Supreme, Anointed;
Lord whose shining crown with terror is laden.
Supreme planet: a seat he has set up
Facing the confined orbit of the red planet [Mars].
Daily within the Lion he is afire;
His light his bright kingships on the lands pronounces.

We can now also understand an enigmatic verse in the New Year's rituals, stating that it was "the constellation Lion that measured the waters of the deep." These statements place the time of the Deluge within a definite framework, for though astronomers nowadays cannot precisely ascertain where the Sumerians set the beginning of a zodiacal house, the following timetable for the ages is considered accurate.

60 B.C. to A.D. 2100—Age of Pisces
2220 B.C. to 60 B.C.—Age of Aries
4380 B.C. to 2220 B.C.—Age of Taurus
6540 B.C. to 4380 B.C.—Age of Gemini
8700 B.C. to 6540 B.C.—Age of Cancer
10,860 B.C. to 8700 B.C.—Age of the Lion

If the Deluge occurred in the Age of the Lion, or sometime between 10,860 B.C. and 8700 B.C., then the date of the Deluge falls well within our timetable: According to modern science, the last ice age ended abruptly in the southern hemisphere some twelve to thirteen thousand years ago, and in the northern hemisphere one or two thousand years later.

The zodiacal phenomenon of precession offers even more comprehensive corroboration of our conclusions. We concluded earlier that the Nefilim landed on Earth 432,000 years (120 *shar's*) before the Deluge, in the Age of Pisces. In terms of the precessional cycle, 432,000 years comprise sixteen full cycles, or Great Years, and more than halfway through another Great Year, into the "age" of the constellation of the Lion.

We can now reconstruct the complete timetable for the events embraced by our findings.

Years Ago	*EVENT*
445,000	The Nefilim, led by Enki, arrive on Earth from the Twelfth Planet. Eridu—Earth Station I—is established in southern Mesopotamia.
430,000	The great ice sheets begin to recede. A hospitable climate in the Near East.
415,000	Enki moves inland, establishes Larsa.
400,000	The great interglacial period spreads globally. Enlil arrives on Earth, establishes Nippur as Mission Control Center.
	Enki establishes sea routes to southern Africa, organizes gold-mining operations.
360,000	The Nefilim establish Bad-Tibira as their metallurgical center for smelting and refining. Sippar, the spaceport, and other cities of the gods are built.
300,000	The Anunnaki mutiny. Man—the "Primitive Worker"—is fashioned by Enki and Ninhursag.
250,000	"Early *Homo sapiens*" multiply, spread to other continents.
200,000	Life on Earth regresses during new glacial period.
100,000	Climate warms again.
	The sons of the gods take the daughters of Man as wives.
77,000	Ubartutu/Lamech, a human of divine parentage, assumes the reign in Shuruppak under the patronage of Ninhursag.

75,000 The "accursation of Earth"—a new ice age—begins. Regressive types of Man roam Earth.

49,000 The reign of Ziusudra ("Noah"), a "faithful servant" of Enki, begins.

38,000 The harsh climatic period of the "seven passings" begins to decimate Mankind. Europe's Neanderthal Man disappears; only Cro-Magnon Man (based in the Near East) survives.

   Enlil, disenchanted with Mankind, seeks its demise.

13,000 The Nefilim, aware of the impending tidal wave that will be triggered by the nearing Twelfth Planet, vow to let Mankind perish.

   The Deluge sweeps over Earth, abruptly ending the ice age.

15

•

KINGSHIP ON EARTH

THE DELUGE, a traumatic experience for Mankind, was no less so for the "gods"—the Nefilim.

In the words of the Sumerian king lists, "the Deluge had swept over," and an effort of 120 *shar*'s was wiped away overnight. The south African mines, the cities in Mesopotamia, the control center at Nippur, the spaceport at Sippar—all lay buried under water and mud. Hovering in their shuttlecraft above devastated Earth, the Nefilim impatiently awaited the abatement of the waters so that they could set foot again on solid ground.

How were they going to survive henceforth on Earth when their cities and facilities were gone, and even their manpower—Mankind—was totally destroyed?

When the frightened, exhausted, and hungry groups of Nefilim finally landed on the peaks of the "Mount of Salvation," they were clearly relieved to discover that Man and beast alike had not perished completely. Even Enlil, at first enraged to discover that his aims had been partly frustrated, soon changed his mind.

The deity's decision was a practical one. Faced with their own dire conditions, the Nefilim cast aside their inhibitions about Man, rolled up their sleeves, and lost no time in imparting to Man the arts of growing crops and cattle. Since survival, no doubt, depended on the speed with which agriculture and animal domestication could be developed to sustain the Nefilim and a rapidly multiplying Mankind, the Nefilim applied their advanced scientific knowledge to the task.

•

Unaware of the information that could be culled from the biblical and Sumerian texts, many scientists who have studied the origins of agriculture have arrived at the conclusion that its "discovery" by Mankind some 13,000 years ago was related to the neothermal ("newly warm") climate that followed the end of the last ice age. Long before modern scholars, however, the Bible also related the beginnings of agriculture to the aftermath of the Deluge.

"Sowing and Harvesting" were described in Genesis as divine gifts granted to Noah and his offspring as part of the post-Diluvial covenant between the Deity and Mankind:

> For as long as the Earth's days shall be,
> There shall not cease
> Sowing and Harvesting,
> Cold and Warmth,
> Summer and Winter,
> Day and Night.

Having been granted the knowledge of agriculture, "Noah as a Husbandman was first, and he planted a vineyard": He became the first post-Diluvial farmer engaged in the deliberate, complicated task of planting.

The Sumerian texts, too, ascribed to the gods the granting to Mankind of both agriculture and the domestication of animals.

Tracing the beginnings of agriculture, modern scholars have found that it appeared first in the Near East, but not in the fertile and easily cultivated plains and valleys. Rather, agriculture began in the mountains skirting the low-lying plains in a semicircle. Why would farmers avoid the plains and limit their sowing and reaping to the more difficult mountainous terrain?

The only plausible answer is that the low-lying lands were, at the time when agriculture began, uninhabitable; 13,000 years ago the low-lying areas were not yet dry enough following the Deluge. Millennia passed before the plains and valleys had dried sufficiently to permit the people to come down from the mountains surrounding Mesopotamia and to settle the low-lying plains. This, indeed, is what the Book of Genesis tells us: Many generations after the Deluge, people arriving "from the East"—from the mountainous areas east of Mesopotamia—"found a plain in the land of Shin'ar [Sumer], and settled there."

The Sumerian texts state that Enlil first spread cereals "in the hill country"—in the mountains, not in the plains—and that he made cultivation possible in the mountains by keeping the floodwaters away. "He barred the mountains as with a door." The name of this mountainous land east of Sumer, E.LAM, meant "house where vegetation germinated." Later, two of Enlil's helpers, the gods Ninazu and Ninmada, extended the cultivation of cereals to the low-lying plains so that, eventually, "Sumer, the land that knew not grain, came to know grain."

Scholars, who have now established that agriculture began with the domestication of wild emmer as a source of wheat and barley, are unable to explain how the earliest grains (like those found at the Shanidar cave) were already uniform and highly specialized. Thousands of generations of genetic selection are needed by nature to acquire even a modest degree of sophistication. Yet the period, time, or location in which such a gradual and very prolonged process might have taken place on Earth are nowhere to be found. There is no explanation for this botanogenetic miracle, unless the process was not one of natural selection but of artificial manipulation.

Spelt, a hard-grained type of wheat, poses an even greater mystery. It is the product of "an unusual mixture of botanic genes," neither a development from one genetic source nor a mutation of one source. It is definitely the result of mixing the genes of several plants. The whole notion that Man, in a few thousand years, changed animals through domestication, is also questionable.

Modern scholars have no answers to these puzzles, nor to the general question of why the mountainous semicircle in the ancient Near East became a continuous source of new varieties of cereals, plants, trees, fruits, vegetables, and domesticated animals.

The Sumerians knew the answer. The seeds, they said, were a gift sent to Earth by Anu from his Celestial Abode. Wheat, barley, and hemp were lowered to Earth from the Twelfth Planet. Agriculture and the domestication of animals were gifts given to Mankind by Enlil and Enki, respectively.

Not only the presence of the Nefilim but also the periodic arrivals of the Twelfth Planet in Earth's vicinity seem to lie behind the three crucial phases of Man's post-Diluvial civilization: agriculture, circa 11,000 B.C., the Neolithic culture, circa 7500 B.C., and the sudden civilization of 3800 B.C. took place at intervals of 3,600 years.

It appears that the Nefilim, passing knowledge to Man in measured doses, did so in intervals matching the periodic returns of the Twelfth Planet to Earth's vicinity. It was as though some on-site inspection, some face-to-face consultation possible only during the "window" period that allowed landings and takeoffs between Earth and the Twelfth Planet, had to take place among the "gods" before another "go ahead" could be given.

The "Epic of Etana" provides a glimpse of the deliberations that took place. In the days that followed the Deluge, it says:

> The great Anunnaki who decree the fate
> sat exchanging their counsels regarding the land.
> They who created the four regions,
> who set up the settlements, who oversaw the land,
> were too lofty for Mankind.

The Nefilim, we are told, reached the conclusion that they needed an intermediary between themselves and the masses of humans. They were, they decided, to be gods—*elu* in Akkadian, meaning "lofty ones." As a bridge between themselves as lords and Mankind, they introduced "Kingship" on Earth: appointing a human ruler who would assure Mankind's service to the gods and channel the teachings and laws of the gods to the people.

A text dealing with the subject describes the situation before either tiara or crown had been placed on a human head, or scepter handed down; all these symbols of Kingship—plus the shepherd's crook, the symbol of righteousness and justice—"lay deposited before Anu in Heaven." After the gods had reached their decision, however, "Kingship descended from Heaven" to Earth.

Both Sumerian and Akkadian texts state that the Nefilim

retained the "lordship" over the lands, and had Mankind first rebuild the pre-diluvial cities exactly where they had originally been and as they had been planned: "Let the bricks of all the cities be laid on the dedicated places, let all the [bricks] rest on holy places." Eridu, then, was first to be rebuilt.

The Nefilim then helped the people plan and build the first royal city, and they blessed it. "May the city be the nest, the place where Mankind shall repose. May the King be a Shepherd."

The first royal city of Man, the Sumerian texts tell us, was Kish. "When Kingship was lowered again from Heaven, the Kingship was in Kish." The Sumerian king lists, unfortunately, are mutilated just where the name of the very first human king was inscribed. We do know, however, that he started a long line of dynasties whose royal abode changed from Kish to Uruk, Ur, Awan, Hamazi, Aksak, Akkad, and then to Ashur and Babylon and more recent capitals.

The biblical "Table of Nations" likewise listed Nimrud —the patriarch of the kingdoms at Uruk, Akkad, Babylon, and Assyria—as descended from Kish. It records the spread of Mankind, its lands and Kingships, as an outgrowth of the division of Mankind into three branches following the Deluge. Descended from and named after the three sons of Noah, these were the peoples and lands of Shem, who inhabited Mesopotamia and the Near Eastern lands; Ham, who inhabited Africa and parts of Arabia; and Japheth, the Indo-Europeans in Asia Minor, Iran, India, and Europe.

These three broad groupings were undoubtedly three of the "regions" whose settlement was discussed by the great Anunnaki. Each of the three was assigned to one of the leading deities. One of these was, of course, Sumer itself, the region of the Semitic peoples, the place where Man's first great civilization arose.

The other two also became sites of flourishing civilizations. Circa 3200 B.C.—about half a millennium after the blooming of the Sumerian civilization—statehood, Kingship, and civilization made their first appearance in the Nile valley, leading in time to the great civilization of Egypt.

Nothing was known until some fifty years ago about the first major Indo-European civilization. But by now it is well established that an advanced civilization, encompassing large cities, a developed agriculture, a flourishing trade, existed in the Indus valley in ancient times. It came into being, scholars believe, some 1,000 years after the Sumerian civilization began. (Fig. 161)

Ancient texts as well as archaeological evidence attest to the close cultural and economic links between these two river-valley civilizations and the older Sumerian one. Moreover, both direct and circumstantial evidence has con-

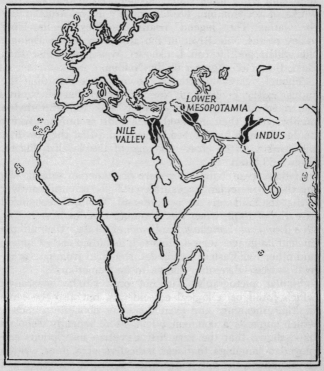

LOWER MESOPOTAMIA

NILE VALLEY

INDUS

Fig. 161

vinced most scholars that the civilizations of the Nile and Indus not only were linked to, but were actually offspring of, the earlier civilization of Mesopotamia.

The most imposing monuments of Egypt, the pyramids, have been found to be, under a stone "skin," simulations of the Mesopotamian ziggurats; and there is reason to believe that the ingenious architect who designed the plans for the great pyramids and supervised their construction was a Sumerian venerated as a god. (Fig. 162)

The ancient Egyptian name for their land was the "Raised Land," and their prehistoric memory was that "a very great god who came forth in the earliest times" found their land lying under water and mud. He undertook great works of reclamation, literally raising Egypt from under the waters. The "legend" neatly describes the low-lying valley of the Nile River in the aftermath of the Deluge; this olden god, it can be shown, was none other than Enki, the chief engineer of the Nefilim.

Though relatively little is known as yet regarding the Indus valley civilization, we do know that they, too, venerated the number twelve as the supreme divine number; that they depicted their gods as human-looking beings wearing horned headdresses; and that they revered the symbol of the cross—the sign of the Twelfth Planet. (Figs. 163, 164)

If these two civilizations were of Sumerian origin, why are their written languages different? The scientific answer is that the languages are not different. This was recognized as early as 1852, when the Reverend Charles Foster (*The One Primeval Language*) ably demonstrated that all the ancient languages then deciphered, including early Chinese and other Far Eastern languages, stemmed from one primeval source—thereafter shown to be Sumerian.

Similar pictographs had not only similar meanings, which could be a logical coincidence, but also the same multiple meanings and even the same phonetic sounds—which suggests a common origin. More recently, scholars have shown that the very first Egyptian inscriptions employed a language that was indicative of a prior written development; the only place where a written language had a prior development was Sumer.

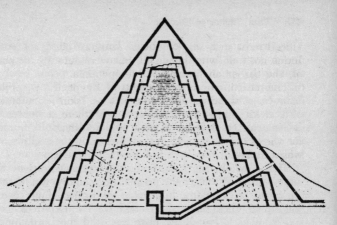

Fig. 162

Fig. 163

Fig. 164

So we have a single written language that for some reason was differentiated into three tongues: Mesopotamian, Egyptian/Hamitic, and Indo-European. Such a differentiation could have occurred by itself over time, distance, and geographical separation. Yet the Sumerian texts claim that it occurred as the result of a deliberate decision of the gods, once again initiated by Enlil. Sumerian stories on the subject are paralleled by the well-known biblical story of the Tower of Babel, in which we are told "that the whole Earth was of one language and of the same words." But after the people settled in Sumer, learned the art of brickmaking, built cities, and raised high towers (ziggurats), they planned to make for themselves a *shem* and a tower to launch it. Therefore "did the Lord mingle the Earth's tongue."

The deliberate raising of Egypt from under the muddy waters, the linguistic evidence, and the Sumerian and biblical texts support our conclusion that the two satellite civilizations did not develop by chance. On the contrary, they were planned and brought about by the deliberate decision of the Nefilim.

Fearing, evidently, a human race unified in culture and purpose, the Nefilim adopted the imperial policy: "Divide and rule." For while Mankind reached cultural levels that included even airborne efforts—after which "anything they shall scheme to do shall no longer be impossible for them"—the Nefilim themselves were a declining lot. By the third millennium B.C., children and grandchildren, to say nothing of humans of divine parentage, were crowding the great olden gods.

The bitter rivalry between Enlil and Enki was inherited by their principal sons, and fierce struggles for supremacy ensued. Even the sons of Enlil—as we have seen in earlier chapters—fought among themselves, as did the sons of Enki. As has happened in recorded human history, overlords tried to keep the peace among their children by dividing the land among the heirs. In at least one known instance, one son (Ishkur/Adad) was deliberately sent away by Enlil to be the leading local deity in the Mountain Land.

As time went on, the gods became overlords, each jealously guarding the territory, industry, or profession over which he had been given dominion. Human kings were the intermediaries between the gods and the growing and spreading humanity. The claims of ancient kings that they went to war, conquered new lands, or subjugated distant peoples "on the command of my god" should not be taken lightly. Text after text makes it clear that this was literally so. The gods retained the powers of conducting foreign affairs, for these affairs involved other gods in other territories. Accordingly, they had the final say in matters of war or peace.

With the proliferation of people, states, cities, and villages, it became necessary to find ways to remind the people who their particular overlord, or "lofty one," was. The Old Testament echoes the problem of having people adhere to *their* god and not "prostitute after other gods." The solution was to establish many places of worship, and to put up in each of them the symbols and likenesses of the "correct" gods.

The age of paganism began.

•

Following the Deluge, the Sumerian texts inform us, the Nefilim held lengthy counsels regarding the future of gods and Man on Earth. As a result of these deliberations, they "created the four regions." Three of them—Mesopotamia, the Nile valley, and the Indus valley—were settled by Man.

The fourth region was "holy"—a term whose original literal meaning was "dedicated, restricted." Dedicated to the gods alone, it was a "pure land," an area that could be approached only with authorization; trespassing could lead to quick death by "awesome weapons" wielded by fierce guards. This land or region was named TIL.MUN (literally, "the place of the missiles"). It was the restricted area where the Nefilim had reestablished their space base after the one at Sippar had been wiped out by the Deluge.

Once again the area was put under the command of Utu/Shamash, the god in charge of the fiery rockets. Ancient heroes like Gilgamesh strove to reach this Land of Living, to be carried by a *shem* or an Eagle to the Heavenly Abode of the Gods. We recall the plea of Gilgamesh to Shamash:

Let me enter the Land, let me raise my *Shem*. . . .
By the life of my goddess mother who bore me,
of the pure faithful king, my father—
my step direct to the Land!

Ancient tales—even recorded history—recall the ceaseless efforts of men to "reach the land," find the "Plant of Life," gain eternal bliss among the Gods of Heaven and Earth. This yearning is central to all the religions whose roots lie deep in Sumer: the hope that justice and righteousness pursued on Earth will be followed by an "afterlife" in some Heavenly Divine Abode.

But where was this elusive land of the divine connection? The question can be answered. The clues are there. But beyond it loom other questions. Have the Nefilim been encountered since? What will happen when they are encountered again?

And if the Nefilim were the "gods" who "created" Man on Earth, *did evolution alone, on the Twelfth Planet, create the Nefilim?*

SOURCES

•

I. Principal sources for biblical texts

A. Genesis through Deuteronomy: *The Five Books of Moses,* new edition, revised by Dr. M. Stern, Star Hebrew Book Company, undated.

B. For latest translation and interpretation based on Sumerian and Akkadian finds: "Genesis," from *The Anchor Bible,* trans. by E. A. Speiser, Garden City, N.Y.: Doubleday & Co., 1964.

C. For "archaic" flavor: *The Holy Bible,* King James Version, Cleveland and New York: The World Publishing Co., undated.

D. For verification of recent interpretations of biblical verses: *The Torah,* new translation of the Holy Scriptures according to the masoretic text, New York: Jewish Publication Society of America, 1962; *The New American Bible,* translation by members of the Catholic Biblical Association of America, New York: P. J. Kenedy & Sons, 1970; and *The New English Bible,* planned and directed by the Church of England, Oxford: Oxford University Press; Cambridge: Cambridge University Press, 1970.

E. For reference on usage comparison and translation aids: *Veteris Testamenti Concordantiae Hebraicae Atque Chaldaicae* by Solomon Mandelkern, Jerusalem: Schocken Books, Inc., 1962; *Encyclopedic Dictionary of the Bible,* a translation and adaptation of the work by A. van den Born, by the Catholic Biblical Association of America, New York: McGraw-Hill Book Co., Inc., 1963; and *Millon-Hatanach* (Hebrew), Hebrew-Aramaic by Jushua Steinberg, Tel Aviv: Izreel Publishing House Ltd., 1961.

II. Principal sources for Near Eastern texts

Barton, George A. *The Royal Inscriptions of Sumer and Akkad.* 1929.

Borger, Riekele. *Babylonisch-Assyrisch Lesestücke.* 1963.

Budge, E. A. Wallis. *The Gods of the Egyptians.* 1904.

Budge, E. A. W., and King, L. W. *Annals of the Kings of Assyria.* 1902.

Chiera, Edward. *Sumerian Religious Texts.* 1924.

Ebeling, E.; Meissner, B.; and Weidner, E. (eds.). *Reallexikon der Assyrologie und Vorderasiatischen Archäology.* 1932-1957.

Ebeling, Erich. *Enuma Elish: die Siebente Tafel des Akkadischen Weltschöpfungsliedes.* 1939.

———. *Tod und Leben nach den Vorstellungen der Babylonier.* 1931.

Falkenstein, Adam, and W. von Soden. *Sumerische und Akkadische Hymnen und Gebete.* 1953.

Falkenstein, Adam. *Sumerische Goetterlieder.* 1959.

Fossey, Charles. *La Magie Syrienne.* 1902.

Frankfort, Henri. *Kingship and the Gods.* 1948.

Gray, John. *The Canaanites.* 1964.

Gordon, Cyrus H. "Canaanite Mythology" in *Mythologies of the Ancient World.* 1961.

Gressman, Hugo. *The Development of the Idea of God in the Old Testament.* 1926.

———. *Altorientalische Texte und Bilder zum alten Testamente.* 1909.

Güterbock, Hans G. "Hittite Mythology" in *Mythologies of the Ancient World.* 1961.

Heidel, Alexander. *The Babylonian Genesis.* 1969.

Hilprecht, Herman V. (ed.). *Reports of the Babylonian Expedition: Cuneiform Texts.* 1893-1914.

Jacobsen, Thorkild. "Mesopotamia" in *The Intellectual Adventure of the Ancient Man.* 1946.

Jastrow, Morris. *Die Religion Babyloniens und Assyriens.* 1905-12.

Jean, Charles-F. *La religion sumerienne.* 1931.

Jensen, P. *Texte zur assyrisch-babylonischen Religion.* 1915.

———. *Die Kosmologie der Babylonier.* 1890.

Jeremias, Alfred. *The Old Testament in the Light of the Ancient Near East.* 1911.

———. *Das Alter der babylonischen Astronomie.* 1908.

———. *Handbuch der Altorientalische Geistkultur.*

Jeremias, Alfred, and Winckler, Hugo. *Im Kampfe um den alten Orient.*

King, Leonard W. *Babylonian Magic and Sorcery, being "The Prayers of the Lifting of the Hand."* 1896.

———. *The Assyrian Language.* 1901.

———. *The Seven Tablets of Creation.* 1902.

———. *Babylonian Religion and Mythology.* 1899.

Kramer, Samuel N. *The Sumerians.* 1963.

———. (ed.): *Mythologies of the Ancient World.* 1961.

———. *History Begins at Sumer.* 1959.

———. *Enmerkar and the Lord of Aratta.* 1952.

———. *From the Tablets of Sumer.* 1956.

———. *Sumerian Mythology.* 1961.

Kugler, Franz Xaver. *Sternkunde und Sterndienst in Babylon.* 1907-1913.

Lambert, W. G., and Millard, A. R. *Atra-Hasis, the Babylonian Story of the Flood.* 1970.

Langdon, Stephen. *Sumerian and Babylonian Psalms.* 1909.

———. *Tammuz and Ishtar.* 1914.

———. (ed.): *Oxford Editions of Cuneiform Text.* 1923.

———. "Semitic Mythology" in *The Mythology of All Races.* 1964.

———. *Enuma Elish: The Babylonian Epic of Creation.* 1923.

———. *Babylonian Penitential Psalms.* 1927.

———. *Die Neu-Babylonischen Königsinschriften.* 1912.

Luckenbill, David D. *Ancient Records of Assyria and Babylonia.* 1926-27.

Neugebauer, O. *Astronomical Cuneiform Texts.* 1955.

Pinches, Theophilus G. "Some Mathematical Tablets in the British Museum" in *Hilprecht Anniversary Volume.* 1909.

Pritchard, James B. (ed.). *Ancient Near Eastern Texts Relating to the Old Testament.* 1969.

Rawlinson, Henry C. *The Cuneiform Inscriptions of Western Asia.* 1861-84.

Sayce, A. H. *The Religion of the Babylonians.* 1888.

Smith, George. *The Chaldean Account of Genesis.* 1876.

Thomas, D. Winton (ed.). *Documents from Old Testament Times.* 1961.

Thompson, R. Campbell. *The Reports of the Magicians and Astrologers of Nineveh and Babylon.* 1900.

Thureau-Dangin, François. *Les Inscriptions de Sumer et Akkad.* 1905.

——. *Die sumerischen und akkadische Königsinschriften.* 1907.

——. *Ritueles accadiens.* 1921.

Virolleaud, Charles. *L'Astronomie Chaldéenne.* 1903-1908.

Weidner, Ernst F. *Alter und Bedeutung der Babylonischer Astronomie und Astrallehre.* 1914.

——. *Handbuch der Babylonischen Astronomie.* 1915.

Witzel, P. Maurus. *Tammuz-Liturgien und Verwandtes.* 1935.

III. Studies and articles consulted in various issues of the following periodicals

Der Alte Orient (Leipzig)

American Journal of Archaeology (Concord, Mass.)

American Journal of Semitic Languages and Literatures (Chicago)

Annual of the American Schools of Oriental Research (New Haven)

Archiv für Keilschriftforschung (Berlin)

Archiv für Orientforschung (Berlin)

Archiv Orientalni (Prague)

Assyrologische Bibliothek (Leipzig)

Assyrological Studies (Chicago)

Das Ausland (Berlin)

Babyloniaca (Paris)

Beiträge zur Assyrologie und semitischen Sprachwissenschaft (Leipzig)

Berliner Beiträge zur Keilschriftforschung (Berlin)

Bibliotheca Orientalis (Leiden)

Bulletin of the American Schools of Oriental Research (Jerusalem and Baghdad)

Deutsches Morgenländische Gesellschaft, Abhandlungen (Leipzig)

Harvard Semitic Series (Cambridge, Mass.)

Hebrew Union College Annual (Cincinnati)
Journal Asiatique (Paris)
Journal of the American Oriental Society (New Haven)
Journal of Biblical Literature and Exegesis (Middletown)
Journal of Cuneiform Studies (New Haven)
Journal of Near Eastern Studies (Chicago)
Journal of the Royal Asiatic Society (London)
Journal of the Society of Oriental Research (Chicago)
Journal of Semitic Studies (Manchester)
Keilinschriftliche Bibliothek (Berlin)
Königliche Museen zu Berlin: Mitteilungen aus der Orientalischen Sammlungen (Berlin)
Leipziger semitische Studien (Leipzig)
Mitteilungen der altorientalischen Gesellschaft (Leipzig)
Mitteilungen des Instituts für Orientforschung (Berlin)
Orientalia (Rome)
Orientalische Literaturzeitung (Berlin)
Proceedings of the American Philosophical Society (Philadelphia)
Proceedings of the Society of Biblical Archaeology (London)
Revue d'Assyrologie et d'archéologie orientale (Paris)
Revue biblique (Paris)
Sacra Scriptura Antiquitatibus Orientalibus Illustrata (Vatican)
Studia Orientalia (Helsinki)
Transactions of the Society of Biblical Archaeology (London)
Untersuchungen zur Assyrologie und vorderasiatischen Archäologie (Berlin)
Vorderasiatische Bibliothek (Leipzig)
Die Welt des Orients (Göttingen)
Wissenschaftliche Veröffentlichungen der deutschen Orient-Gesellschaft (Berlin)
Zeitschrift für Assyrologie und verwandte Gebiete (Leipzig)
Zeitschrift für die alttestamentliche Wissenschaft (Berlin, Gissen)
Zeitschrift der deutschen morgenländischen Gesellschaft (Leipzig)
Zeitschrift für Keilschriftforschung (Leipzig)

INDEX

•

Turn the page
for a revealing sneak preview of

THE
END *of* DAYS

by
Zecharia Sitchin,
the 7th and concluding book of
THE EARTH CHRONICLES
now available in hardcover
from William Morrow,
an imprint of HarperCollins Publishers

FOREWORD: THE PAST, THE FUTURE

"When will they return?"

I have been asked this question countless times by people who have read my books, the "they" being the Anunnaki—the extraterrestrials who had come to Earth from their planet Nibiru and were revered in antiquity as gods. Will it be when Nibiru in its elongated orbit returns to our vicinity, and what will happen then? Will there be darkness at noon and the Earth shall shatter? Will it be Peace on Earth, or Armageddon? A Millennium of trouble and tribulations, or a messianic Second Coming? Will it happen in 2012, or later, or not at all?

These are profound questions that combine people's deepest hopes and anxieties with religious beliefs and expectations, questions compounded by current events: Wars in lands where the entwined affairs of gods and men began; the threats of nuclear holocausts; the alarming ferocity of natural disasters. They are questions that I dared not answer all these years—but now are questions the answers to which cannot—must not—be delayed.

Questions about the Return, it ought to be realized, are not new; they have inexorably been linked in the past—as they are today—to the expectation and the apprehension of the Day of the Lord, the End of Days, Armageddon. Four millennia ago, the Near East witnessed a god and his son promising Heaven on Earth. More than three millennia ago, king and people in Egypt yearned for a messianic time. Two millennia ago, the people of Judea wondered whether the Messiah had appeared, and we are still seized with the mysteries of those events. Are prophecies coming true?

We shall deal with the puzzling answers that were given, solve ancient enigmas, decipher the origin and meaning of symbols—the Cross, the Fishes, the Chalice. We shall describe the role of space-related sites in historic events, and show why Past, Present, and Future converge in Jerusalem, the place of the "Bond Heaven-Earth." And we shall ponder why it is that our current twenty-first century A.D. is so similar to the twenty-first century B.C. Is history repeating itself? Is it destined to repeat itself? Is it all guided by a Messianic Clock? Is the time at hand?

More than two millennia ago, Daniel of Old Testament fame repeatedly asked the angels: *When?* When will be the End of Days,

The End of Days

the End of Time? More than three centuries ago the famed Sir Isaac Newton, who elucidated the secrets of celestial motions, composed treatises on the Old Testament's Book of Daniel and the New Testament's Book of Revelation; his recently found handwritten calculations concerning the End of Days will be analyzed, along with more recent predictions of The End.

Both the Hebrew Bible and the New Testament asserted that the secrets of the Future are imbedded in the Past, that the destiny of Earth is connected to the Heavens, that the affairs and fate of Mankind are linked to those of God and gods. In dealing with what is yet to happen, we cross over from history to prophecy; one cannot be understood without the other, and we shall report them both.

It is with that as our guide, let us look at what is to come through the lens of what had been. The answers will be certain to surprise.

Zecharia Sitchin
New York, August 2006

CHAPTER I: THE MESSIANIC CLOCK

Wherever one turns, humankind appears seized with Apocalyptic trepidation, Messianic fervor, and End of Time anxiety.

Religious fanaticism manifests itself in wars, rebellions, and the slaughter of "infidels." Armies amassed by Kings of the West are warring with Kings of the East. A Clash of Civilizations shakes the foundations of traditional ways of life. Carnage engulfs cities and towns; the high and the mighty seek safety behind protective walls. Natural calamities and ever-intensifying catastrophes leave people wondering: Has Mankind sinned, is it witnessing Divine Wrath, is it due for another annihilating Deluge? Is this the Apocalypse? Can there be—will there be—Salvation? Are Messianic times afoot?

The time—the twenty-first century A.D., or was it the twenty-first century B.C.?

The correct answer is Yes and Yes, both in our own time as well as in those ancient times. It is the condition of the present time, as well as at a time more than four millennia ago; and the amazing similarity is due to events in the middle-time in-between—the period associated with the messianic fervor at the time of Jesus.

Those three cataclysmic periods for Mankind and its planet—two in the recorded past (circa 2100 B.C. and when B.C. changed to A.D.), one in the nearing future—are interconnected; one has led to the other, one can be understood only by understanding the other. The Present stems from the Past, the Past is the Future. Essential to all three is **Messianic Expectation**; and linking all three is **Prophecy**.

How the present time of troubles and tribulations will end—what the Future portends—requires entering the realm of Prophecy. Ours will not be a melange of newfound predictions whose main magnet is fear of doom and End, but a reliance upon unique ancient records that documented the Past, predicted the Future, and recorded previous Messianic expectations—prophesying the future in antiquity, and, one believes, the Future that is to come.

In all three apocalyptic instances—the two that had occurred, the one that is about to happen—the physical and spiritual relationship between Heaven and Earth was and remains pivotal for the events. The physical aspects were expressed by the existence on Earth of actual sites that linked Earth with the heavens—sites that were deemed crucial, that were focuses of the events; the spiritual aspects have been expressed in what we call Religion. In all three instances, a changed relationship between Man and God was central; except that when, circa 2100 B.C., Mankind faced the first of these three epochal upheavals, the relationship was between men and *gods*, in the plural. Whether that relationship has really changed, the reader will soon discover.

The story of the gods, the **Anunnaki** ("Those who from heaven to Earth came"), as the Sumerians called them, begins with their coming to Earth from **Nibiru** in need of gold. The story of their planet was told in antiquity in the *Epic of Creation*, a long text on seven tablets; it is usually considered to be an allegorical myth, the product of primitive minds that spoke of planets as living gods combating each other. But as I have shown in my book *The Twelfth Planet*, the ancient text is in fact a sophisticated cosmogony that tells how a stray planet, passing by our solar system, collided with a planet called Tiamat; the collision resulted in the creation of Earth and its Moon, of the Asteroid Belt and comets, and in the capture of the invader itself in a great elliptical orbit that takes about 3,600 Earth years to complete (Fig. 1 in the hardcover edition).

It was, Sumerian texts tell, 120 such orbits—432,000 Earth years—prior to the Deluge (the "Great flood") that the Anunnaki came to Earth. How and why they came, their first cities in the E.DIN (the biblical Eden), their fashioning of the Adam and the reasons for it, and the events of the catastrophic Deluge—have all been told in *The Earth Chronicles* series of my books and will not be repeated here. But before we time-travel to the momentous twenty-first century B.C., some pre-Diluvial and post-Diluvial landmark events need to be recalled.

The biblical tale of the Deluge, starting in chapter 6 of Genesis, ascribes its conflicting aspects to a sole deity, Yahweh, who at first is determined to wipe Mankind off the face of the Earth, and then goes out of his way to save it through Noah and the Ark. The earlier Sumerian sources of the tale ascribe the disaffection with Mankind to the god **Enlil** and the counter-effort to save Mankind to the god **Enki**. What the Bible glossed over for the sake of Monotheism was not just the disagreement between the Enlil and Enki, but a rivalry and a conflict between two clans of Anunnaki that dominated the course of subsequent events on Earth.

The Messianic Clock

That conflict between the two and their offspring, and the Earth regions allocated to them after the Deluge, need to be kept in mind to understand all that happened thereafter.

The two were half-brothers, sons of Nibiru's ruler **Anu**; their conflict on Earth had its roots on their home planet, Nibiru. Enki—then called E.A ("He whose home is Water")—was Anu's firstborn son, but not by the official spouse, Antu. When Enlil was born to Anu by Antu—a half-sister of Anu—Enlil became the Legal Heir to Nibiru's throne though he was not the firstborn son. The unavoidable resentment on the part of Enki and his maternal family was exacerbated by the fact that Anu's accession to the throne was problematic to begin with: Having lost out in a succession struggle to a rival named Alalu, he later usurped the throne in a coup-d'etat, forcing Alalu to flee Nibiru for his life. That not only backtracked Ea's resentments to the days of his forebears, but also brought about other challenges to the leadership of Enlil, as told in the epic *Tale of Anzu*. (For the tangled relationships of Nibiru's royal families and the ancestries of Anu and Antu, Enlil and Ea, see *The Lost Book of Enki*).

The key to unlocking the mystery of the gods' succession (and marriage) rules was my realization that these rules also applied to the people chosen by them to serve as their proxies to Mankind. It was the biblical tale of the Patriarch Abraham explaining (*Genesis* 20:12) that he did not lie when he had presented his wife Sarah as his sister: "Indeed, she is my sister, the daughter of my father, but not the daughter of my mother, and she became my wife." Not only was marrying a half-sister from a different mother permitted, but a son by her—in this case Isaac—became the Legal Heir and dynastic successor, rather the Firstborn Ishmael, the son of the handmaiden Hagar. (How such succession rules caused the bitter feud between Ra's divine descendants in Egypt, the half-brothers Osiris and Seth who married the half-sisters Isis and Nephtys, is explained in *The Wars of Gods and Men*).

Though those succession rules appear complex, they were based on what those who write about royal dynasties call "bloodlines"— what we now should recognize as sophisticated DNA genealogies that also distinguished between general DNA inherited from the parents as well as the mtDNA that is inherited by females only from the mother. The complex yet basic rule was this: Dynastic lines continue through the male line; the Firstborn son is next in succession; a half-sister could be taken as wife *if she had a different mother*; and if a son by such a half-sister is later born, that son—though not Firstborn—became the Legal Heir and the dynastic successor.

The rivalry between the two half-brothers Ea/Enki and Enlil in matters of the throne was complicated by personal rivalry in matters of the heart. They both coveted their half-sister **Ninmah**, whose mother was yet another concubine of Anu. She was Ea's true love, but he was not permitted to marry her. Enlil then took over and had a son by her—**Ninurta**. Though born without wedlock, the succesion rules made Ninurta Enlil's uncontested heir, being both his Firstborn son and one born by a royal half-sister.

Ea, as related in *The Earth Chronicles* books, was the leader of the first group of fifty Anunnaki to come to Earth to obtain the gold needed to protect Nibiru's dwindling atmosphere. When the initial plans failed, his half-brother Enlil was sent to Earth with more Anunnaki for an expanded Mission Earth. If that was not enough to create a hostile atmosphere, Ninmah too arrived on Earth to serve as chief medical officer . . .

A long text known as the *Atrahasis Epic* begins the story of gods and men on Earth with a visit by Anu to Earth to settle once and for all (he hoped) the rivalry between his two sons that was ruining the vital mission; he even offered to stay on Earth and let one of the half-brothers assume the regency on Nibiru. With that in mind, the ancient text tells us, lots were drawn to determine who shall stay on Earth and who shall sit on Nibiru's throne:

> The gods clasped hands together,
> had cast lots and had divided:
> Anu went up [back] to heaven,
> [For Enlil] the Earth was made subject;
> The seas, enclosed as with a loop,
> to Enki the prince were given.

The result of drawing lots, then, was that Anu returned to Nibiru as its king. Ea, given dominion over the seas and waters (in later times, "Poseidon" to the Greeks and "Neptune" to the Romans), was granted the epithet EN.KI ("Lord of Earth") to soothe his feelings; but it was EN.LIL ("Lord of the Command") who was put in overall charge: "To him the Earth was made subject." Resentful or not, Ea/Enki could not defy the rules of succession or the results of the drawing of lots; and so the resentment, the anger at justice denied, and a consuming determination to avenge injustices to his father and forefathers and thus to himself, led Enki's son **Marduk** to take up the fight.

Several texts describe how the Anunnaki set up their settlemernts in the E.DIN (The post-Diluvial Sumer), each with a specific function, and all laid out in accordance with a master plan. The crucial space connection—the ability to constantly stay in communication with the home planet and with the shuttlecraft and spacecraft—was maintained from Enlil's command post in **Nippur**, the heart of which was a dimly lit chamber called the DUR.AN.KI, "The Bond Heaven-Earth." Another vital facility was a spaceport, located at Sippar ("Bird City"). Nippur lay at the center of concentric circles at which the other "cities of the gods" were located; all together they shaped out, for an arriving spacecraft, a landing corridor whose focal point was the Near East's most visible topographic feature—the twin peaks of Mount Ararat (Fig. 2 in the hardcover edition).

And then the Deluge "swept over the earth," obliterated all the cities of the gods with their Mission Control Center and Spaceport, and buried the Edin under millions of tons of mud and silt. Everything had to be done all over again—but much could no longer be the same. First and foremost, it was necessary to create a new spaceport facility, with a new Mission Control Center and new beacon-sites for a Landing Corridor. The new landing path was anchored again on the prominent twin peaks of Ararat; the other components were all new: The actual spaceport in the Sinai Peninsula, on the 30th parallel north; artificial twin peaks as beacon-sites, the Giza pyramids; and a new Mission Control Center at a place called Jerusalem (Fig. 3 in the hardcover edition). It was a layout that played a crucial role in post-Diluvial events.

The Deluge was a watershed (both literally and figuratively) in the affairs of both gods and men, and in the relationship between the two: The Earthlings, who were fashioned to serve and work for the gods, were henceforth treated as junior partners on a devastated planet.

The new relationship between men and gods was formulated, sanctified and codified when Mankind was granted its first high civilization, in Mesopotamia, circa 3800 B.C. The momentous event followed a state visit to Earth by Anu, not just as Nibiru's ruler but also as the head of the pantheon, on Earth, of the ancient gods.

The End of Days

Another (and probably the main) reason for his visit was the establishment and affirmation of peace among the gods themselves—a live-and-let-live arrangement by dividing the lands of the Old World among the two principal Anunnaki clans—that of Enlil and that of Enki; for the new post-Diluvial circumstances and the new location of the space facilities required a new territorial division among the gods.

It was a division that was reflected in the biblical Table of Nations (*Genesis*, chapter 10), in which the spread of Mankind, emanating from the three sons of Noah, was recorded by nationality and geography: Asia to the nations/lands of Shem, Europe to the descendants of Japhet, Africa to the nation/lands of Ham. The historical records show that the parallel division among the gods allotted the first two to the Enlilites, the third one to Enki and his sons. The connecting Sinai peninsula, where the vital post-Diluvial spaceport was located, was set aside as a neutral Sacred Region.

While the Bible simply listed the lands and nations according to their Noahite division, the earlier Sumerian texts recorded the fact that the division was a deliberate act, the result of deliberations by the leadership of the Anunnaki. A text known as the *Epic of Etana* tells us that

> The great Anunnaki who decree the fates
> sat exchanging their counsels regarding the Earth.
> They created the four regions,
> set up the settlements.

In the First Region, the lands between the two rivers Euphrates and Tigris (Mesopotamia), Man's first known high civilization, that of Sumer, was established. Where the pre-Diluvial cities of the gods had been, Cities of Man arose, each with its sacred precinct where a deity resided in his or her ziggurat—Enlil in Nippur, Ninmah in Shuruppak, Ninurta in Lagash, **Nannar/Sin** in Ur, **Inanna/Ishtar** in Uruk, **Utu/Shamash** in Sippar, and so on. In each such urban center an EN.SI, a "Righteous Shepherd"—initially a chosen demigod—was selected to govern the people in behalf of the gods; his main assignment was to promulgate codes of justice and morality. In the sacred precinct, a priesthood overseen by a high priest served the god and his spouse, supervised the holiday celebrations, and handled the rites of offerings, sacrifices and prayers to the gods. Art and sculpture, music and dance, poetry and hymns, and above all writing and recordkeeping flourished in the temples and extended to the royal palace.

From time to time one of those cities was selected to serve as the land's capital; there the ruler was king, LU.GAL ("Great man"). Initially and for a long time thereafter this person, the most powerful man in the land, served as both king and high priest. He was carefully chosen, for his role and authority, and all the physical symbols of Kingship, were deemed to have come to Earth directly from Heaven, from Anu on Nibiru. A Sumerian text dealing with the subject stated that before the symbols of Kingship (tiara/crown and scepter) and of Righteousness (the shepherd's staff) were granted to an earthly king, they "lay deposited before Anu in heaven." Indeed, the Sumerian word for Kingship was *Anuship*.

This aspect of "Kingship" as the essence of civilization, just behavior and a moral code for Mankind, was explicitly expressed in the statement, in the Sumerian King Lists, that after the Deluge *"Kingship was brought down from Heaven."* It is a profound statement that must be borne in mind as we progress in this book to the messianic expectations—in the words of the New Testament, for the **Return of the "Kingship of Heaven" to Earth.**

Circa 3100 B.C. a similar yet not identical civilization was established in the Second Region in Africa, that of the river Nile (Nubia and Egypt). Its history was not as harmonious as that among the Enlilites, for rivalry and contention continued among Enki's six sons, to whom not cities but whole land domains were allocated. Paramount was an ongoing conflict between Enki's firstborn **Marduk** (*Ra* in Egypt) and **Ningishzidda** (*Thoth* in Egypt), a conflict that led to the exile of Thoth and a band of African followers to the New World (where he became known as *Quetzaloatl*, the Winged Serpent). Marduk/Ra himself was punished and exiled when, opposing the marriage of his young brother Dumuzi to Enlil's granddaughter Inanna/Ishtar, he caused his brother's death. It was as compensation to Inanna/Ishtar that she was granted dominion over the Third Region of civilization, that of the Indus Valley, circa 2900 B.C. It was for good reason that the three civilizations—as was the spaceport in the sacred region—were all centered on the 30th parallel north (Fig. 4 in the hardcover edition).

According to Sumerian texts, the Anunnaki established Kingship—civilization and its institutions, as most clearly exemplified in Mesopotamia—as a new order in their relationships with Mankind, with kings/priests serving both as a link and a separator between gods and men. But as one looks back on that seemingly "golden age" in the affairs of gods and men, it becomes evident that the affairs of the gods constantly dominated and determined the affairs of Men and the fate of Mankind. Overshadowing all was the determination of Marduk/Ra to undo the injustice done to his father Ea/Enki, when under the succession rules of the Anunnaki not Enki but Enlil was declared the Legal Heir of their father Anu, the ruler on their home planet Nibiru.

In accord with the sexagesimal ("base sixty") mathematical system that the gods granted the Sumerians, the twelve great gods of the Sumerian pantheon were given numerical ranks in which Anu held the supreme Rank of Sixty; the Rank of Fifty was granted to Enlil; that of Enki was 40, and so farther down, alternating between male and female deities (Fig. 5 in the hardcover edition). Under the succession rules, Enlil's son Ninurta was in line for the rank of 50 on Earth, while Marduk held a nominal rank of 10; and initially, these two successors-in-waiting were not yet part of the twelve "Olympians."

And so, the long, bitter and relentness struggle by Marduk that began with the Enlil-Enki feud focused later on Marduk's contention with Enil's son Ninurta for the succession to the Rank of Fifty, and then extended to Enlil's granddaughter Inanna/Ishtar whose marriage to Dumuzi, Enki's youngest son, was so opposed by Marduk that it ended with Dumuzi's death. In time Marduk/Ra faced conflicts even with other brothers and half-brothers of his, in addition to the conflict with Thoth that we have already mentioned—principally with Enki's son Nergal who married a granddaughter of Enlil named Ereshkigal.

In the course of these struggles, the conflicts at times flared up to full-fledged wars between the two divine clans; some of those wars are called "The Pyramid Wars" in my book *The Wars of Gods and Men*. In one notable instance the fighting led to the burying alive of Marduk inside the Great Pyramid; in another, it led to its capture by Ninurta. Marduk was also exiled, more than once—both as punishment and as a self-imposed absence. His persistent efforts to attain the status to which he believed he was entitled included the event recorded in the Bible as the Tower of Babel incident; but in the end, after numerous frustrations, success came only when Earth and Heaven were aligned with the **Messianic Clock**.

Indeed, the first cataclysmic set of events, in the twenty-first century B.C., and the Messianic expectations that accompanied it, is principally the story of Marduk; it also brought to center stage his son **Nabu**—a deity, the son of a god, but whose mother was an Earthling.

Throughout the history of Sumer that spanned almost two thousand years, its royal capital shifted—from the first one, Kish (Ninurta's first city) to Uruk (the city that Anu granted to Inanna/ Ishtar) to Ur (Sin's seat and center of worship); then to others and then back to the initial ones; and finally, for the third time, back to Ur. But at all times Enlil's city Nippur, his "cult center" as scholars are wont to call it, remained the religious center of Sumer and the Sumerian people; it was there that the annual cycle of worshipping the gods was determined.

The twelve "Olympians" of the Sumerian pantheon, each with his or her celestial counterpart among the twelve members of the Solar System (Sun, Moon and ten planets, including Nibiru), were also honored with one month each in the annual cycle of a twelve-month year. The Sumerian term for "month," EZEN, actually meant holiday, festival; and each such month was devoted to celebrating the worship-festival of one of the twelve supreme gods. It was the need to determine the exact time when each such month began and ended (and not in order to enable peasants to know when to sow or harvest, as schoolbooks explain) that led to the introduction of *Mankind's first calendar* in **3760** B.C. It is known as the **Calendar of Nippur** because it was the task of its priests to determine the calendar's intricate timetable and to announce, for the whole land, the time of the religious festivals. That calendar is still in use to this day as the Jewish religious calendar which, in A.D. 2006, numbered the year as 5766.

In pre-Diluvial times Nippur served as Mission Control Center, Enlil's command post where he set up the DUR.AN.KI, the "Bond Heaven-Earth" for the communications with the home planet Nibiru and with the spacecraft connecting them. (After the Deluge, these functions were relocated to a place later known as Jerusalem). Its central position, equidistant from the other functional centers in the E.DIN (*see* Fig. 2), was also deemed to be equidistant from the "four corners of the Earth" and gave it the nickname *"Navel of the Earth."* A hymn to Enlil referred to Nippur and its functions thus:

The End of Days

Enlil,
When you marked off divine settlements on Earth,
Nippur you set up as your very own city...
You founded the Dur-An-Ki
In the center of the four corners of the Earth.

(The term "the Four Corners of the Earth" is also found in the Bible; and when Jerusalem replaced Nippur as Mission Control Center after the Deluge, it too was nicknamed the Navel of the Earth).

In Sumerian the term for the four regions of the Earth was UB, but it also is found as AN.UB—the heavenly, the *celestial* four "corners"—in this case an astronomical term connected with the calendar. It is taken to refer to the four points in the Earth-Sun annual cycle that we nowadays call the Summer Solstice, the Winter Solstice, and the two crossing of the equator—once as the Spring Equinox and then as the Autumnal Equinox. In the Calendar of Nippur, the year began on the day of the Spring Equinox and it has so remained in the ensuing calendars of the ancient Near East. That determined the time of the most important festival of the year—the New Year festival, an event that lasted ten days during which detailed and canonized rituals had to be followed.

Determining calendrical time by Heliacal Rising entailed the observation of the skies at dawn, when the sun just begins to rise on the eastern horizon but the skies are still dark enough to show the stars in the background. The day of the equinox having been determined by the fact that on it daylight and nighttime were precisely equal, the position of the sun at heliacal rising was then marked by the erection of a stone pillar to guide future observations—a procedure that was followed, for example, later on at Stonehenge in Britain; and, as at Stonehenge, long term observations revealed that the group of stars ("constellation") in the background has not remained the same (Fig. 6 in the hardcover edition); there, the alignment stone called the "Heel Stone" that points to sunrise on solstice day nowadays pointed originally to sunrise circa 2000 B.C.

The phenomenon, called Precession of the Equinoxes or just Precession, results from the fact that as the Earth completes one annual orbit around the Sun, it does not return to the same exact celestial spot. There is a slight, very slight retardation; it amounts to one degree (out of 360 in the circle) in 72 years. It was Enki who first grouped the stars observable from Earth into "constellations," and divided the heavens in which the Earth circled the sun into twelve parts—what has since been called the Zodiacal Circle of constellations (Fig. 7 in the hardcover edition). Since each twelfth part of the circle occupied 30 degrees of the celestial arc, the retardation or Precessional shift from one Zodiacal House to another lasted (mathematically) **2160** years (72 x 30), and a complete zodiacal cycle lasted 25,920 years (2160 x 12). The approximate dates of the **Zodiacal Ages**—following the equal twelve-part division and not actual astronomical observations—have been added here for the readere's guidance.

That this was the achievement from a time preceding Mankind's civilizations is attested by the fact that a zodiacal calendar was applied to Enki's first stays on Earth (when the first two zodiacal houses were named in his honor); that this was not the achievement of a Greek astronomer (Hipparchus) in the third century B.C. (as most textbooks still suggest) is attested by the fact that the twelve zodiacal houses were known to the Sumerians millennia earlier by names (Fig. 8 in the hardcover edition) and depictions (Fig. 9 in the hardcover edition) that we use to this day.

In *When Time Began* the calendrical timetables of gods and men were discussed at length. Having come from Nibiru, whose orbital period, the SAR, meant 3,600 (Earth-) years, that unit was naturally the first calendrical yardstick of the Anunnaki even on the fast-orbiting Earth. Indeed, the texts dealing with their early days on Earth, such as the Sumerian King Lists, designated the periods of this or that leader's time on Earth in terms of Sars. I termed this **Divine Time**. The calendar granted to Mankind, one based on the orbital aspects of the Earth (and its Moon), was named **Earthly Time**. Pointing out that the 2160-year zodiacal shift (less than a year for the Anunnaki) offered them a better ratio—the "golden ratio" of 10:6—between the two extremes; I called this time unit **Celestial Time**.

As Marduk discovered, that Celestial Time was the "clock" by which his destiny was to be determined.

But which was **Mankind's Messianic Clock**, determining its fate and destiny—*Earthly Time*, such as the count of fifty-year Jubilees, a count in centuries, or the Millennium? Was it *Divine Time*, geared to Nibiru's orbit? Or was it—is it—*Celestial Time* that follows the slow rotation of the zodiacal clock?

The quandary, as we shall see, baffled Mankind in antiquity; it still lies at the core of the current Return issue. The question that

The End of Days

is posed has been asked before—by Babylonian and Assyrian star-gazing priests, by biblical Prophets, in the Book of Daniel, in the Revelation of St. John the Divine, by the likes of Sir Isaac Newton, by all of us today.

The answer will be astounding. Let us embark on the painstaking quest.

BESTSELLING AUTHOR
ZECHARIA
SITCHIN
The Earth Chronicles

Startling documentary evidence of the extraterrestrial gods who changed the course of human development

BOOK I: THE 12TH PLANET
978-0-06-137913-0

BOOK II: THE STAIRWAY TO HEAVEN
978-0-06-137920-8

BOOK III: THE WARS OF GODS AND MEN
978-0-06-137927-7

BOOK IV: THE LOST REALMS
978-0-06-137925-3

BOOK V: WHEN TIME BEGAN
978-0-06-137928-4

BOOK VI: THE COSMIC CODE
978-0-06-137924-6

And Don't Miss the Companion Volumes:

GENESIS REVISITED
Is Modern Science Catching Up with Ancient Knowledge?
978-0-380-76159-3

DIVINE ENCOUNTERS
A Guide to Visions, Angels and Other Emissaries
978-0-380-78076-1

THE END OF DAYS
Armageddon and Prophecies of the Return
978-0-06-123921-2

SIT 0108